QMC
a30213 007212072b
D0421446

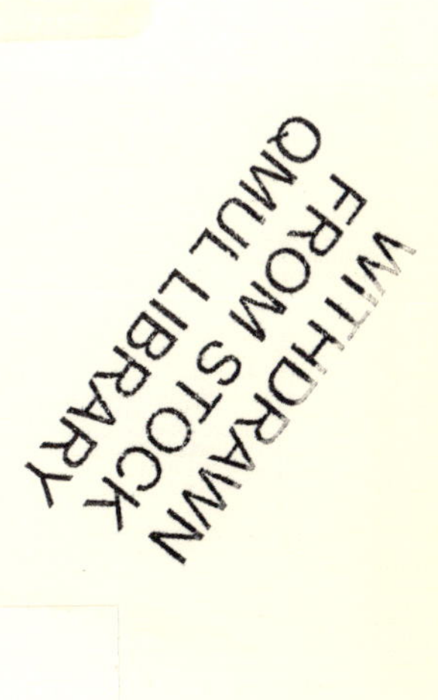
WITHDRAWN
FROM STOCK
QMUL LIBRARY

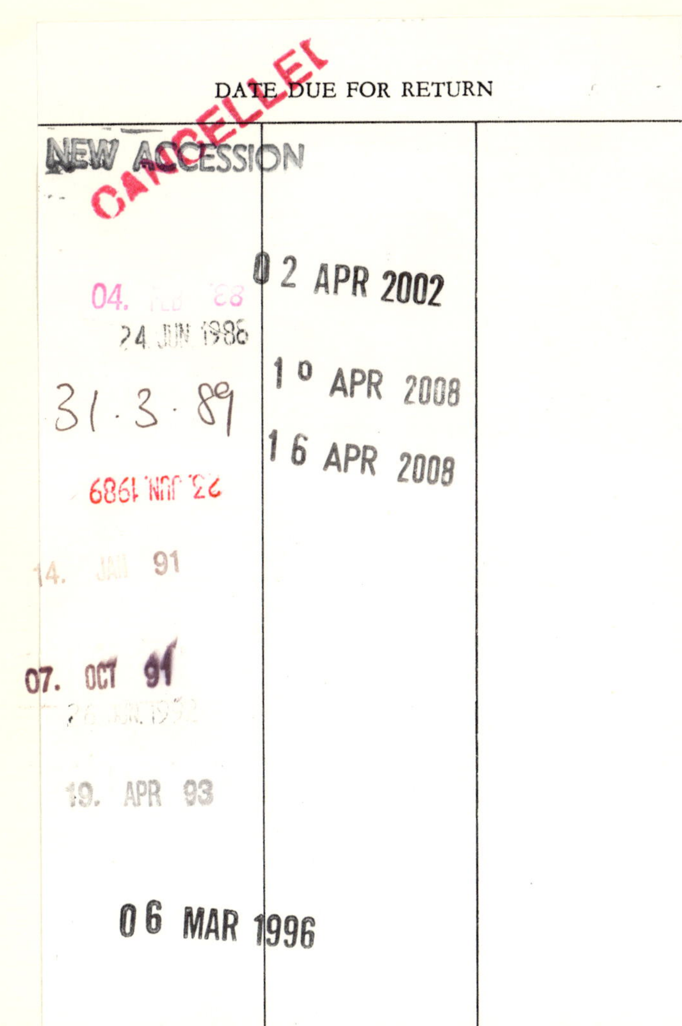
DATE DUE FOR RETURN
NEW ACCESSION
CANCELLED
02 APR 2002
24 JUN 1986
10 APR 2008
31.3.89
16 APR 2008
22 JUN 1989
91
07. OCT 91
19. APR 93
06 MAR 1996

Texts and Monographs in Physics

M. Chaichian N.F. Nelipa

# Introduction to Gauge Field Theories

With 75 Figures

Springer-Verlag
Berlin Heidelberg New York Tokyo 1984

Professor Dr. Masud Chaichian

University of Helsinki, Department of High Energy Physics, Siltavuorenpenger 20 C, SF-00170 Helsinki 17, Finland

Professor Dr. Nikolai F. Nelipa

Institute for Nuclear Physics, Moscow State University, Leninskije gory, SU-117 234 Moscow, USSR

*Translator*

Dr. Juri Estrin

Lehmkuhlskamp 13, D-2110 Buchholz/N, Fed. Rep. of Germany



ISBN 3-540-13008-X Springer-Verlag Berlin Heidelberg New York Tokyo
ISBN 0-387-13008-X Springer-Verlag New York Heidelberg Berlin Tokyo

Library of Congress Cataloging in Publication Data. Chaichian, M. (Masud), 1941–. Introduction to gauge field theories. (Texts and monographs in physics). 1. Gauge fields (Physics) I. Nelipa, N. F. (Nikolaĭ Fedorovich) II. Title. III. Series. QC793.3.F5C484 1984 530.1'43 84-5318



Typesetting: K + V Fotosatz, Beerfelden.
Offset printing and bookbinding: Konrad Triltsch, Graphischer Betrieb, Würzburg
2153/3130-543210

# Preface

In recent years, gauge fields have attracted much attention in elementary particle physics. The reason is that great progress has been achieved in solving a number of important problems of field theory and elementary particle physics by means of the quantum theory of gauge fields. This refers, in particular, to constructing unified gauge models and theory of strong interactions between the elementary particles.

This book expounds the fundamentals of the quantum theory of gauge fields and its application for constructing unified gauge models and the theory of strong interactions.

In writing the book, the authors' aim was three-fold: firstly, to outline the basic ideas underlying the unified gauge models and the theory of strong interactions; secondly, to discuss the major unified gauge models, the theory of strong interactions and their experimental implications; and, thirdly, to acquaint the reader with a rather special mathematical approach (path-integral method) which has proved to be well suited for constructing the quantum theory of gauge fields.

Gauge fields are a vigorously developing area. In this book, we have selected for presentation the more or less traditional and commonly accepted material. There also exist a number of different approaches which are presently being developed. The most important of them are touched upon in the Conclusion.

The book is intended for those who are familiar with the basic facts of elementary particle physics and with the fundamentals of relativistic quantum mechanics. In order to enable the reader to avoid the necessity of consulting different sources, we present in Chap. 1 the basic facts about the global groups of the space-time and the internal symmetry, about the Lagrangians invariant under these groups as well as about the operator form of the quantum field theory.

During the preparation of this book the authors have benefited from discussing various questions with many of their colleagues. It is a pleasure to express gratitude to all of them and to acknowledge the stimulating discussions and their useful advice.

Our special thanks go to Mrs. Christine Pendl from Springer-Verlag for her considerable help in improving the manuscript.

Helsinki, Moscow, February 1984 *M. Chaichian, N. F. Nelipa*

# Contents

# Part IV
# Gauge Theory of Strong Interactions

# Introduction

In recent years, the physics of elementary particles has undergone substantial changes with regard to its experimental as well as to its theoretical aspects. More than two hundred particles have been discovered to date, and their number keeps growing.

All known elementary particles can be divided into three groups, viz. *gauge bosons, leptons* and *hadrons*. The first group consists of the photons, the $W$- and $Z$-bosons; the second group comprises the electrons, the muons and the $\tau$-particles as well as their neutrinos; the third group includes the mesons, the baryons and the meson and baryon resonances.

Besides the gravitational interaction, three types of interactions between the elementary particles are known: the *strong*, the *electromagnetic*, and the *weak interactions*. Each of these interactions is characterized by a certain coupling constant; the coupling constants $g$, $e$ and $G$ of the strong, electromagnetic and weak interactions have the following orders of magnitude: $g \sim 1-10$, $e \sim 0.1$, and $G \sim 10^{-5}$, respectively.

The weak interaction occurs between all the particles, with the exception of photons; the electromagnetic interaction occurs between the charged particles and photons; the strong interaction occurs between all hadrons.

It has been established experimentally that the hadrons form definite *multiplets of particles*. First, the isotopic multiplets of hadrons have been discovered followed by the observation of the unitary and – relatively recently – charmed multiplets. These multiplets can be connected with the representations of the $SU_2$-, $SU_3$-, and $SU_4$-group, respectively.

The existence of the multiplets is accounted for in a simple way by the *quark model of hadrons*. According to this model, all the hadrons can be constructed from the quarks whose number is equal to the dimension of the lowest representation of the group. The bosons are composed of a quark and an anti-quark, while the fermions are composed of various combinations of either three quarks or three anti-quarks. In this scheme, the leptons are considered as non-structured particles with the electromagnetic and the weak interactions existing between them. It is assumed that between the quarks, there exist the strong, the electromagnetic, and the week interactions. In the framework of the symmetries corresponding to the *global groups*, all three types of the interactions are considered separately. It is thus only natural that there emerged the idea of constructing models which would unify these various types of particle interactions.

Using the symmetries corresponding to the *local groups* opens interesting possibilities of the realization of this idea. The point is that, to construct locally invariant theories, new fields have to be introduced which are referred to as the *gauge fields.* It may be assumed that all the three types of interactions are mediated by the same fields, namely, by the vector gauge fields. Then all interaction types have a common basis, and there appears a possibility of unifying them.

As is well-known, the interactions can be of *long-range* nature (e.g., the electromagnetic interaction) or of *short-range* nature (e.g., the weak and the strong interactions). The long-range interactions are mediated by massless, and the short-range interactions by massive particles. The gauge fields can be both massless and massive. They are massless when the Lagrangian and the vacuum state are invariant under a given symmetry group. If the Lagrangian is invariant, while the vacuum state is noninvariant, i.e., if the symmetry is *spontaneously broken*, the gauge fields may acquire mass. That is to say, the gauge fields can mediate both long-range and short-range interactions. Furthermore, the gauge field theories with spontaneous symmetry-breaking have been found to be *renormalizable*, as distinct from the ordinary massive vector fields.

The three ideas mentioned, namely, the *gauge invariance,* the *spontaneous symmetry-breaking*, and the *renormalizability*, are the groundwork of the models which unify the different interactions. Hence, the problem of constructing the unified models of particle interactions is that of constructing renormalizable gauge-invariant theories with spontaneous symmetry-breaking.

The first step in constructing a gauge-invariant theory is to find the *gauge-invariant Lagrangians.* This is dealt with in Part I of the book. In Chap. 1, we give a brief account of the fundamentals of the global invariance and the globally invariant Lagrangians. In Chap. 2, a general method of constructing the gauge-invariant Lagrangians is given, along with the basic properties of the gauge fields. Chapter 3 deals with the spontaneous symmetry-breaking and with finding the expressions for the Lagrangians with spontaneously broken symmetry.

After obtaining the expressions for the Lagrangian, we pass, in Part II, to constructing the *quantum theory of gauge fields.* In Chap. 4 the expressions for the transition amplitude are found in terms of the path integral. The path-integral technique proves to be most suited for treating the gauge fields. In Chap. 5, the covariant perturbation theory for some models is presented.

Part III deals with the gauge theory of the *electroweak* interactions of particles. In Chap. 6, the Lagrangians for the standard model of the electroweak interaction of leptons as well as of quarks are obtained. In Chap. 7 the electromagnetic interaction of leptons is considered. In Chap. 8 the weak interaction of particles is treated in the framework of the standard model using the covariant perturbation theory are considered. The subject of Chap. 9 is the problem of divergences in higher orders of perturbation theory and of removing these divergences by means of renormalization.

In the framework of gauge fields, the problem of the *strong interactions* of particles appears in a new light. Firstly, it is possible to formulate a new theory – *quantum chromodynamics* – which is generally believed to be a candidate theory of strong interactions. Secondly, due to the fact that the strong, electromagnetic and weak interactions are mediated by the same fields, constructing a unified theory of all three interaction types – *the grand unification* – becomes possible. The gauge theory of strong interactions is discussed in Part IV where the basic ideas of the theory of the strong interactions as well as quantum chromodynamics (perturbative, and on a lattice) and the grand unification are presented.

In some cases, the gauge fields make it possible to introduce in the theory objects of a new nature, the so-called *solitons*. The solitons are believed to be promising in the theory of elementary particles. The classical and the quantum theory of solitons is presented in Chap. 15.

## Units and Notation

In this book, the system of units is used for which $\hbar = c = 1$.

The letters set in lightface (e.g., $p$, $q$, etc.) denote the 4-vectors; the boldfaced letters (e.g., $\boldsymbol{p}$, $\boldsymbol{q}$, etc.) stand for the 3-vectors.

The scalar product of two 4-vectors is written as

$$(ab) = a_\mu b_\mu = a_0 b_0 - a_1 b_1 - a_2 b_2 - a_3 b_3 = a_0 b_0 - \boldsymbol{a}\boldsymbol{b} \,.$$

The operations are denoted as follows: * refers to the complex conjugation, $T$ means transposed, and † the Hermitian conjugation.

The transformation properties of the wave functions (the representations) with respect to the Lorentz group are characterized by Greek indices and with respect to the internal symmetry groups by Latin indices; $u_i(x)$ denotes an arbitrary (scalar, vector, etc.) wave function with $i$ components.

# Part I

# Invariant Lagrangians

# 1. Global Invariance

The symmetry properties of elementary particles can be divided into two classes: *space-time* properties and *internal* properties. These properties are described by the spatial symmetry groups and the internal symmetry groups, respectively. Of the spatial symmetry groups, the Lorentz group and the Poincaré group are of most interest to us, while of the internal symmetry groups, those of interest are the unitary groups $U_1$, $SU_2$, $SU_3$, and $SU_4$.

The transformation groups can be *global* or *local*. The global transformations are identical for all points of the space-time whereas the local ones depend on the space-time coordinates.

A most general *requirement* to be met by a physical theory is formulated as that *of its invariance* with respect to a certain transformation group. Since one of the fundamental characteristics of a theory is the Lagrangian, the problem usually consists in formulating a Lagrangian invariant under the transformations of a given group.

This chapter briefly gives the basic information concerning the global groups of interest, as well as the Lagrangians invariant with respect to these groups. The local symmetry groups and the Lagrangians invariant with respect to them will be considered in the next chapter. The reader who is familiar with the fundamentals of relativistic quantum field theory can skip this chapter.

## 1.1 The Global Lorentz Group. Relativistic Invariance

### 1.1.1 The Lorentz Group

Consider a four-dimensional pseudo-Euclidean space-time with coordinates $x_0, x_1, x_2, x_3$ and with the metric $s^2 = x_0^2 - \boldsymbol{x}^2$. An arbitrary rotation around a fixed point $O$ in this space (to be referred to as the origin of the coordinate system) can be represented as a product of six successive rotations in the planes $(x_1x_2)$, $(x_1x_3)$, $(x_2x_3)$, $(x_1x_0)$, $(x_2x_0)$, $(x_3x_0)$. For example, the rotations in the plane $(x_1x_2)$ by the angle $\beta$ and in the plane $(x_0x_3)$ by the angle $u$ are described as follows:

$$\begin{aligned} x_1' &= x_1\cos\beta + x_2\sin\beta \qquad & x_3' &= x_3 \\ x_2' &= -x_1\sin\beta + x_2\cos\beta \qquad & x_0' &= x_0\,, \end{aligned} \tag{1.1.1}$$

$$\begin{aligned} x_1' &= x_1 \quad & x_3' &= x_3 \operatorname{ch} u + x_0 \operatorname{sh} u \\ x_2' &= x_2 \quad & x_0' &= x_3 \operatorname{sh} u + x_0 \operatorname{ch} u \,. \end{aligned} \tag{1.1.1}$$

Such rotations can be represented with the aid of matrices

$$a_{12} = \begin{pmatrix} \cos\beta & \sin\beta & 0 & 0 \\ -\sin\beta & \cos\beta & 0 & 0 \\ 0 & 0 & 1 & 0 \\ 0 & 0 & 0 & 1 \end{pmatrix}, \qquad a_{03} = \begin{pmatrix} 1 & 0 & 0 & 0 \\ 0 & 1 & 0 & 0 \\ 0 & 0 & \operatorname{ch} u & \operatorname{sh} u \\ 0 & 0 & \operatorname{sh} u & \operatorname{ch} u \end{pmatrix}. \tag{1.1.2}$$

The different form of the transformations is due to the pseudo-Euclidean, rather than Euclidean, space in which they are performed. The matrices representing rotations in other planes have a similar form. To obtain a matrix corresponding to an arbitrary rotation, the product of all six matrices representing the respective rotations in different planes should be taken.

The set of all rotations in the four-dimensional space-time has the following properties:

i) Two successive rotations (referred to as the product of two rotations) are a rotation as well; the product of the rotations is represented by the product of their matrices which, in turn, is a matrix of the same type.

ii) Among the rotations, there is one transforming the space into itself (referred to as the unit rotation); this rotation is represented by a unit matrix.

iii) For each rotation $a$ there exists an inverse rotation $a^{-1}$; e.g., the inverse rotation in the $(x_1, x_2)$ plane is given by the angle $(-\beta)$. The product of a given rotation and its inverse is equivalent to the unit rotation: $aa^{-1} = I$. The matrix $a^{-1}$ corresponding to the inverse rotation is inverse to the matrix of the original rotation.

A set of rotations possessing the above three properties by definition constitutes a *group*; the matrices of rotations form a group as well.

The group of rotations in the four-dimensional pseudo-Euclidean space considered above is referred to as the *Lorentz group*. In a general form the transformations of coordinates $x_\mu$ as given by the Lorentz group read

$$x_\mu \to x_\mu' = a_{\mu\nu} x_\nu \,,$$

where $a_{\mu\nu}$ is the product of all the matrices corresponding to rotations in different planes.

The group of transformations in the four-dimensional pseudo-Euclidean space that includes, besides the rotations, also translations along all four coordinate axes is called the *Poincaré group*. With respect to this group, the coordinates are transformed as follows

$$x_\mu \to x_\mu' = a_{\mu\nu} x_\nu + a_\mu \,,$$

where $a_\mu$ is an arbitrary constant four-vector.

The number of independent parameters which determine a group is called the *dimension of the group*. The Lorentz group has the dimension six. Continuous groups of finite dimension (i.e., those with a finite number of parameters) are called the *Lie groups.* The Lorentz group is an example of a Lie group.

A group is said to be *Abelian* if all its elements are commutable and to be *non-Abelian* otherwise. The Lorentz group is non-Abelian.

### 1.1.2 The Lie Algebra of the Lorentz Group

The matrices given by (1.1.2) represent a rotation in the four-dimensional space-time by a finite angle.

It is convenient to investigate rotations by infinitesimal angles $\varepsilon_{\alpha\beta}$. Then, by taking a Taylor expansion of each element of the infinitesimal rotation matrix with respect to the corresponding angle and by retaining the first-order terms one gets, e.g., for the matrices $a_{12}$ and $a_{03}$

$$(a_{12}(\varepsilon_{12}))_{\mu\nu} = \delta_{\mu\nu} + (l_{12})_{\mu\nu}\varepsilon_{12},$$

$$(a_{03}(\varepsilon_{03}))_{\mu\nu} = \delta_{\mu\nu} + (l_{03})_{\mu\nu}\varepsilon_{03},$$

where $\mu$ and $\nu$ are the matrix indices and

$$l_{12} = \frac{\partial}{\partial\beta} a_{12}\bigg|_{\beta=0} = \begin{pmatrix} 0 & 1 & 0 & 0 \\ -1 & 0 & 0 & 0 \\ 0 & 0 & 0 & 0 \\ 0 & 0 & 0 & 0 \end{pmatrix},$$

$$l_{03} = \frac{\partial}{\partial u} a_{03}\bigg|_{u=0} = \begin{pmatrix} 0 & 0 & 0 & 0 \\ 0 & 0 & 0 & 0 \\ 0 & 0 & 0 & 1 \\ 0 & 0 & 1 & 0 \end{pmatrix}.$$

The matrices found in this way which correspond to infinitesimal coordinate transformations are referred to as the *generators of a group*; thus $l_{12}$, $l_{03}$ are examples of generators of the Lorentz group.

The number of generators of a group is equal to the number of the independent parameters of the group. It can be proven by straightforward calculations that the six generators $l_{\alpha\beta}$ of the Lorentz group obey the following commutation rules:

$$[l_{\mu\nu}, l_{\varrho\sigma}]_{-} = g_{\mu\varrho} l_{\nu\sigma} + g_{\nu\sigma} l_{\mu\varrho} - g_{\mu\sigma} l_{\nu\varrho} - g_{\nu\varrho} l_{\mu\sigma}, \qquad (1.1.3)$$

where $g_{\mu\nu}$ is the metric tensor.

The set of generators of the Lorentz group has the following properties: the sum of two generators of the group as well as the product of a generator and an arbitrary number belong to the same set; the commutator of two generators is expressed as a combination of the generators of this group. The set of generators possessing these properties is defined as the *Lie algebra* of the group. The set of generators $l_{\alpha\beta}$ thus constitutes the Lie algebra of the Lorentz group.

Among the generators of a group, some can turn out to be commutable. Their maximum number defines the *rank* of the group. The Lorentz group has two commuting generators; its rank is thus equal to 2.

### 1.1.3 Representations of the Lorentz Group

A particle with spin $s$ and a non-zero mass is described by a wave function with $2s+1$ components or by a $2s+1$-component multiplet. The components of a wave function also transform under coordinate transformations. For example, the transformation of the components of a four-component wave function, in a general form, reads

$$u_1'(x') = A_{11}u_1(x) + A_{12}u_2(x) + A_{13}u_3(x) + A_{14}u_4(x)\,,$$
$$u_2'(x') = A_{21}u_1(x) + A_{22}u_2(x) + A_{23}u_3(x) + A_{24}u_4(x)\,,$$
$$u_3'(x') = A_{31}u_1(x) + A_{32}u_2(x) + A_{33}u_3(x) + A_{34}u_4(x)\,,$$
$$u_4'(x') = A_{41}u_1(x) + A_{42}u_2(x) + A_{43}u_3(x) + A_{44}u_4(x)\,.$$

This transformation of the components is characterized (like the coordinate transformation) by a four-row matrix

$$T_\mathrm{a} = \begin{pmatrix} A_{11} & A_{12} & A_{13} & A_{14} \\ A_{21} & A_{22} & A_{23} & A_{24} \\ A_{31} & A_{32} & A_{33} & A_{34} \\ A_{41} & A_{42} & A_{43} & A_{44} \end{pmatrix}.$$

A wave function with a different number of components will transform in a similar way, the dimension of the matrix being equal to this number.

The matrices $T_\mathrm{a}$ characterize the rotation of the components of the wave functions by a finite angle. It is convenient to investigate infinitesimal transformations of wave functions and the corresponding matrices, i.e., the *generators of a representation* of the group. Let $M_{\alpha\beta}$ be the generators of a representation of the group. Then

$$(T_\mathrm{a})_{\mu\nu} = \delta_{\mu\nu} + (M_{\alpha\beta})_{\mu\nu}\varepsilon_{\alpha\beta} + O(\varepsilon^2_{\alpha\beta})\,.$$

Each generator $l_{\alpha\beta}$ of the group can be put in correspondence with a generator $M_{\alpha\beta}$ of infinitesimal transformations of the components of the wave function. Let us require that the generators $M_{\alpha\beta}$ obey the same Lie algebra as the generators $l_{\alpha\beta}$ of the group itself.

$$[M_{\mu\nu}, M_{\varrho\sigma}]_- = g_{\mu\varrho}M_{\nu\sigma} + g_{\nu\sigma}M_{\mu\varrho} - g_{\mu\sigma}M_{\nu\varrho} - g_{\nu\varrho}M_{\mu\sigma}\,. \tag{1.1.4}$$

The generators $M_{\alpha\beta}$ then become the representation of the Lorentz group. More precisely, we put in correspondence with each space-time rotation matrix $a$ a transformation matrix $T_\mathrm{a}$ for the wave functions in such a manner that the product of two matrices, $T_{\mathrm{a}_1\mathrm{a}_2} = T_{\mathrm{a}_1}T_{\mathrm{a}_2}$, corresponds to the product

$a_1 a_2$ of the space-time rotation matrices (i.e., to two successive rotations). By definition, the matrices of wave function transformation obeying this condition are said to constitute a *representation* of the Lorentz group.

The Lie algebra being given, the problem consists in finding the generators of the representations of the group, i.e., in the case of the Lorentz group, in finding the set of the matrices $M_{\alpha\beta}$ satisfying the commutation relations (1.1.4).

Let us present the simplest representations of the Lorentz group together with the wave functions that transform according to these representations.

i) Associate with any rotation of the four-dimensional space-time a unit operator acting on the wave functions. On rotating the four-dimensional space-time, a wave function then transforms into itself. Such a one-component wave function $\phi(x)$ is called *scalar.* It describes a particle of zero spin.

ii) Attribute to each coordinate transformation matrix $l_{\alpha\beta}$ an identical matrix acting on four-component wave functions. Since $l_{\alpha\beta} = M_{\alpha\beta}$, the condition (1.1.4) is automatically fulfilled. In this case the representation of the Lorentz group is comprised by the generators of the group themselves. A wave function $u_\alpha(x)$ whose components transform by means of the generators $l_{\alpha\beta}$ is said to be a *vector wave function*. It describes spin-1 particles (provided that a certain additional condition is imposed on its components).

iii) Bring each element of the Lorentz group into correspondence with the matrix $T_{\alpha\beta,\gamma\delta} = a_{\alpha\gamma} a_{\beta\delta}$. The wave function $u_{\alpha\beta}$ transforms then as follows:

$$u'_{\alpha\beta}(x') = T_{\alpha\beta,\gamma\delta} u_{\gamma\delta}(x) .$$

Such a representation of the Lorentz group is called a *tensor* one, the function $u_{\alpha\beta}(x)$ being the second-rank tensor.

iv) Attribute to each element of the Lorentz group a matrix

$$\sigma_{\mu\nu} = \frac{1}{2\mathrm{i}}(\gamma_\mu \gamma_\nu - \gamma_\nu \gamma_\mu) ,$$

where $\gamma_\mu$ are the *Dirac matrices*

$$\gamma_0 = \begin{pmatrix} 1 & 0 & 0 & 0 \\ 0 & 1 & 0 & 0 \\ 0 & 0 & -1 & 0 \\ 0 & 0 & 0 & -1 \end{pmatrix} , \quad \gamma_1 = \begin{pmatrix} 0 & 0 & 0 & 1 \\ 0 & 0 & 1 & 0 \\ 0 & -1 & 0 & 0 \\ -1 & 0 & 0 & 0 \end{pmatrix} ,$$

$$\gamma_2 = \begin{pmatrix} 0 & 0 & 0 & -\mathrm{i} \\ 0 & 0 & \mathrm{i} & 0 \\ 0 & \mathrm{i} & 0 & 0 \\ -\mathrm{i} & 0 & 0 & 0 \end{pmatrix} , \quad \gamma_3 = \begin{pmatrix} 0 & 0 & 1 & 0 \\ 0 & 0 & 0 & -1 \\ -1 & 0 & 0 & 0 \\ 0 & 1 & 0 & 0 \end{pmatrix} .$$

The wave function

$$\psi(x) = \begin{pmatrix} \psi_1(x) \\ \psi_2(x) \\ \psi_3(x) \\ \psi_4(x) \end{pmatrix}$$

whose components transform via the matrices $\sigma_{\mu\nu}$ under rotations of the four-dimensional space-time

$$\psi'(x') = \exp(\mathrm{i}\,\varepsilon_{\mu\nu}\sigma_{\mu\nu})\,\psi(x)$$

is called a *spinor*. It describes spin-1/2 particles.

The spinor

$$\bar{\psi}(x) = \psi^{\dagger}(x)\,\gamma_0 = (\psi_1^*(x),\ \psi_2^*(x),\ -\psi_3^*(x),\ -\psi_4^*(x))$$

is called the *Dirac conjugate* spinor.

The number of independent components in the multiplet is referred to as the *dimension of the representation*.

A representation whose dimension is equal to the order of the group is said to be *adjoint*. An adjoint representation of the Lorentz group is thus a representation of dimension 6. The representation of minimum dimension (excluding the unit one) is called *fundamental*. All other representations of a group can be obtained from the fundamental one by means of multiplication. The fundamental representation of the Lorentz group (with the space inversion) is the matrix $\sigma_{\mu\nu}$.

The transformation law for wave functions under infinitesimal rotations by the angles $\varepsilon_{\mu\nu}$ is explicitly written as

i) for a scalar wave function

$$\phi(x) \to \phi'(x') = \phi(x)\,, \tag{1.1.5}$$

ii) for a vector wave function

$$u_\mu(x) \to u'_\mu(x') = (g_{\mu\nu} + \varepsilon_{\mu\nu})\,u_\nu(x) + O(\varepsilon^2)\,, \tag{1.1.6}$$

iii) for a spinor wave function

$$\psi(x) \to \psi'(x') = (1 + \mathrm{i}\,\varepsilon_{\mu\nu}\sigma_{\mu\nu})\,\psi(x) + O(\varepsilon^2)\,. \tag{1.1.7}$$

Transformations of the wave functions with a larger number of components have a similar form.

Hence, the Lorentz group has representations of any dimension or, in other words, it has wave functions with any number of components; the transformation matrices for the components of the wave functions of a given dimension are determined by the Lorentz group as well.

### 1.1.4 Reducible and Irreducible Representations

Consider a wave function which is, e.g., a second-rank tensor (having sixteen components). It can always be expressed as the sum of a symmetric tensor and an antisymmetric one:

$$u_{\mu\nu}(x) = u^{\mathrm{s}}_{\mu\nu}(x) + u^{\mathrm{a}}_{\mu\nu}(x)\,,$$

so that

$$u^{\mathrm{s}}_{\mu\nu}(x) = \tfrac{1}{2}[u_{\mu\nu}(x) + u_{\nu\mu}(x)]\,, \qquad u^{\mathrm{a}}_{\mu\nu}(x) = \tfrac{1}{2}[u_{\mu\nu}(x) - u_{\nu\mu}(x)]\,.$$

Obviously, the components of the symmetric and the antisymmetric tensors will independently transform upon rotations of the four-dimensional space, without any mixing. In other words, the sixteen components of the tensor $u_{\mu\nu}(x)$ decompose, under four-dimensional rotations, into two "independent" subsets: a six-dimensional one, $u^{\mathrm{a}}_{\mu\nu}(x)$, and a ten-dimensional one, $u^{\mathrm{s}}_{\mu\nu}(x)$. The ten components $u^{\mathrm{s}}_{\mu\nu}(x)$ decompose, in their turn, into two independent subsets: a one-dimensional (scalar) invariant which is the sum of the diagonal elements (the trace) and the remaining nine components (that constitute a matrix with a zero trace).

Hence, the sixteen-component, second-rank tensor $u_{\mu\nu}(x)$ decomposes under four-dimensional rotations into three so-called invariant subspaces: a one-dimensional, a six-dimensional, and a nine-dimensional one.

Accordingly, the representation $T_{\alpha\beta,\,\mu\nu}$, i.e., the matrix performing the transformation of a tensor wave function $u_{\mu\nu}(x)$ in four-dimensional rotations

$$u'_{\alpha\beta} = T_{\alpha\beta,\,\mu\nu}u_{\mu\nu}\,,$$

will be made out of three independent matrices of smaller dimensions.

A representation that decomposes into the sum of independent matrices of lower dimensions is said to be *reducible*, otherwise it is called *irreducible*.

As is seen, the tensor representation $T_{\alpha\beta,\,\mu\nu}$ is reducible: it decomposes into three irreducible representations.

### 1.1.5 Relativistically Invariant Quantities

Let us consider the wave functions to form a representation of a symmetry group. Specifying the group then uniquely determines the transformation of the wave functions, cf., e.g., (1.1.5–7). Hence, combinations $Q(x)$ of wave functions can be formed which do not change under transformations of the Poincaré group: $Q'(x') = Q(x)$. Such combinations are called *relativistic invariants*. Here are several examples of relativistic invariants:

| | |
|---|---|
| $\phi(x)\,\phi(x)$ | scalar field, |
| $u_\mu(x)\,u_\mu(x)$ | vector field, |
| $\bar{\psi}(x)\,\gamma_\mu\partial_\mu\psi(x)$ | spinor field, $\gamma_\mu$ denoting the Dirac matrices, |
| $\bar{\psi}(x)\,\gamma_\mu\psi(x)\,u_\mu(x)$ | spinor and vector fields, |
| $\bar{\psi}(x)\,\psi(x)\,\phi(x)$ | spinor and scalar fields. |

The invariance of these combinations can be directly proven by means of (1.1.5–7).

### 1.1.6 The Lagrangian Formalism

There are two ways of constructing quantum field theory, namely, the *Lagrangian formalism* and the *Hamiltonian formalism.*

The Lagrangian formalism is concerned with the Lagrangian density $L(x_0, \boldsymbol{x})$: a function defined at a given point of the field. The integral of the Lagrangian density $L(x_0, \boldsymbol{x})$ over a certain volume $\Omega$ of the four-dimensional space-time

$$I = \int_{\Omega} d\boldsymbol{x}\, dx_0 L(x_0, \boldsymbol{x}) \equiv \int_{\Omega} dx\, L(x) \tag{1.1.8}$$

is called the action. Henceforth, $L(x)$ will be simply referred to as the *Lagrangian.*

The Lagrangian must fulfil a number of requirements. First of all, it has to be a relativistic invariant. Furthermore, we shall assume (i) that the Lagrangian is a real function; (ii) that it contains functions and their derivatives referring to one point only, i.e., that it describes local interactions; (iii) that it depends only on the wave functions $u_i(x)$ and their derivatives $\partial_\mu u_i(x)$; and (iv) that it does not explicitly depend on the coordinate $x$.

Thus, the Lagrangian has the form

$$L(x) = L(u_i(x), \partial_\mu u_i(x)) \,. \tag{1.1.9}$$

Several Lagrangians which fulfil the above requirements are presented here:
i) for a real scalar field $\phi(x)$ ($m$ denoting the mass)

$$L_0(\phi, \partial_\mu \phi) = \tfrac{1}{2}\partial_\mu \phi(x)\partial_\mu \phi(x) - \tfrac{1}{2}m^2\phi^2(x); \tag{1.1.10}$$

ii) for a complex scalar field $\phi(x)$

$$L_0(\phi, \partial_\mu \phi) = \partial_\mu \phi^*(x)\partial_\mu \phi(x) - m^2\phi^*(x)\phi(x); \tag{1.1.11}$$

iii) for an electromagnetic field

$$L_0(A_\mu, \partial_\mu A_\nu) = -\tfrac{1}{4}F_{\mu\nu}F_{\mu\nu}, \qquad F_{\mu\nu} = \partial_\mu A_\nu - \partial_\nu A_\mu; \tag{1.1.12}$$

iv) for a complex spinor field $\psi(x)$ ($M$ denoting the mass)

$$L_0(\psi, \partial_\mu \psi) = \frac{\mathrm{i}}{2}[\bar{\psi}(x)\gamma_\mu \partial_\mu \psi(x) - \partial_\mu \bar{\psi}(x)\gamma_\mu \psi(x)] - M\bar{\psi}(x)\psi(x) \,. \tag{1.1.13}$$

Here the symbol $\gamma_\mu$ denotes the Dirac matrices.

The relativistic invariance of these Lagrangians can be proven in a straightforward way by making use of (1.1.5 – 7).

The Lagrangians given by (1.1.10 – 13) correspond to free fields. Interaction between fields is described by an *interaction Lagrangian* $L_\mathrm{I}$. For instance,

i) $L_\mathrm{I}(\phi) = \dfrac{f}{4!}\phi^4(x)$ for the self-interaction of a scalar field,

ii) $L_{\mathrm{I}}(\psi, A_\mu) = -e\bar{\psi}\gamma_\mu\psi A_\mu$ for the interaction between spinor and electromagnetic fields.

The *total Lagrangian L* of a system of interacting fields is written as the sum of the Lagrangians of the free field, $L_0$, and of the interaction Lagrangian $L_{\mathrm{I}}$:

$$L = L_0 + L_{\mathrm{I}}\,.$$

Field equations are obtained by using the *principle of least action* according to which those motions will realize for which the action (1.1.8) is minimal. By taking the variation $\delta I$ and setting it equal to zero the field equation is obtained:

$$\frac{\partial L}{\partial u_i} - \partial_\mu \frac{\partial L}{\partial(\partial_\mu u_i)} = 0\,. \tag{1.1.14}$$

Considering the Lagrangians (1.1.10–13) we get the following equations:
i) for a real scalar field

$$(\Box + m^2)\,\phi(x) = 0\,, \quad \text{where}$$
$$\Box = \partial_\mu\partial_\mu\,;$$

ii) for a complex scalar field

$$(\Box + m^2)\,\phi(x) = 0\,, \quad (\Box + m^2)\,\phi^*(x) = 0\,;$$

iii) for an electromagnetic field

$$\Box A_\mu(x) = 0\,;$$

iv) for a complex spinor field

$$(\mathrm{i}\gamma_\mu\partial_\mu - M)\,\psi(x) = 0\,.$$

### 1.1.7 The Hamiltonian Formalism

In the Hamiltonian formalism, a system is characterized by the generalized coordinates $Q_i(t)$ and the generalized momenta $P_i(t)$ defined as

$$P_i = \partial L/\partial \dot{Q}_i\,, \tag{1.1.15}$$

where $\dot{Q}_i = \dfrac{\partial Q_i}{\partial t}$ [note that in the case of fields, $Q_i$ coincides with $u_i(x)$].

*The Hamiltonian* $H(Q_i, P_i)$ depends on the canonical variables $Q_i$, $P_i$, while the canonical equations of motion have the form

$$\dot{Q}_i = \frac{\partial H(Q_i, P_i)}{\partial P_i}\,, \quad \dot{P}_i = -\frac{\partial H(Q_i, P_i)}{\partial Q_i}\,. \tag{1.1.16}$$

The Hamiltonian and the Lagrangian are related by the equation

$$H(Q_i, P_i) = \sum_i P_i \dot{Q}_i - L(Q_i, P_i) \,. \tag{1.1.17}$$

### 1.1.8 The Operator Form of the Quantum Field Theory

Various methods are used to develop the quantum field theory. One of them is based on passing from classical fields to quantized fields by replacing classical quantities (e.g., field functions, conserved quantities) with the corresponding *operators.*

i) Let us quantize, for example, a real scalar field which is described by the Klein-Gordon equation:

$$(\Box + m^2)\,\phi(x) = 0\,, \qquad \Box = \partial_\mu \partial_\mu \,.$$

This equation has two solutions corresponding to positive and negative energy:

$$\phi(x) = \phi^{(+)}(x) + \phi^{(-)}(x)\,, \qquad \text{where}$$

$$\phi^{(+)}(x) = \frac{1}{(2\pi)^{3/2}} \int d\boldsymbol{q}\, \frac{1}{\sqrt{2q_0}}\, \phi^{(+)}(\boldsymbol{q})\, \mathrm{e}^{iqx}\,,$$

$$\phi^{(-)}(x) = \frac{1}{(2\pi)^{3/2}} \int d\boldsymbol{q}\, \frac{1}{\sqrt{2q_0}}\, \phi^{(-)}(\boldsymbol{q})\, \mathrm{e}^{-iqx}\,.$$

Here $(q_0, \boldsymbol{q})$ is the four-momentum of the scalar field.

Let us replace the functions $\phi^{(\pm)}(x)$ by the operators $\hat{\phi}^{(\pm)}(x)$. The hat denotes an operator. These satisfy certain commutation relations. For example, in the coordinate representation they have the form

$$[\hat{\phi}^{(-)}(x), \hat{\phi}^{(+)}(y)]_- = \frac{1}{(2\pi)^3} \int d\boldsymbol{q}\, \frac{\exp[-iq(x-y)]}{2q_0} = \frac{1}{\mathrm{i}}\, \mathscr{D}^{(-)}(x-y)\,.$$

The operators $\hat{\phi}^{(+)}(\boldsymbol{q})$ and $\hat{\phi}^{(-)}(\boldsymbol{q})$ describe creation and annihilation of a scalar particle with the momentum $\boldsymbol{q}$, respectively.

Let us define the *vacuum* state $\Phi_0$ as a state where particles are absent. That is to say, the energy and the momentum of the vacuum are zero. Acting on the vacuum with the creation operators gives rise to states with a certain number of particles. For instance, we find for the vector of the state $\Phi_2$ with two scalar mesons whose momenta are $\boldsymbol{q}_1$ and $\boldsymbol{q}_2$

$$\Phi_2 = \hat{\phi}^{(+)}(\boldsymbol{q}_1)\hat{\phi}^{(+)}(\boldsymbol{q}_2)\, \Phi_0 \,.$$

Other fields (vector fields, spinor fields, etc.) can also be quantized in a similar way.

ii) A system of interacting particles can transform from one state into another. In what follows we shall consider the scattering of particles. A general formulation of the problem is as follows. An initial system of non-interacting particles is given, at $t = -\infty$, by the state vector $\Phi(-\infty)$. One has to find the final state of the system at $t = +\infty$, which is described by the state vector $\Phi(+\infty)$, resulting from the interaction.

In solving this problem we shall proceed from the equations for the state vector $\Phi(t)$ in the interaction representation

$$\mathrm{i}\,\partial\Phi(t)/\partial t = H_{\mathrm{I}}(t)\,\Phi(t)\,, \tag{1.1.18}$$

where $H_{\mathrm{I}}$ is the field interaction Hamiltonian.

Let us express the state vector $\Phi(t)$ as

$$\Phi(t) = \hat{S}(t, t_0)\,\Phi(t_0)\,, \tag{1.1.19}$$

where $\Phi(t_0)$ is the value of the state vector $\Phi(t)$ at an initial moment $t = t_0$ and $\hat{S}(t, t_0)$ is some unknown operator.

Substituting (1.1.19) into (1.1.18) gives the equation for the operator $\hat{S}(t, t_0)$:

$$\mathrm{i}\,\partial\hat{S}(t, t_0)/\partial t = H_{\mathrm{I}}(t)\hat{S}(t, t_0)\,; \tag{1.1.20}$$

$\hat{S}(t, t_0)$ satisfies the initial condition $\hat{S}(t_0, t_0) = 1$.

Let us look for a solution of (1.1.20) as a series in the coupling constant $g$:

$$\hat{S}(t, t_0) = \sum_{n=0}^{\infty} g^n \hat{S}_n(t, t_0)\,. \tag{1.1.21}$$

Substituting (1.1.21) into (1.1.20) we find

$$\hat{S}_n(t, t_0) = \frac{(-\mathrm{i})^n}{n!} \int_{t_0}^{t} dt_1 \int_{t_0}^{t} dt_2 \ldots \int_{t_0}^{t} dt_n\, T(H_{\mathrm{I}}(t_1) H_{\mathrm{I}}(t_2) \ldots H_{\mathrm{I}}(t_n))\,. \tag{1.1.22}$$

Here $T$ is the *chronological ordering* operator which arranges the factors in such a way that their temporal arguments increase from right to left.

It follows from (1.1.19) that

$$\Phi(\infty) = \hat{S}(\infty, -\infty)\,\Phi(-\infty)\,.$$

The operator $\hat{S}(\infty, -\infty)$ is called the *scattering S-matrix.* This operator, by acting on the vector of state of the initial system given in the infinite past ($t = -\infty$), yields the vector of state of the system in the infinite future ($t = \infty$).

In order to obtain the scattering $S$-matrix in *perturbation theory,* the finite integration limits should be replaced in (1.1.22) by the infinite ones:

$$\hat{S}_n(\infty, -\infty) = \frac{(-\mathrm{i})^n}{n!} \int_{-\infty}^{+\infty} dx_1 \int_{-\infty}^{+\infty} dx_2 \ldots \int_{-\infty}^{+\infty} dx_n\, T(H_{\mathrm{I}}(x_1) H_{\mathrm{I}}(x_2) \ldots H_{\mathrm{I}}(x_n))\,. \tag{1.1.23}$$

The probability of transition of the system from the initial state $\Phi_{\mathrm{i}}$ into the final state $\Phi_{\mathrm{f}}$ is described by the quantity

$$S_{\mathrm{fi}} = \Phi_{\mathrm{f}}(\infty)\hat{S}(\infty, -\infty)\Phi_{\mathrm{i}}(-\infty)\,, \tag{1.1.24}$$

which is called the *matrix element*, or the *amplitude*.

iii) Let us consider the quantum electrodynamics as an example. The corresponding interaction Hamiltonian is

$$H_{\mathrm{I}} = e\bar{\psi}\gamma_\mu\psi A_\mu\,, \tag{1.1.25}$$

where $e$ is the electromagnetic interaction constant.

Substituting this formula into (1.1.24) leads to an expression for the matrix element for electrodynamic processes in any order of perturbation theory. For example, the matrix element for the process of scattering of the photon $\gamma$ on the electron $e^-$,

$$\gamma(k) + e^-(p) \to \gamma(k') + e^-(p')\,,$$

reads, in the second order of perturbation theory,

$$S_{\mathrm{fi}}^{(2)} = \Phi_{\mathrm{f}}^\dagger \hat{S}_2 \Phi_{\mathrm{i}}\,. \tag{1.1.26}$$

Here

$$\hat{S}_2 = \frac{(-\mathrm{i})^2}{2!}\int dx_1 dx_2 T(H_{\mathrm{I}}(x_1)H_{\mathrm{I}}(x_2))\,, \tag{1.1.27}$$

$$\begin{aligned}\Phi_{\mathrm{i}} &= a_\lambda^{(+)}(\boldsymbol{k})a_r^{(+)\dagger}(\boldsymbol{p})\,\Phi_0\,,\\ \Phi_{\mathrm{f}} &= a_{\lambda'}^{(+)}(\boldsymbol{k}')a_{r'}^{(+)\dagger}(\boldsymbol{p}')\,\Phi_0\,,\end{aligned} \tag{1.1.28}$$

$T$ is the chronological ordering operator, $a_\lambda^{(+)}(\boldsymbol{k})$ and $a_r^{(+)\dagger}(\boldsymbol{p})$ are the creation operators of photons and electrons and $\lambda$ and $r$ are the photon and electron spin projections on the three momenta of particles.

By substituting (1.1.27) and (1.1.28) into (1.1.26) we obtain an expression for the matrix element which is the vacuum expectation value of the product of the field operators. Making use of Wick's theorem, we find from this expression

$$S_{\mathrm{fi}}^{(2)} = (\mathrm{i}e)^2\bar{v}_{r'}^{(+)}(\boldsymbol{p}')\varepsilon_\mu^{\lambda'}\gamma_\mu\frac{\not{f}_1 + m}{f_1^2 - m^2}\varepsilon_\nu^\lambda\gamma_\nu v_r^{(-)}(\boldsymbol{p})(2\pi)^8\delta(p + k - p' - k')\,. \tag{1.1.29}$$

(The cross-term has been omitted.) Here $v_r^{(-)}(\boldsymbol{p})$ and $\bar{v}_{r'}^{(+)}(\boldsymbol{p}')$ are the electron wave functions in the initial and the final states, respectively, $\varepsilon_\mu^\lambda$ is the photon wave function, and $\gamma_\mu$ are the Dirac matrices;

$$f_1 = p + k, \not{f}_1 = f_{1\mu}\gamma_\mu\,.$$

iv) Analytic expressions for the matrix elements can be presented in a graphical form. For that, correspondence rules have to be established between the analytic expressions and the graphs. The graph corresponding to the matrix

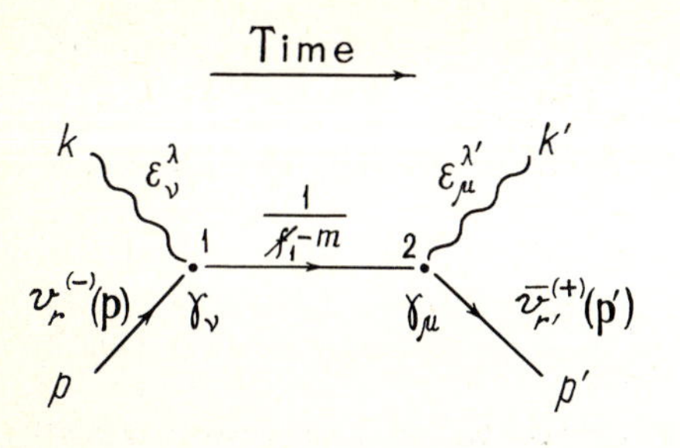


**Fig. 1.1.** Diagram for the Compton scattering on the electron (without the cross term)

element (1.1.29) is shown in Fig. 1.1. The incident electron is described by the line entering the point and the emerging electron by the line originating from the point. The incident and the emerging photons are represented by undirected wavy lines. The motion of the electron from point 1 to point 2 (the propagator) is represented by a directed line and an electromagnetic interaction vertex by a point.

Graphical representation of the matrix elements has been proposed by Feynman, and these graphs are referred to as the *Feynman diagrams.*

The outlined form of quantum field theory is known as the *operator form.* The path integral formulation of quantum field theory will be considered below (Chaps. 4 and 5).

## 1.2 Global Groups of Internal Symmetry. Unitary Symmetry

### 1.2.1 Internal Symmetry Properties

So far we have considered such properties of physical systems which are invariant under transformations of the four-dimensional space-time. Besides, there exist transformations leaving the space-time coordinates unchanged, $x \to x' = x$, but changing only the wave functions, $\Phi_a(x) \to \Phi'_a(x)$. Such transformations are related to the internal properties of fields and corresponding elementary particles and are referred to as *internal transformations.* The properties of physical systems that are invariant under the internal transformations are called the *internal properties.* The latters are described by symmetry groups. We shall consider in more detail the unitary groups.

#### The Unitary Group $SU_n$

Let us consider transformations of the wave function components by quadratic complex $n \times n$ matrices, $V$ satisfying the condition

$$VV^{\dagger} = 1 . \tag{1.2.1}$$

Such transformations form a group referred to as a *unitary* group.

It follows from the unitarity of a matrix that the absolute value of its determinant is equal to unity since

$$\det VV^{\dagger} = \det V \det V^{\dagger} = |\det V|^2 = 1 ,$$

yielding

$$|\det V| = 1 . \tag{1.2.2}$$

By expressing $V$ in the form $V = U\exp(\mathrm{i}\alpha)$, where $U$ is a unitary matrix whose determinant is equal to unity (referred to as a unimodular matrix), we get, instead of (1.2.1),

$$UU^{\dagger} = I, \ \det U = 1 . \tag{1.2.3}$$

Consequently, one unitary transformation can be subdivided into two. The first one consists of multiplying by $\exp(\mathrm{i}\alpha)$ and forms the group of phase transformations. The second one is carried out by the matrix $U$. The set of such transformations gives rise to the group $SU_n$.

Let us count the number of independent real parameters of the group $SU_n$. The group transformation is given by a matrix $a_{ik}$ with $n^2$ complex or $2n^2$ real numbers. The requirement of unitarity,

$$\sum_k a^{\dagger}_{ik} a_{kj} = \delta_{ij} ,$$

places $n^2$ conditions, and the requirement of unimodularity, $\det a_{ik} = 1$, means one more condition. Thus, a unimodular matrix contains $n^2 - 1$ independent real parameters. Consequently, the Lie algebra of the group $SU_n$ consists of $n^2 - 1$ generators.

## The $U_1$-Symmetry

Consider the group of *phase transformations* $U_1$:

$$u_i(x) \to u_i'(x) = \mathrm{e}^{-\mathrm{i}g\varepsilon} u_i(x) , \qquad u_i^*(x) \to u_i^{*\prime}(x) = \mathrm{e}^{\mathrm{i}g\varepsilon} u_i^*(x) . \tag{1.2.4}$$

Infinitesimal ($\varepsilon \ll 1$) phase transformations are expressed as follows:

$$u_i(x) \to u_i'(x) = u_i(x) - \mathrm{i}g\varepsilon u_i(x) , \qquad u_i^*(x) \to u_i^{*\prime}(x) = u_i^*(x) + \mathrm{i}g\varepsilon u_i^*(x) . \tag{1.2.5}$$

The group $U_1$ is one-parametric and Abelian.

The previously introduced relativistic Lagrangians (1.1.11, 13) are invariant also with respect to the group of phase transformations.

## The $SU_2$-Symmetry

The group $SU_2$ describes the *isotopic* properties of particles. This group is closely related (homomorphic) to the group of rotations in the three-dimensional isotopic space. The group $SU_2$ is three-parametric: its three generators $M_i$ form the following Lie algebra

$$[M_i, M_k]_- = \varepsilon_{ikn} M_n \,. \tag{1.2.6}$$

As seen, the group $SU_2$ is non-Abelian, because its generators are non-commuting.

The group $SU_2$ has tensor as well as spinor representations. The simplest ones are

i) a scalar representation: a unit matrix acting on a one-component wave function; it describes one particle (isotopic singlet);

ii) an isovector representation: three-row matrices obeying the relations (1.2.6); these act on the three-component wave function

$$u^a(x) = \begin{pmatrix} u^1(x) \\ u^2(x) \\ u^3(x) \end{pmatrix}, \qquad a = 1, 2, 3,$$

which describes an isotopic triplet of particles. The generators can be chosen, e.g., in the form $\omega_i/\mathrm{i}$, where

$$\omega_1 = \begin{pmatrix} 0 & 0 & 0 \\ 0 & 0 & -\mathrm{i} \\ 0 & \mathrm{i} & 0 \end{pmatrix}, \quad \omega_2 = \begin{pmatrix} 0 & 0 & \mathrm{i} \\ 0 & 0 & 0 \\ -\mathrm{i} & 0 & 0 \end{pmatrix}, \quad \omega_3 = \begin{pmatrix} 0 & -\mathrm{i} & 0 \\ \mathrm{i} & 0 & 0 \\ 0 & 0 & 0 \end{pmatrix}; \tag{1.2.7}$$

iii) an isodoublet representation: two-row matrices acting on the two-component wave function

$$\psi^a(x) = \begin{pmatrix} \psi^1(x) \\ \psi^2(x) \end{pmatrix}, \qquad a = 1, 2$$

describing an isotopic doublet of particles; as the generators obeying (1.2.6) the matrices $\tau_i/2i$ are usually chosen, $\tau_i$ denoting the *Pauli matrices*

$$\tau_1 = \begin{pmatrix} 0 & 1 \\ 1 & 0 \end{pmatrix}, \quad \tau_2 = \begin{pmatrix} 0 & -\mathrm{i} \\ \mathrm{i} & 0 \end{pmatrix}, \quad \tau_3 = \begin{pmatrix} 1 & 0 \\ 0 & -1 \end{pmatrix}. \tag{1.2.7'}$$

The transformations of the simplest isotopic wave functions under infinitesimal rotations by the angles $\varepsilon_n$ in the isotopic space are written as follows:

i) for an isoscalar

$$\phi(x) \to \phi'(x) = \phi(x) \,, \tag{1.2.8}$$

ii) for an isovector

$$u_\mu^a(x) \to u_\mu'^a(x) = (\delta_{ab} - \mathrm{i}\varepsilon_n \omega_n^{ab}) u_\mu^b(x) + O(\varepsilon^2) \,, \tag{1.2.9}$$

iii) for an isodoublet

$$\psi^a(x) \to \psi^{a'}(x) = \left(\delta_{ab} - \frac{\mathrm{i}}{2}\varepsilon_n \tau_n^{ab}\right)\psi^b(x) + O(\varepsilon^2)\,. \tag{1.2.10}$$

The Lagrangian must be invariant under the Lorentz group as well as under the group $SU_2$, i.e., only relativistic and isotopic scalars can enter the Lagrangian. This is the case, e.g., for the following Lagrangians:

$$L(\phi^a, \partial_\mu \phi^a) = (\partial_\mu \phi^{a*}(x))(\partial_\mu \phi^a(x)) - m^2 \phi^{a*}(x)\phi^a(x)\,, \tag{1.2.11}$$

$$L(\psi^a, \partial_\mu \psi^a) = \mathrm{i}\,\bar{\psi}^a(x)\gamma_\mu \partial_\mu \psi^a(x) - M\bar{\psi}^a(x)\psi^a(x)\,, \tag{1.2.12}$$

whose invariance under the group $SU_2$ can be explicitly proven with the aid of (1.2.8 – 10).

The index $a$ characterizes the dimensionality of the isotopic multiplet: $a = 1$ corresponds to a singlet, $a = 1, 2$ to a doublet, $a = 1, 2, 3$ to a triplet, etc. Set $a = 1, 2$. The Lagrangian (1.2.11) then describes a scalar field with respect to the Lorentz group and an isodoublet field with respect to the group $SU_2$ while the Lagrangian (1.2.12) describes a spinor field with respect to the Lorentz group and an isodoublet field with respect to the group $SU_2$.

### The $SU_3$-Symmetry

The group $SU_3$ describes the *unitary* properties of particles. This group is eight-parametric. Usually the generators of its fundamental representation are chosen in the form $\lambda_k/2\mathrm{i}$, where $\lambda_k$ denote the *Gell-Mann matrices*:

$$\lambda_1 = \begin{pmatrix} 0 & 1 & 0 \\ 1 & 0 & 0 \\ 0 & 0 & 0 \end{pmatrix}, \quad \lambda_2 = \begin{pmatrix} 0 & -\mathrm{i} & 0 \\ \mathrm{i} & 0 & 0 \\ 0 & 0 & 0 \end{pmatrix}, \quad \lambda_3 = \begin{pmatrix} 1 & 0 & 0 \\ 0 & -1 & 0 \\ 0 & 0 & 0 \end{pmatrix},$$

$$\lambda_4 = \begin{pmatrix} 0 & 0 & 1 \\ 0 & 0 & 0 \\ 1 & 0 & 0 \end{pmatrix}, \quad \lambda_5 = \begin{pmatrix} 0 & 0 & -\mathrm{i} \\ 0 & 0 & 0 \\ \mathrm{i} & 0 & 0 \end{pmatrix}, \quad \lambda_6 = \begin{pmatrix} 0 & 0 & 0 \\ 0 & 0 & 1 \\ 0 & 1 & 0 \end{pmatrix},$$

$$\lambda_7 = \begin{pmatrix} 0 & 0 & 0 \\ 0 & 0 & -\mathrm{i} \\ 0 & \mathrm{i} & 0 \end{pmatrix}, \quad \lambda_8 = \frac{1}{\sqrt{3}}\begin{pmatrix} 1 & 0 & 0 \\ 0 & 1 & 0 \\ 0 & 0 & -2 \end{pmatrix}. \tag{1.2.13}$$

The matrices $\lambda_k$ obey the following commutation relations (the Lie algebra):

$$[\lambda_k, \lambda_j]_- = 2\mathrm{i} f_{kjn}\lambda_n, \quad \mathrm{Tr}\{\lambda_k \lambda_j\} = 2\delta_{kj}\,. \tag{1.2.14}$$

The constants $f_{kjn}$ are totally antisymmetric with respect to interchange of indices. The independent constants which are different from zero have the values

$$f_{123}=1, f_{147}=f_{246}=f_{345}=f_{516}=f_{257}=f_{637}=\frac{1}{2}, f_{458}=f_{678}=\frac{\sqrt{3}}{2}.$$

The rank of the group $SU_3$ is equal to 2.

The simplest irreducible representations of the group $SU_3$ are
i) the scalar representation: a unit matrix acting on the one-component function describing a unitary singlet;
ii) the triplet contravariant representation: three-row matrices obeying (1.2.14) and acting on the three-component function

$$\psi^a(x)=\begin{pmatrix}\psi^1(x)\\ \psi^2(x)\\ \psi^3(x)\end{pmatrix},$$

which describes a unitary triplet of particles. These matrices are usually chosen in the form of (1.2.13);
iii) the triplet covariant representation: three-row matrices of the representation conjugate to (1.2.13); these matrices act on the three-component function

$$\psi_a(x)=(\psi_1(x),\ \psi_2(x),\ \psi_3(x))$$

which describes a unitary triplet of anti-particles.

The infinitesimal transformations of unitary wave functions look as follows:
i) for a scalar

$$\phi(x)\to\phi'(x)=\phi(x), \tag{1.2.15}$$

ii) for a contravariant triplet

$$\psi^a(x)\to\psi'^a(x)=\left(1-\frac{\mathrm{i}}{2}\varepsilon_n\lambda_n^{ab}\right)\psi^b(x)+O(\varepsilon^2), \tag{1.2.16}$$

iii) for a covariant triplet

$$\psi_a(x)\to\psi'_a(x)=\left(1+\frac{\mathrm{i}}{2}\varepsilon_n(\lambda_n^T)^{ab}\right)\psi_b(x)+O(\varepsilon^2). \tag{1.2.17}$$

Examples of Lagrangians invariant under both the Lorentz group and the group $SU_3$ are given by

$$L(\phi^a,\partial_\mu\phi^a)=[\partial_\mu\phi^{a*}(x)][\partial_\mu\phi^a(x)]-m^2\phi^{a*}(x)\phi^a(x),\quad a=1,2,3, \tag{1.2.18}$$

$$L(\psi^a,\partial_\mu\psi^a)=\mathrm{i}\bar\psi^a(x)\gamma_\mu\partial_\mu\psi^a(x)-M\bar\psi^a(x)\psi^a(x). \tag{1.2.19}$$

The invariance of these Lagrangians under the group $SU_3$ can be proven with the help of relations (1.2.15 – 17). The index $a$ characterizes the dimensionality of the unitary multiplet: $a=1$ corresponds to a singlet, $a=1, 2, 3$ to a triplet, etc. Take $a = 1, 2, 3$. The Lagrangian (1.2.18) then describes a scalar field with respect to the Lorentz group and a triplet field with respect to the group $SU_3$ while the Lagrangian (1.2.19) describes a spinor field with respect to the Lorentz group and a triplet field with respect to the group $SU_3$.

### 1.2.2 Condition for the Invariance of the Lagrangian

Let us formulate in a general form the condition for the invariance of the Lagrangian under the transformations of an arbitrary global group $G$ of internal symmetry. In the case of groups of internal symmetry, only the wave functions (or fields) undergo the transformations; consequently,

$$x_\mu \to x'_\mu = x_\mu . \tag{1.2.20}$$

The group being given, one has for the infinitesimal transformations of the wave functions $u_i(x)$

$$u_i(x) \to u'_i(x) = u_i(x) + \delta u_i(x) , \tag{1.2.21}$$

$$\delta u_i(x) = T^k_{ij} \varepsilon_k u_j(x) . \tag{1.2.22}$$

Here $T^k_{ij}$ and $\varepsilon_k$ are the generators and the infinitesimal parameters of the group, respectively.

The Lie algebra of the group $G$ is given by the relation

$$[T_i, T_k]_- = f_{ikl} T_l . \tag{1.2.23}$$

The constants $f_{ikl}$ are called the *structure constants.* They have the following properties

$$f_{ikl} f_{lmn} + f_{kml} f_{lin} + f_{mil} f_{lkn} = 0; \quad f_{ikl} = -f_{kil} . \tag{1.2.24}$$

The invariance of the Lagrangian under the transformations of the group $G$ implies that

$$\int_\Omega dx\, L(u'_i(x), \partial_\mu u'_i(x)) - \int_\Omega dx\, L(u_i(x), \partial_\mu u_i(x)) = 0 , \tag{1.2.25}$$

where $\Omega$ is the region of integration. For infinitesimal transformations, (1.2.25) can be rewritten as

$$\delta I = 0 . \tag{1.2.26}$$

Thus, the invariance of action under the transformations of the group $G$ means that the *variation of action* must vanish under these transformations.

Finding an explicit expression for the variation of action we have

$$\delta I = \int dx \left[ \frac{\partial L}{\partial u_i} \delta u_i + \frac{\partial L}{\partial(\partial_\mu u_i)} \delta(\partial_\mu u_i) \right] = 0 .$$

Due to the arbitrariness of the region of integration, we obtain from this

$$\frac{\partial L}{\partial u_i}\delta u_i + \frac{\partial L}{\partial(\partial_\mu u_i)}\delta(\partial_\mu u_i) = 0\,, \tag{1.2.27}$$

or, taking into account (1.2.22) and the arbitrariness of the quantities $\varepsilon_k$,

$$\frac{\partial L}{\partial u_i} T^k_{ij} u_j + \frac{\partial L}{\partial(\partial_\mu u_i)} T^k_{ij}\partial_\mu u_j = 0\,. \tag{1.2.28}$$

These identities express the necessary and sufficient *conditions for the Lagrangian to be invariant* under the transformations of an arbitrary global group $G$ of internal symmetry.

### 1.2.3 Classification of Groups

A Lie group is essentially uniquely determined by its Lie algebra. The Lie algebra can be defined by a basis $X_i$ $(i = 1, \ldots, d)$ which obeys the commutation relations,

$$[X_i, X_j]_- = c_{ijk} X_k\,,$$

where the structure constants $c_{ijk}$ obey (1.2.24). One class of Lie groups related to the set of the so called *simple* Lie algebras is of special interest in physics. This class consists of only four series [(i – iv) below] which are called classical groups, and of additional five exceptional groups (v):

i) unitary groups $SU_n$, $n \geqslant 2$;
ii) orthogonal groups of odd dimension $O_{2n+1}$, $n \geqslant 2$;
iii) symplectic groups $Sp_{2n}$, $n \geqslant 3$;
iv) orthogonal groups of even dimension $O_{2n}$, $n \geqslant 4$;
v) exceptional groups $E_6$, $E_7$, $E_8$, $F_4$, $G_2$.

In the case of simple (or semi-simple) compact groups, the structure constants $f_{ijk}$ are totally antisymmetric:

$$f_{ijk} = -f_{jik} = -f_{ikj}\,. \tag{1.2.29}$$

We have already considered some unitary groups.

Each simple Lie group has a definite set of multiplet dimensions characteristic of this group (for example, the Lorentz group has multiplets of any dimension, while the group $SU_3$ has multiplets only of certain dimensions: 1, 3, 6, 8, 10, etc.).

# 2. Local (Gauge) Invariance

In the preceding chapter we have considered the groups of global transformations and the Lagrangian invariant under these groups. A *globally* invariant Lagrangian can, however, be non-invariant under a certain group of local transformations. To obtain a *locally* invariant Lagrangian, new fields have to be introduced. These are called *gauge fields.*

Both space-time groups and internal symmetry groups can be local. In this book we confine ourselves to the case when the space-time symmetry group (the Lorentz group, the Poincaré group) is global while the internal symmetry groups are localized.

In this chapter we will show in a general form how a locally invariant, under an internal symmetry group, Lagrangian can be obtained in a simple way from the corresponding globally invariant Lagrangian. Furthermore, the major properties of the gauge fields will be clarified. It turns out that for solving these problems the requirement of invariance under the group of local transformations is sufficient. The general results will be illustrated by considering two examples of specific local groups.

## 2.1 Locally (Gauge-) Invariant Lagrangians

### 2.1.1 The Group of Local Transformations

A group of global transformations is characterized by the parameter $\varepsilon$ in (1.2.22) being independent of the coordinate. Suppose now that the parameters of the group are coordinate dependent. The functions of the field then transform according to

$$\delta u_i(x) = T^k_{ij}\varepsilon_k(x) u_j(x) \,. \tag{2.1.1}$$

The group of such local transformations is called the *local* or *gauge* group.

### 2.1.2 Gauge Fields

It can be easily proven that a globally invariant Lagrangian may be non-invariant under the group of local transformations (2.1.1). Indeed, by taking into account the relations

$$\delta(\partial_\mu u_i(x)) = T^k_{ij}\varepsilon_k(x)\,\partial_\mu u_j(x) + T^k_{ij}u_j(x)\,\partial_\mu\varepsilon_k(x)$$

that follow from (2.1.1), one gets for the variation of the Lagrangian

$$\delta L = \frac{\partial L}{\partial u_i}\,\delta u_i(x) + \frac{\partial L}{\partial(\partial_\mu u_i)}\,\delta(\partial_\mu u_i(x)) = \frac{\partial L}{\partial u_i}\,T^k_{ij}\varepsilon_k(x)\,u_j(x)$$

$$+\frac{\partial L}{\partial(\partial_\mu u_i)}\,T^k_{ij}\varepsilon_k(x)\,\partial_\mu u_j(x) + \frac{\partial L}{\partial(\partial_\mu u_i)}\,T^k_{ij}u_j(x)\,\partial_\mu\varepsilon_k(x)\,. \qquad (2.1.2)$$

Due to the global invariance the relations (1.2.28) are satisfied and the sum of the two first terms on the right-hand side of (2.1.2) is zero, so that it reads

$$\delta L = \frac{\partial L}{\partial(\partial_\mu u_i)}\,T^k_{ij}u_j\partial_\mu\varepsilon_k \neq 0\,. \qquad (2.1.3)$$

Thus, the variation of $L(x)$ does not vanish, i.e., $L(x)$ is not invariant under the local transformations (2.1.1). To achieve the invariance of the Lagrangian under the transformations (2.1.1), a new field

$$A^l_l(x)\,, \quad l = 1, 2, \ldots M$$

can be added to the fields $u_i(x)$ to compensate the right-hand side of (2.1.3) and to result in a new Lagrangian, invariant under the transformations (2.1.1). The fields thus introduced are called *compensating* fields; in the modern usage they are referred to as *gauge* fields.

Let us demonstrate how, by using only the requirement that the Lagrangian be invariant under an arbitrary local group of internal symmetry, the following questions can be answered:

i) How do the gauge fields $A^l_l(x)$ transform under the Lorentz group and under the gauge group $G$, i.e., what is the meaning of the index $l$ [cf. (2.1.10)]?
ii) What is the form of the interaction between the fields $u_i(x)$ and the gauge fields [cf. (2.1.12)]?
iii) How can the locally invariant Lagrangian be deduced from the globally invariant Lagrangian [cf. (2.1.13) and (2.2.7)]?
iv) What is the explicit form of the transformation of the gauge fields under the gauge group $G$ [cf. (2.2.7)]?
v) What is the invariant Lagrangian for the gauge fields (2.2.21 – 23)?

### 2.1.3 Conditions for the Local Invariance of the Lagrangian

Suppose that the new Lagrangian $\mathscr{L}(x)$ contains only the gauge fields $A^l_l(x)$ and not their derivatives: $\mathscr{L} = \mathscr{L}(u_i, \partial_\mu u_i, A^l_l)$. The infinitesimal field transformations take the form

$$\delta u_i(x) = T^k_{ij}\varepsilon_k(x)\,u_j(x)\,, \qquad (2.1.4)$$

$$\delta A^l_l(x) = P^k_{li}A^l_i(x)\,\varepsilon_k(x) + R^k_{l\mu}\partial_\mu\varepsilon_k(x)\,. \qquad (2.1.5)$$

Here $P$ and $R$ are some unknown constant matrices to be found later. The second term in (2.1.5) has been introduced to compensate the right-hand side of (2.1.3).

The condition of *local invariance* of the Lagrangian $\mathscr{L}(u_i, \partial_\mu u_i, A_l')$ reads

$$\delta\mathscr{L} = \frac{\partial\mathscr{L}}{\partial u_i}\delta u_i + \frac{\partial\mathscr{L}}{\partial(\partial_\mu u_i)}\delta(\partial_\mu u_i) + \frac{\partial\mathscr{L}}{\partial A_l'}\delta A_l' = 0$$

or, after substituting (2.1.4, 5),

$$\left(\frac{\partial\mathscr{L}}{\partial u_i}T_{ij}^k u_j + \frac{\partial\mathscr{L}}{\partial(\partial_\mu u_i)}T_{ij}^k\partial_\mu u_j + \frac{\partial\mathscr{L}}{\partial A_l'}P_{lm}^k A_m'\right)\varepsilon_k(x)$$
$$+\left(\frac{\partial\mathscr{L}}{\partial(\partial_\mu u_i)}T_{ij}^k u_j + \frac{\partial\mathscr{L}}{\partial A_l'}R_{l\mu}^k\right)\partial_\mu\varepsilon_k(x) = 0\,. \quad (2.1.6)$$

The arbitrariness of $\varepsilon_k(x)$ and $\partial_\mu\varepsilon_k(x)$ implies that their coefficients in (2.1.6) are zero:

$$\frac{\partial\mathscr{L}}{\partial u_i}T_{ij}^k u_j + \frac{\partial\mathscr{L}}{\partial(\partial_\mu u_i)}T_{ij}^k\partial_\mu u_j + \frac{\partial\mathscr{L}}{\partial A_l'}P_{lm}^k A_m' = 0\,, \quad (2.1.7)$$

$$\frac{\partial\mathscr{L}}{\partial(\partial_\mu u_i)}T_{ij}^k u_j + \frac{\partial\mathscr{L}}{\partial A_l'}R_{l\mu}^k = 0\,. \quad (2.1.8)$$

### 2.1.4 Connection Between the Globally and the Locally Invariant Lagrangians

Now we are going to demonstrate that the identity (2.1.8) makes it possible to elucidate the meaning of the index $l$ in $A_l'$ and to explicitly find the dependence of the new Lagrangian $\mathscr{L}$ on the gauge field.

i) The set of equations (2.1.8) consists of $4n$ equations, since $\mu = 0, 1, 2, 3$; $k = 1, 2 \dots n$. For the single-valuedness of the dependence of $\mathscr{L}$ on $A_l'$, the number of components $A_l'$ (where $l = 1, 2, \dots, M$) has to equal the number of equations of the set (2.1.8), i.e., $M = 4n$. Suppose also that the matrices $R$ are non-singular and that their inverse matrices exist, which are determined by

$$(R_{l\mu}^k)^{-1}R_{m\mu}^k = \delta_{lm}, \quad (R_{l\mu}^k)^{-1}R_{l\nu}^i = \delta_{ki}g_{\mu\nu}\,. \quad (2.1.9)$$

Then the gauge field can be expressed as

$$A_\mu^k = (R_{l\mu}^k)^{-1}A_l'\,. \quad (2.1.10)$$

Hence, under the Lorentz group, the gauge field $A_\mu^k(x)$ transforms as a 4-vector; the index $k$ determines the number of components with respect to the gauge group. The question of how these components transform will be clarified later [cf. (2.2.7)].

ii) With the help of (2.1.9, 10) the set of equations (2.1.8) is rewritten as

$$\frac{\partial \mathscr{L}}{\partial(\partial_\mu u_i)} T^k_{ij} u_j + \frac{\partial \mathscr{L}}{\partial A^k_\mu} = 0 \,. \tag{2.1.11}$$

This set of equations is complete and Jacobian. For the Lagrangian to satisfy this set of equations, the gauge fields $A^k_\mu$ should enter $\mathscr{L}$ as a combination

$$\nabla_\mu u_i \equiv \partial_\mu u_i - T^k_{ij} u_j A^k_\mu \,, \tag{2.1.12}$$

which is commonly called the *covariant derivative*. Hence, a local Lagrangian should have the form

$$\mathscr{L}(u_i, \partial_\mu u_i, A^l_l) = \mathscr{L}'(u_i, \nabla_\mu u_i) \equiv L(u_i, \nabla_\mu u_i) \,. \tag{2.1.13}$$

The following relations are then fulfilled:

$$\begin{aligned}
\frac{\partial \mathscr{L}}{\partial u_i} &= \left.\frac{\partial \mathscr{L}'}{\partial u_i}\right|_{\nabla u_i = \text{const}} - \left.\frac{\partial \mathscr{L}'}{\partial \nabla_\mu u_j}\right|_{u_i = \text{const}} T^k_{ji} A^k_\mu \,, \\
\frac{\partial \mathscr{L}}{\partial(\partial_\mu u_i)} &= \left.\frac{\partial \mathscr{L}'}{\partial \nabla_\mu u_i}\right|_{u_i = \text{const}} , \\
\frac{\partial \mathscr{L}}{\partial A^l_l} &= - \left.\frac{\partial \mathscr{L}'}{\partial \nabla_\mu u_i}\right|_{u_i = \text{const}} T^k_{ij} u_j (R^k_{\mu l})^{-1} \,.
\end{aligned} \tag{2.1.14}$$

The requirement of the local invariance thus results in replacing the derivatives in the global Lagrangian by the covariant derivatives (2.1.12). The second term in the covariant derivative determines the interaction of the fields $u_i$, which naturally can be called the *matter fields*, with the gauge fields $A^k_\mu$. It should be emphasized that the number of gauge fields $A^k_\mu$ is equal to the number of the generators of the gauge group.

## 2.2 Gauge Fields

Consider now the main properties of the gauge fields.

### 2.2.1 Transformations of the Gauge Fields

First we concern ourselves with the transformations of the gauge fields $A^k_\mu$. Taking into account (2.1.5, 9), and (2.1.10) we get

$$\delta A^k_\mu = (C^k_\mu)^{jm}_\nu A^m_\nu \varepsilon_j(x) + \partial_\mu \varepsilon_k(x) \,, \tag{2.2.1}$$

where

$$(C^k_\mu)^{jm}_\nu = (R^k_{i\mu})^{-1}(P^j_{il})(R^m_{l\nu})$$

is an unknown matrix. To find an explicit expression for the matrix $(C^k_\mu)^{jm}_\nu$ we use the identities (2.1.7) together with (2.1.14). By substituting (2.1.14) into (2.1.7) and by making use of (2.1.12) and (2.2.1) we obtain

$$\frac{\partial \mathscr{L}'}{\partial u_i} T^k_{ij} u_j + \frac{\partial \mathscr{L}'}{\partial \nabla_\mu u_i} T^k_{ij} \nabla_\mu u_j + \frac{\partial \mathscr{L}'}{\partial \nabla_\mu u_i}$$

$$\cdot (-T^l_{ij} T^k_{jn} A^l_\mu u_n + T^k_{ij} T^l_{jn} u_n A^l_\mu - (C^l_\mu)^{km}_\nu T^l_{ij} u_j A^m_\nu) = 0 \, . \tag{2.2.2}$$

For the Lagrangian $\mathscr{L}'(u_i, \nabla_\mu u_i)$ the condition of invariance, analogous to (1.2.28), holds:

$$\frac{\partial \mathscr{L}'(u_i, \nabla_\mu u_i)}{\partial u_i} T^k_{ij} u_j + \frac{\partial \mathscr{L}'(u_i, \nabla_\mu u_i)}{\partial \nabla_\mu u_i} T^k_{ij} \nabla_\mu u_j = 0 \, . \tag{2.2.3}$$

Consequently, the first two terms in (2.2.2) vanish and the remaining ones, in view of (1.2.23), give

$$\frac{\partial \mathscr{L}'(u_i, \nabla_\mu u_i)}{\partial \nabla_\mu u_i} (f_{kml} T^l_{ij} u_j A^m_\mu - (C^l_\mu)^{km}_\nu T^l_{ij} u_j A^m_\nu) = 0 \, . \tag{2.2.4}$$

It follows that

$$f_{kml} T^l_{ij} u_j A^m_\mu - (C^l_\mu)^{km}_\nu T^l_{ij} u_j A^m_\nu = 0 \quad \text{or} \tag{2.2.5}$$

$$(C^l_\mu)^{km}_\nu = f_{kml} g_{\mu\nu} \, . \tag{2.2.6}$$

By substituting (2.2.6) into (2.2.1) we obtain the desired expression for the infinitesimal transformation of gauge fields

$$\delta A^k_\mu = A'^k_\mu - A^k_\mu = f_{lmk} A^m_\mu \varepsilon_l(x) + \partial_\mu \varepsilon_k(x) \, . \tag{2.2.7}$$

Hence, the locally invariant Lagrangian is obtained from the globally invariant Lagrangian by replacing the ordinary derivative by the covariant derivative; it should be kept in mind that the gauge field transforms according to (2.2.7).

To find the expressions for the infinitesimal transformations of the covariant derivatives, we make use of (1.2.23), (2.1.12) and (2.2.7). This yields

$$\delta(\nabla_\mu u_i) = T^k_{ij} \varepsilon_k(x) \nabla_\mu u_j \, . \tag{2.2.8}$$

A comparison of (2.1.4) and (2.2.8) shows that $\nabla_\mu u_i$ transforms under the gauge group in the same way as $u_i$, i.e., $\nabla_\mu u_i$ is a covariant quantity.

### 2.2.2 The Lagrangian for Gauge Fields

The Lagrangian (2.1.13) is made up of the free Lagrangian for the matter fields $u_i$ and the interaction Lagrangian for the matter fields with the gauge fields $A^k_\mu$.

Let us find the expression for the Lagrangian of the gauge fields which is invariant under the local group of internal symmetry. This Lagrangian depends on the gauge fields as well as on their derivatives: $\mathscr{L}_0(A^k_\mu, \partial_\mu A^k_\nu)$. Let us make use of the condition of invariance for $\mathscr{L}_0$ which is written as

$$\delta \mathscr{L}_0 = \frac{\partial \mathscr{L}_0}{\partial A^k_\mu} \delta A^k_\mu + \frac{\partial \mathscr{L}_0}{\partial(\partial_\nu A^k_\mu)} \delta(\partial_\nu A^k_\mu) = 0 \tag{2.2.9}$$

or, with the account of (2.2.7)

$$\left(\frac{\partial \mathscr{L}_0}{\partial A^k_\mu} f_{lmk} A^m_\mu + \frac{\partial \mathscr{L}_0}{\partial(\partial_\nu A^k_\mu)} f_{lmk} \partial_\nu A^m_\mu\right) \varepsilon_l(x) + \left(\frac{\partial \mathscr{L}_0}{\partial A^l_\nu} + \frac{\partial \mathscr{L}_0}{\partial(\partial_\nu A^k_\mu)} f_{lmk} A^m_\mu\right) \partial_\nu \varepsilon_l(x) + \frac{\partial \mathscr{L}_0}{\partial(\partial_\nu A^k_\mu)} \partial_\nu \partial_\mu \varepsilon_k(x) = 0\,. \tag{2.2.10}$$

Due to the arbitrariness of the functions $\varepsilon_k(x)$, (2.2.10) leads to the following identities:

$$\frac{\partial \mathscr{L}_0}{\partial A^k_\mu} f_{lmk} A^m_\mu + \frac{\partial \mathscr{L}_0}{\partial(\partial_\nu A^k_\mu)} f_{lmk} \partial_\nu A^m_\mu \equiv 0\,, \qquad l = 1, 2, \dots, n\,, \tag{2.2.11}$$

$$\frac{\partial \mathscr{L}_0}{\partial A^l_\nu} + \frac{\partial \mathscr{L}_0}{\partial(\partial_\nu A^k_\mu)} f_{lmk} A^m_\mu = 0\,, \qquad l = 1, 2, \dots, n\,, \tag{2.2.12}$$

$$\frac{\partial \mathscr{L}_0}{\partial(\partial_\mu A^k_\nu)} + \frac{\partial \mathscr{L}_0}{\partial(\partial_\nu A^k_\mu)} = 0\,, \qquad k = 1, 2, \dots, n\,. \tag{2.2.13}$$

In deriving the last identity, the relation

$$\frac{\partial \mathscr{L}_0}{\partial(\partial_\nu A^k_\mu)} \partial_\nu \partial_\mu \varepsilon_k(x) = \frac{1}{2}\left(\frac{\partial \mathscr{L}_0}{\partial(\partial_\mu A^k_\nu)} + \frac{\partial \mathscr{L}_0}{\partial(\partial_\nu A^k_\mu)}\right) \partial_\mu \partial_\nu \varepsilon_k(x) \tag{2.2.14}$$

has been used.

Let us demonstrate that the identities (2.2.11 – 13) determine the explicit form of the Lagrangian. From (2.2.13) it follows that the derivative of the field $A^k_\mu$ can only enter the Lagrangian as a combination

$$A^k_{\mu\nu} = \partial_\mu A^k_\nu - \partial_\nu A^k_\mu \tag{2.2.15}$$

in which

$$A^k_{\mu\nu} = -A^k_{\nu\mu}\,. \tag{2.2.16}$$

Taking into account that

$$\frac{\partial \mathscr{L}_0}{\partial(\partial_\nu A^k_\mu)} = -2\frac{\partial \mathscr{L}_0}{\partial A^k_{\mu\nu}}, \qquad \frac{\partial \mathscr{L}_0}{\partial(\partial_\mu A^k_\nu)} = 2\frac{\partial \mathscr{L}_0}{\partial A^k_{\mu\nu}} \tag{2.2.17}$$

and using the new variables (2.2.15) we obtain, instead of (2.2.11, 12),

$$\frac{\partial \mathscr{L}_0}{\partial A^k_\mu} f_{lmk} A^m_\mu - 2\frac{\partial \mathscr{L}_0}{\partial A^k_{\mu\nu}} f_{lmk}\,\partial_\nu A^m_\mu = 0\,, \tag{2.2.18}$$

$$\frac{\partial \mathscr{L}_0}{\partial A^l_\mu} - 2\frac{\partial \mathscr{L}_0}{\partial A^k_{\nu\mu}} f_{lmk} A^m_\nu = 0\,. \tag{2.2.19}$$

Consider the set of equations (2.2.19). Like the set of equations (2.1.8), it gets divided into four independent sets ($\mu = 0, 1, 2, 3$). At a fixed $\mu$ the set (2.2.19) consists of $n$ equations and depends on the variables $A^k_\mu$ and $A^k_{\mu\nu}$, $k = 1, \ldots, n$, $\nu = 0, 1, 2, 3$; $\nu \neq \mu$. This set is complete and Jacobian. For the Lagrangian $\mathscr{L}_0$ to satisfy the set of equations (2.2.19), the fields $A^k_\mu$ and $A^k_{\mu\nu}$ should enter $\mathscr{L}_0$ through the combination

$$F^k_{\mu\nu} = A^k_{\mu\nu} - \tfrac{1}{2} f_{lmk}(A^l_\mu A^m_\nu - A^l_\nu A^m_\mu), \qquad k = 1, \ldots, n;\ \nu \neq \mu\,, \tag{2.2.20}$$

i.e.,

$$\mathscr{L}_0(A^k_\mu, \partial_\nu A^k_\mu) = \mathscr{L}'_0(F^k_{\mu\nu})\,. \tag{2.2.21}$$

From this it follows that

$$\frac{\partial \mathscr{L}_0}{\partial(\partial_\mu A^k_\nu)} = 2\frac{\partial \mathscr{L}'_0}{\partial F^k_{\mu\nu}}, \qquad \frac{\partial \mathscr{L}_0}{\partial A^k_\mu} = 2\frac{\partial \mathscr{L}'_0}{\partial F^l_{\mu\nu}} f_{mkl} A^m_\nu\,. \tag{2.2.21'}$$

Considering finally the set of identities (2.2.18) and taking into account (1.2.24) and (2.2.21′) we arrive at one more condition imposed on the Lagrangian of gauge fields:

$$\frac{\partial \mathscr{L}'_0}{\partial F^k_{\mu\nu}} f_{lmk} F^l_{\mu\nu} = 0\,, \qquad m = 1, \ldots, n\,. \tag{2.2.22}$$

Thus, a locally invariant *Lagrangian for the gauge fields* is a function of $F^k_{\mu\nu}$ only and obeys condition (2.2.22). The choice of the Lagrangian satisfying these requirements is not unique. The simplest Lagrangian, quadratic in $F^k_{\mu\nu}$, has been proposed by Yang and Mills:

$$\mathscr{L}_{\mathrm{YM}} = -\tfrac{1}{4} F^k_{\mu\nu} F^k_{\mu\nu}\,, \qquad \text{where} \tag{2.2.23}$$

$$F^k_{\mu\nu} = \partial_\mu A^k_\nu - \partial_\nu A^k_\mu - \tfrac{1}{2} f_{lmk}(A^l_\mu A^m_\nu - A^l_\nu A^m_\mu)\,. \tag{2.2.24}$$

This Lagrangian satisfies the condition (2.2.22), taking into account (1.2.29).

By using (2.2.7) and (2.2.21) the following expression for the infinitesimal transformations of the tensor $F^k_{\mu\nu}$ is obtained:

$$\delta F^k_{\mu\nu} = f_{lmk}\varepsilon_l(x) F^m_{\mu\nu} . \tag{2.2.25}$$

*The full Lagrangian* $\mathscr{L}$ of the system of the matter fields $u_i(x)$ and of gauge fields will be given by the sum of the Lagrangian $\mathscr{L}'_0$ of the gauge fields and of the local Lagrangian $\mathscr{L}'$ of the matter fields (which contains the Lagrangian of the matter fields as well as the interaction Lagrangian between the matter and the gauge fields):

$$\mathscr{L} = \mathscr{L}'_0 + \mathscr{L}' . \tag{2.2.26}$$

### 2.2.3 Conserved Currents

According to the Noether theorem, the invariance of the Lagrangian under a group of continuous transformations implies the conservation of some quantity. The invariance of the Lagrangian under local transformations is connected with conserved currents. To obtain an expression for these currents, we start from the full Lagrangian. The condition of its invariance under the local transformation group is expressed as

$$\delta\mathscr{L} = \frac{\partial\mathscr{L}}{\partial A^k_\mu}\delta A^k_\mu + \frac{\partial\mathscr{L}}{\partial(\partial_\nu A^k_\mu)}\delta(\partial_\nu A^k_\mu) + \frac{\partial\mathscr{L}}{\partial u_i}\delta u_i + \frac{\partial\mathscr{L}}{\partial(\partial_\mu u_i)}\delta(\partial_\mu u_i) = 0 . \tag{2.2.27}$$

With the field equations for $u_i(x)$ and $A^k_\mu(x)$ in the form (1.1.14), Eq. (2.2.27) can be rewritten as

$$\delta\mathscr{L} = \partial_\mu\left(\frac{\partial\mathscr{L}}{\partial(\partial_\mu A^k_\nu)}\delta A^k_\nu + \frac{\partial\mathscr{L}}{\partial(\partial_\mu u_i)}\delta u_i\right) = 0 . \tag{2.2.28}$$

With the new variables $u_i$, $\nabla_\mu u_i$, $A^k_\mu$, and $F^k_{\mu\nu}$ and with the use of (2.1.14) and (2.2.21′) as well as (2.1.1) and (2.2.7), Eq. (2.2.28) yields

$$\begin{aligned}\delta\mathscr{L} = {} & \partial_\mu\left(\frac{\partial\mathscr{L}}{\partial\nabla_\mu u_i}T^k_{ij}u_j + 2\frac{\partial\mathscr{L}}{\partial F^m_{\mu\nu}}f_{klm}A^l_\nu\right)\varepsilon^k(x) \\ & + \left[\frac{\partial\mathscr{L}}{\partial\nabla_\mu u_i}T^k_{ij}u_j + 2\frac{\partial\mathscr{L}}{\partial F^m_{\mu\nu}}f_{klm}A^l_\nu + 2\partial_\nu\left(\frac{\partial\mathscr{L}}{\partial F^k_{\nu\mu}}\right)\right]\partial_\mu\varepsilon^k(x) \\ & + \left(\frac{\partial\mathscr{L}}{\partial F^k_{\mu\nu}} + \frac{\partial\mathscr{L}}{\partial F^k_{\nu\mu}}\right)\partial_\mu\partial_\nu\varepsilon^k(x) = 0 .\end{aligned} \tag{2.2.29}$$

Since the functions $\varepsilon^k(x)$ are arbitrary, this gives

$$\partial_\mu\left(\frac{\partial\mathscr{L}}{\partial\nabla_\mu u_i}T^k_{ij}u_j + 2\frac{\partial\mathscr{L}}{\partial F^m_{\mu\nu}}f_{klm}A^l_\nu\right) = 0 , \tag{2.2.30}$$

$$\frac{\partial \mathscr{L}}{\partial \nabla_\mu u_i} T^k_{ij} u_j + 2 \frac{\partial \mathscr{L}}{\partial F^m_{\mu\nu}} f_{klm} A^l_\nu + \frac{\partial \mathscr{L}}{\partial A^k_\mu} = 0 . \tag{2.2.31}$$

In deriving (2.2.31), it has been taken into account that, according to (2.2.21′) and (1.1.14),

$$\partial_\nu \frac{\partial \mathscr{L}}{\partial F^k_{\nu\mu}} = \frac{1}{2} \partial_\nu \frac{\partial \mathscr{L}}{\partial(\partial_\nu A^k_\mu)} = \frac{1}{2} \frac{\partial \mathscr{L}}{\partial A^k_\mu} .$$

Let the quantity

$$J^k_\mu(x) = \frac{\partial \mathscr{L}}{\partial A^k_\mu(x)} \tag{2.2.32}$$

be called the *current*. Then from (2.2.31) the expression for the current follows:

$$J^k_\mu = -\frac{\partial \mathscr{L}}{\partial \nabla_\mu u_i} T^k_{ij} u_j - 2 \frac{\partial \mathscr{L}}{\partial F^m_{\mu\nu}} f_{klm} A^l_\nu , \tag{2.2.33}$$

while (2.2.30) results in the conservation law for the current:

$$\partial_\mu J^k_\mu = 0 , \quad k = 1, 2, \ldots, n . \tag{2.2.34}$$

Let us illustrate the general results obtained above on the examples of two concrete local groups, namely $U_1$ and $SU_2$.

## 2.3 The Abelian Group $U_1$. The Electromagnetic Field

Let the Lagrangian for a spinor field $\psi$ with mass $M$ be given:

$$L = \frac{\mathrm{i}}{2} (\bar\psi \gamma_\mu \partial_\mu \psi - \partial_\mu \bar\psi \gamma_\mu \psi) - M \bar\psi \psi . \tag{2.3.1}$$

This Lagrangian is invariant under the global Abelian group of phase transformations

$$\psi \to \psi' = \mathrm{e}^{-\mathrm{i}g\varepsilon} \psi , \quad \bar\psi \to \bar\psi' = \bar\psi \mathrm{e}^{\mathrm{i}g\varepsilon} , \tag{2.3.2}$$

where $\varepsilon$ is a (constant) parameter of the group and $g$ is the coupling constant (as will be seen below).

From (2.3.2) it follows that

$$\delta\psi = -\mathrm{i}\varepsilon g \psi , \quad \delta\bar\psi = \mathrm{i}\varepsilon g \bar\psi . \tag{2.3.3}$$

By comparing (2.1.1) and (2.3.3) we find that

$$T_{11} = -\mathrm{i}g , \; T_{22} = \mathrm{i}g , \; T_{21} = T_{12} = 0 . \tag{2.3.4}$$

(The indices 1 and 2 refer to $\psi$ and $\bar{\psi}$, respectively.) Besides, for the group $U_1$, the structure constants obey the condition $f_{klm} = 0$.

Consider the local group of phase transformations. Then $\varepsilon(x)$ is a function of the coordinate $x$. The Lagrangian for a spinor field invariant under the local group of gauge transformations can be obtained from (2.3.1) by replacing the derivatives by covariant derivatives as defined by (2.1.12):

$$\partial_\mu \psi \to \nabla_\mu \psi = \partial_\mu \psi + \mathrm{i} g \psi A_\mu ,$$

$$\partial_\mu \bar{\psi} \to \nabla_\mu \bar{\psi} = \partial_\mu \bar{\psi} - \mathrm{i} g \bar{\psi} A_\mu .$$

In other words, the role of the gauge field in the case under consideration is played by the electromagnetic field $A_\mu(x)$, and the interaction Lagrangian $\mathscr{L}_\mathrm{I}$ for the spinor and gauge fields is given by

$$\mathscr{L}_\mathrm{I} = -g \bar{\psi} \gamma_\mu \psi A_\mu . \tag{2.3.5}$$

The Lagrangian for the gauge field $A_\mu$ is expressed, according to (2.2.23, 24), as follows:

$$\mathscr{L}_0 = -\tfrac{1}{4} F_{\mu\nu} F_{\mu\nu} , \quad \text{where} \tag{2.3.6}$$

$$F_{\mu\nu} = \partial_\mu A_\nu - \partial_\nu A_\mu$$

is the electromagnetic field tensor, i.e., (2.3.6) gives the Lagrangian for the free electromagnetic field.

An infinitesimal transformation of the field $A_\mu$ is defined by (2.2.7):

$$\delta A_\mu = \partial_\mu \varepsilon(x) .$$

The full locally invariant Lagrangian

$$\mathscr{L} = \frac{\mathrm{i}}{2} (\bar{\psi} \gamma_\mu \partial_\mu \psi - \partial_\mu \bar{\psi} \gamma_\mu \psi) - M \bar{\psi} \psi - \tfrac{1}{4} F_{\mu\nu} F_{\mu\nu} - g \bar{\psi} \gamma_\mu \psi A_\mu$$

coincides with the Lagrangian of quantum electrodynamics. According to (2.2.33), the conserved current

$$J_\mu = -g \bar{\psi} \gamma_\mu \psi , \tag{2.3.7}$$

is associated with it. This current coincides with the usual electromagnetic one. The constant $g$ can be identified with the electromagnetic interaction constant $e$.

## 2.4 The Non-Abelian Group $SU_2$. The Yang-Mills Field

Consider an $SU_2$ isodoublet of spinor fields

$$\psi^a = \begin{pmatrix} \psi^1 \\ \psi^2 \end{pmatrix} , \tag{2.4.1}$$

where $\psi^1$ and $\psi^2$ describe, e.g., a proton and a neutron (of mass $M$), respectively. The free Lagrangian for such a doublet is given by

$$L = \frac{\mathrm{i}}{2}(\bar{\psi}^a \gamma_\mu \partial_\mu \psi^a - \partial_\mu \bar{\psi}^a \gamma_\mu \psi^a) - M \bar{\psi}^a \psi^a . \tag{2.4.2}$$

It is invariant under the global non-Abelian group $SU_2$ under which the functions $\psi$ transform as follows:

$$\begin{aligned} \psi^a \to \psi'^a &= \left[\exp\left(-\frac{\mathrm{i}}{2} g\, \varepsilon_k \tau_k\right)\right]_{ab} \psi^b , \\ \bar{\psi}^a \to \bar{\psi}'^a &= \bar{\psi}^b \left[[\exp\left(\frac{\mathrm{i}}{2} g\, \varepsilon_k \tau_k\right)\right]_{ba} , \end{aligned} \tag{2.4.3}$$

where $\varepsilon_k$ denotes the (constant) parameters of the group, $\tau_k$ denotes the matrices defined by (1.2.7′), and $g$ is a constant.

According to (2.4.3), one has for infinitesimal transformations of the functions

$$\begin{aligned} \delta \psi^a &= -\frac{\mathrm{i}g}{2} \varepsilon_k (\tau_k)_{ab} \psi^b , \\ \delta \bar{\psi}^a &= \frac{\mathrm{i}g}{2} \varepsilon_k \bar{\psi}^b (\tau_k)_{ba} , \end{aligned} \tag{2.4.4}$$

and, consequently, for the generators of the transformations

$$T^k_{ab} = -\frac{\mathrm{i}g}{2}(\tau_k)_{ab} . \tag{2.4.5}$$

From the relation

$$[T^k, T^l]_- = -\frac{g^2}{4}[\tau_k, \tau_l]_- = g\, \varepsilon_{klm} T_m ,$$

it follows that

$$f_{klm} = g\, \varepsilon_{klm} ,$$

where $\varepsilon_{klm}$ is the totally antisymmetric tensor ($\varepsilon_{123} = 1$).

Let us turn to the group of local transformations. The Lagrangian (2.4.2) will become gauge invariant, provided that the substitution

$$\partial_\mu \psi^a \to \nabla_\mu \psi^a = \partial_\mu \psi^a + \frac{\mathrm{i}g}{2}(\tau_k)_{ab} \psi^b A^k_\mu , \quad k = 1, 2, 3 \tag{2.4.6}$$

is made, according to (2.1.12). As can be seen, in this case a triplet of vector fields $A^k_\mu$ is the gauge field (the number of the gauge fields being equal to the number of generators of the group).

The Lagrangian for the gauge fields has, according to (2.2.23, 24), the form

$$\mathscr{L}_{\text{YM}} = -\tfrac{1}{4} F^k_{\mu\nu} F^k_{\mu\nu}\,, \quad \text{where} \tag{2.4.7}$$

$$F^k_{\mu\nu} = \partial_\mu A^k_\nu - \partial_\nu A^k_\mu - \frac{g}{2}\varepsilon_{klm}(A^l_\mu A^m_\nu - A^l_\nu A^m_\mu)$$

is the tensor of the Yang-Mills field.

Equation (2.4.7) contains, besides the quadratic terms, the cubic and the quartic terms in the fields $A^k_\mu$, i.e., the Yang-Mills field is *self-interacting*.

The gauge fields $A^k_\mu$ transform, according to (2.2.7), as follows:

$$\delta A^k_\mu = g\,\varepsilon_{klm} A^m_\mu \varepsilon_l(x) + \partial_\mu \varepsilon_k(x)\,. \tag{2.4.8}$$

For the full locally invariant Lagrangian we obtain

$$\mathscr{L} = \frac{\mathrm{i}}{2}(\bar{\psi}^a \gamma_\mu \partial_\mu \psi^a - \partial_\mu \bar{\psi}^a \gamma_\mu \psi^a) - M\bar{\psi}^a \psi^a + \mathscr{L}_{\text{YM}} - \frac{g}{2}\bar{\psi}^a \gamma_\mu (\tau_k)_{ab} \psi^b A^k_\mu\,. \tag{2.4.9}$$

The constant $g$ plays the role of a coupling constant for the gauge field with the spinor field and with itself.

As follows from (2.2.33), the Lagrangian (2.4.9) implies the conservation of the current

$$J^k_\mu = -\frac{g}{2}\bar{\psi}^a \gamma_\mu (\tau_k)_{ab} \psi^b$$
$$- g\,\varepsilon_{klm} A^m_\nu \left[\partial_\mu A^l_\nu - \partial_\nu A^l_\mu - \frac{g}{2}\varepsilon_{lij}(A^i_\mu A^j_\nu - A^i_\nu A^j_\mu)\right].$$

# 3. Spontaneous Symmetry-Breaking

As we have seen in Chap. 2, the gauge-invariant theories include massless gauge fields. From the viewpoint of physical applications both *massless* and *massive* gauge fields are relevant. Just adding a mass term to the Lagrangian for the gauge field is not allowed since it would lead to the violation of the gauge-invariance of the Lagrangian. Therefore, a different approach has been proposed in which gauge fields acquire a mass by breaking of the gauge-invariance of the vacuum, while the Lagrangian of the gauge field remains gauge invariant (spontaneous symmetry-breaking). Symmetry-breaking of the vacuum may be incomplete. A part of the gauge fields then remains massless. This makes it possible to build theories including both massive and massless gauge fields, a circumstance used in unifying the *short-range* and the *long-range* interactions (e.g., the weak and the electromagnetic interactions), whose *mediators* are massive intermediate bosons and massless photons, respectively.

First we shall explain what *spontaneous symmetry-breaking* is and then consider the *mechanism* of spontaneous breaking of the global and the local invariance emphasizing their specific features. Finally, we shall dwell on the *residual symmetry.*

## 3.1 Degeneracy of the Vacuum States and Symmetry-Breaking

Consider a quantum-mechanical system. Let it be described by the Lagrangian $L$ or the Hamiltonian $H$. The system can be in various energy states $E_n$ determined by the equation

$$H\psi_n = E_n\psi_n .$$

Each state is specified by a certain value of the energy $E_n$ and by the wave function $\psi_n$. The state of minimum energy $E_0$ described by the wave function $\psi_0$ is called a vacuum one. If a single vacuum state corresponds to the value $E_0$ it is called a *non-degenerate* vacuum state, otherwise it is called *degenerate.*

Let a definite transformation group $G$ be given. The *vacuum* state is *invariant* under the group $G$ if it transforms into itself and *non-invariant* otherwise.

In the framework of local relativistic quantum field theory there exists a connection between the invariance of the vacuum state under a group of trans-

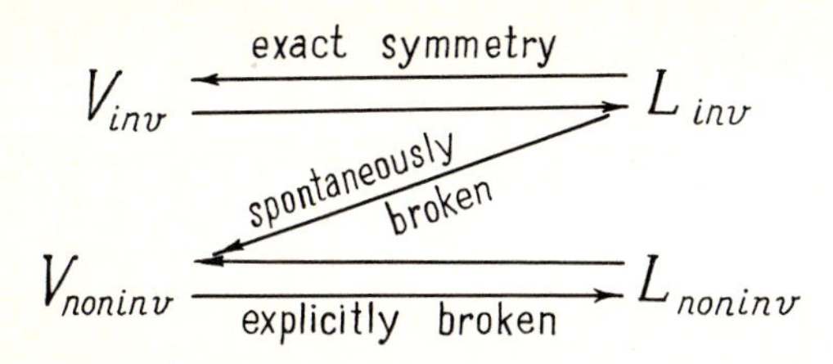


**Fig. 3.1.** Connection between the vacuum states and the Lagrangians

formations and the invariance of the Lagrangian under the same group (the Coleman theorem):

i) If the vacuum state is invariant the Lagrangian must necessarily be invariant, too ("the invariance of the vacuum state is the invariance of the Universe"). The case of the vacuum state and the Lagrangian both being invariant is referred to as that of *exact symmetry* (Fig. 3.1).

ii) If the vacuum state is non-invariant the Lagrangian may be either non-invariant or invariant. In both cases the symmetry as a whole is broken. In the case of non-invariance of both the vacuum state and the Lagrangian we speak of *explicit* symmetry-breaking (Fig. 3.1). If the vacuum state is non-invariant whereas the Lagrangian is invariant, the symmetry breaking is called *spontaneous* (Fig. 3.1).

It can be shown that the case of spontaneous symmetry-breaking necessarily leads to the occurrence of zero-mass particles. This statement is known as Goldstone's theorem. Accordingly, the massless particles are called *goldstones*. As yet, such particles have not been experimentally observed.

The non-invariance of the vacuum state gives rise to quantities which enable one to convert at least a part of the massless gauge fields into massive ones.

Let us consider several simple examples of spontaneous symmetry-breaking to get insight into the mechanism for the generation of the particle masses and to rationalize the occurrence of goldstones as well as the means of eliminating them in the case of gauge theories.

## 3.2 Spontaneous Breaking of Global Symmetry

### 3.2.1 Exact Symmetry

Consider a model described by the Lagrangian

$$L = (\partial_\mu \phi^*)(\partial_\mu \phi) - m^2 \phi^* \phi - \tfrac{1}{4} f (\phi^* \phi)^2 , \qquad (3.2.1)$$

where $\phi(x)$ is a complex scalar field, $f$ is the coupling constant of the scalar fields, $f > 0$, and $m$ is the mass of the scalar particle, $m^2 > 0$.

This Lagrangian is invariant under the global group $U_1$ of the phase transformations

$$\phi(x) \to \phi'(x) = \mathrm{e}^{-\mathrm{i}g\varepsilon} \phi(x) , \qquad \phi^*(x) \to \phi'^*(x) = \mathrm{e}^{\mathrm{i}g\varepsilon} \phi^*(x) . \qquad (3.2.2)$$

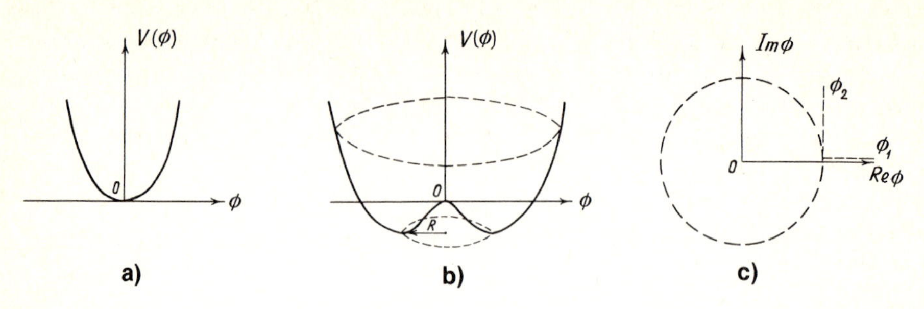


**Fig. 3.2a – c.** Vacuum states: **(a)** non-degenerate; **(b)** degenerate; **(c)** fixed

The energy of an arbitrary model with the Lagrangian $L$ is determined by (1.1.17). In the case under consideration

$$E = (\partial_0\phi^*)(\partial_0\phi) + (\partial_i\phi^*)(\partial_i\phi) + m^2\phi^*\phi + \tfrac{1}{4}f(\phi^*\phi)^2 . \tag{3.2.3}$$

The minimum, or the *vacuum* value of the energy, is defined by the conditions

$$\frac{\partial E}{\partial\phi} = m^2\phi^* + \frac{f}{2}\phi^{*2}\phi = 0 , \qquad \frac{\partial E}{\partial\phi^*} = m^2\phi + \frac{f}{2}\phi^*\phi^2 = 0 , \tag{3.2.4}$$

which show that the vacuum state of this model is associated with the point $\phi(x) = \phi^*(x) = 0$ (cf. Fig. 3.2a).

Hence, the vacuum state of the model considered is non-degenerate and invariant under the transformations (3.2.2). The Lagrangian (3.2.1) is also invariant under the transformations of the group $U_1$. The model thus has the exact $U_1$-symmetry. The vacuum state corresponds to the minimum of energy and is a stable state. The *vacuum expectation value* of the field $\phi(x)$ is zero:

$$\phi_v(x) = \phi_v^*(x) = 0 . \tag{3.2.5}$$

### 3.2.2 Spontaneous Symmetry-Breaking

Now consider a model described by the Lagrangian

$$L_1 = (\partial_\mu\phi^*)(\partial_\mu\phi) + m^2\phi^*\phi - \tfrac{1}{4}f(\phi^*\phi)^2 . \tag{3.2.6}$$

This Lagrangian differs from that given by (3.2.1) by the sign of the second term but is still invariant under the group of the global transformations (3.2.2).

1) The energy of the system with the Lagrangian $L_1$ is expressed as

$$E = (\partial_0\phi^*)(\partial_0\phi) + (\partial_i\phi^*)(\partial_i\phi) - m^2\phi^*\phi + \tfrac{1}{4}f(\phi^*\phi)^2 ,$$

and the condition of a minimum of the energy reads

$$\frac{\partial E}{\partial\phi} = -m^2\phi^* + \frac{f}{2}\phi^{*2}\phi = 0 , \qquad \frac{\partial E}{\partial\phi^*} = -m^2\phi + \frac{f}{2}\phi^*\phi^2 = 0 .$$

It follows that the energy is a minimum at $\phi^*(x)\phi(x) = 2m^2/f$ or $|\phi(x)| = \sqrt{2}m/\sqrt{f}$. The system has an infinite set of vacuum states, each of them corresponding to a point on the circumference of a circle of radius $R = \sqrt{2}m/\sqrt{f}$ on the complex plane $\phi$ (Fig. 3.2b). That is to say, the vacuum states are infinitely degenerate.

2) The transformations (3.2.2) convert a certain vacuum state (a point of the circumference) into any other vacuum state (into another point of the circumference). This means that the vacuum states are not invariant under the transformations (3.2.2).

3) However, the Lagrangian $L_1$ still remains invariant under the transformations (3.2.2). The system described by the Lagrangian $L_1$ thus has a spontaneously broken $U_1$-symmetry.

Since the vacuum states are non-invariant, the vacuum expectation value of the field $\phi_v(x)$ is not zero; it is equal to

$$|\phi_v(x)| = \sqrt{2}m/\sqrt{f}. \tag{3.2.7}$$

4) In order to construct a theory, a definite stable vacuum state, i.e., a definite point on the circumference (Fig. 3.2b) has to be chosen. One should bear in mind that different degenerate vacuum states are not related to each other and no superposition can be formed out of them (there is no such physical state). In other words, to different vacuum states correspond different worlds, and only one of them can be used.

This can be explained by the following reasoning. The probability of tunneling between the two minima decreases with increasing the number of degrees of freedom, and for an infinite number of degrees of freedom (which is the case for a field) this probability vanishes. In fact, for a field in a finite volume $\Omega$ the Lagrangian of a system is $L' \sim \int d^3x L \sim L\Omega$, the kinetic energy is $\sim \dot{\phi}^2\Omega$, and the potential energy is $\sim V\Omega$, where

$$V = |\partial\phi|^2 - m^2\phi^*\phi + \tfrac{1}{4}f(\phi^*\phi)^2.$$

The problem reduces to calculating the quantum-mechanical probability of passing a barrier of width $\sim\phi_v$ and height $\sim\Omega m^4/f$ by a particle of mass $\sim m\Omega$. This probability is proportional to $\exp(-\Omega m^3/f)$ and tends to zero as $\Omega \to \infty$, i.e., transitions between two vacuum states are not possible, indeed.

5) Due to the fact that any point on the circumference can be converted by means of transformation (3.2.2) into any other such point, all the degenerate vacuum states (all the points on the circumference) can be chosen, on equal terms, as the ground state. Commonly such a vacuum state is chosen to which there corresponds the intercept of the circumference with the real axis of the plane $\phi(x)$ (Fig. 3.2c). The coordinate system is then shifted along the real axis. In the initial coordinate system, $\phi(x)$ can be written as

$$\phi(x) = \frac{1}{\sqrt{2}}\left(\frac{2m}{\sqrt{f}} + \phi_1(x) + i\phi_2(x)\right), \tag{3.2.8}$$

where $\phi_1(x)$ and $\phi_2(x)$ are the real and the imaginary parts of the function $\phi(x)$, respectively, in the shifted coordinate system.

6) Substituting (3.2.8) into (3.2.6) we find

$$L_1(\phi) \to L_2(\phi_1, \phi_2) = \tfrac{1}{2}(\partial_\mu \phi_1)^2 - \tfrac{1}{2} m_1^2 \phi_1^2 + \tfrac{1}{2}(\partial_\mu \phi_2)^2 - \frac{f}{16}(\phi_1^4 + 2\phi_1^2\phi_2^2 + \phi_2^4) - \frac{m\sqrt{f}}{2}(\phi_1^2 + \phi_2^2)\phi_1, \tag{3.2.9}$$

where $m_1 = \sqrt{2}m$ is the mass of the particle $\phi_1(x)$.

The Lagrangian $L_2$ does not contain a term proportional to $\phi_2^2(x)$, i.e., the scalar particle described by the function $\phi_2(x)$ is massless. It emerged as a result of spontaneous symmetry-breaking (and is a goldstone).

Thus, the model with the Lagrangian $L_1$ has a spontaneously broken symmetry. As a consequence, the initial complex field got converted into a goldstone and a real scalar field of mass $\sqrt{2}m$. Hence, spontaneous breaking of global symmetry leads to the rise of goldstones. It should be stressed that the Lagrangians $L_1$ and $L_2$ are completely equivalent, both describing the dynamics of the same system.

7) A comparison of the two examples considered clearly shows the essence of spontaneous symmetry-breaking: (i) There exists a *critical* point (in the considered case $m^2 = 0$) which determines whether spontaneous symmetry-breaking occurs. (ii) In the case of negative $m^2$-term, there exists a stable non-degenerate and invariant vacuum state of the system, and $\phi_v(x) = 0$. (iii) In the case of positive $m^2$-terms, there appear other stable vacuum states of the system; these are degenerate and non-invariant; in this case $\phi_v(x) \neq 0$.

It should be noted that there is a large number of physical systems both in classical and in quantum physics, which exhibit spontaneous symmetry-breaking (for example, ferromagnets, superconductors, etc.).

8) It should also be noticed that, at first glance, a change of sign of the $m^2$-term in the Lagrangian (3.2.6) means introducing unphysical particles with imaginary mass. However, this is not the case. As a matter of fact, the term $m^2\phi^*\phi$ in the Lagrangian corresponds to the mass term only if the state $\phi = 0$ is the stable equilibrium position, i.e., corresponds to the minimum of the potential energy. In the case under consideration, the potential energy

$$V(\phi) = |\partial\phi|^2 - m^2\phi^*\phi + \tfrac{1}{4}f(\phi^*\phi)^2$$

has a local maximum, rather than a minimum, at $\phi = 0$ (Fig. 3.2b). Accordingly, the term $m^2\phi^*\phi$ in (3.2.6) is not a mass term. At $\phi = 0$ the system is unstable whereas at $|\phi| = \sqrt{2}m/\sqrt{f}$ it is stable. To determine the physical masses of the particles, one has to consider a series expansion of the potential energy in the vicinity of the true minimum. According to (3.2.8), the first term of this expansion will contain the field $\phi_1$. The mass of the field $\phi_1$, which is equal to $\sqrt{2}m$, is real, the term $\frac{1}{2}m^2\phi_1^2$ in the Lagrangian (3.2.9) is a mass term and has the usual (minus) sign. Hence, in the case of the Lagrangian (3.2.6) it is the field $\phi_1$ rather than the field $\phi$ which is a physical field.

## 3.3 Spontaneous Breaking of Local Symmetry

### 3.3.1 Exact Symmetry

Consider a model described by the Lagrangian (3.2.1). Suppose this model is also invariant under local $U_1$-transformations. The corresponding locally invariant Lagrangian can be written, with the use of the equations of Sect. 2.3, as follows:

$$\mathscr{L}_1 = -\frac{1}{4}F_{\mu\nu}F_{\mu\nu}+(\partial_\mu\phi^*-\mathrm{i}gA_\mu\phi^*)(\partial_\mu\phi+\mathrm{i}gA_\mu\phi)-m^2\phi^*\phi-\frac{f}{4}(\phi^*\phi)^2. \tag{3.3.1}$$

This Lagrangian is invariant under the transformations

$$\begin{aligned}
&\phi(x)\to\phi'(x)=\mathrm{e}^{-\mathrm{i}g\varepsilon(x)}\phi(x)\,,\\
&\phi^*(x)\to\phi^{*\prime}(x)=\mathrm{e}^{\mathrm{i}g\varepsilon(x)}\phi^*(x)\,,\\
&A_\mu(x)\to A'_\mu(x)=A_\mu(x)+\partial_\mu\varepsilon(x)\,.
\end{aligned} \tag{3.3.2}$$

A stable point $\phi_\mathrm{v}=\phi_\mathrm{v}^*=0$ corresponds to the vacuum. The model thus has an exact local $U_1$-symmetry.

### 3.3.2 Spontaneously Broken Symmetry

Consider the cases of Abelian and non-Abelian symmetry groups.

**Abelian Group**

In the case of $f>0$ and a positive $m^2$-term, a circle of radius $R$ (Fig. 3.2b) corresponds to the vacuum state, like in the case of the Lagrangian (3.2.6). This implies spontaneous symmetry-breaking. Taking as the vacuum state the point given by (3.2.7) and substituting (3.2.8) into (3.3.1) yields

$$\begin{aligned}
&\mathscr{L}_1(\phi, A_\mu)\to\mathscr{L}_2(\phi_1,\phi_2,A_\mu)\\
&=-\frac{1}{4}F_{\mu\nu}F_{\mu\nu}+\frac{2g^2m^2}{f}A_\mu A_\mu+\frac{1}{2}\partial_\mu\phi_1\partial_\mu\phi_1-m^2\phi_1^2\\
&\quad+\frac{1}{2}\partial_\mu\phi_2\partial_\mu\phi_2+\frac{2mg}{\sqrt{f}}A_\mu\partial_\mu\phi_2+\mathscr{L}_\mathrm{I}\,,
\end{aligned} \tag{3.3.3}$$

where $\mathscr{L}_\mathrm{I}$ is the interaction Lagrangian for the fields $A_\mu(x)$, $\phi_1(x)$, $\phi_2(x)$:

$$\begin{aligned}
\mathscr{L}_\mathrm{I}=&\,gA_\mu(\phi_1\partial_\mu\phi_2-\phi_2\partial_\mu\phi_1)+\frac{2g^2m}{\sqrt{f}}A_\mu^2\phi_1+\frac{g^2}{2}A_\mu^2(\phi_1^2+\phi_2^2)+\frac{m^4}{f}\\
&-\frac{1}{16}f(\phi_1^4+\phi_2^4+2\phi_1^2\phi_2^2)-\frac{1}{2}m\sqrt{f}(\phi_1^2+\phi_2^2)\phi_1\,.
\end{aligned}$$

The Lagrangian (3.3.3) is invariant under the gauge transformations obtained by substituting (3.2.8) into (3.3.2):

$$\phi_1(x) \to \phi_1'(x) = \cos[g\varepsilon(x)]\left[\phi_1(x) + \frac{2m}{\sqrt{f}}\right] + \sin[g\varepsilon(x)]\,\phi_2(x) - \frac{2m}{\sqrt{f}}$$

$$\phi_2(x) \to \phi_2'(x) = \cos[g\varepsilon(x)]\,\phi_2(x) - \sin[g\varepsilon(x)]\left[\phi_1(x) + \frac{2m}{\sqrt{f}}\right], \qquad (3.3.4)$$

$$A_\mu(x) \to A_\mu'(x) = A_\mu(x) + \partial_\mu \varepsilon(x)\,.$$

A remarkable feature of the Lagrangian (3.3.3) is that it contains a *massive* vector particle with mass $2gm/\sqrt{f}$. A direct introduction of a term proportional to $A_\mu A_\mu$ in the Lagrangian (3.3.1) is not allowed because of its non-invariance under the transformations (3.3.2). However, due to breaking the invariance of the vacuum state the vacuum expectation value given by (3.2.7) emerged. This brought about a mass term of the vector field in the Lagrangian (3.3.3) which remains invariant under the transformations (3.3.4).

Another consequence of spontaneous symmetry-breaking is, as expected, the appearance of the massless real field $\phi_2(x)$ (goldstone) in the Lagrangian.

The free Lagrangian in (3.3.3) is non-diagonal due to the term $(2mg/\sqrt{f})\,A_\mu \partial_\mu \phi_2$ which is the product of two different fields. To determine the mass spectrum, we diagonalize the free Lagrangian in (3.3.3). To that end we carry out the following field transformation

$$A_\mu(x) = B_\mu(x) - \beta\partial_\mu\phi_2(x)\,, \qquad (3.3.5)$$

where $\beta$ is an unknown parameter. It can be found by substituting (3.3.5) into (3.3.3) and equating to zero the coefficient of the term $B_\mu \partial_\mu \phi_2$. The result reads

$$A_\mu(x) = B_\mu(x) - \frac{\sqrt{f}}{2mg}\,\partial_\mu\phi_2\,.$$

With this formula, the Lagrangian (3.3.3) can be rewritten as

$$\mathcal{L}_2 \to \mathcal{L}_3 = -\frac{1}{4}(\partial_\mu B_\nu - \partial_\nu B_\mu)^2 + \frac{2g^2m^2}{f}B_\mu B_\mu + \frac{1}{2}(\partial_\mu\phi_1)(\partial_\mu\phi_1)$$
$$- m^2\phi_1^2 + \mathcal{L}_\mathrm{I}\,, \qquad (3.3.6)$$

where $\mathcal{L}_\mathrm{I}$ is the interaction Lagrangian of the fields $B_\mu(x)$, $\phi_1(x)$, $\phi_2(x)$.

Only two massive particles, namely a vector one, $B_\mu(x)$, and a real scalar one, $\phi_1(x)$, remain in the free Lagrangian (3.3.6). To eliminate the goldstones in the Lagrangian (3.3.6), one has to transform, besides the field $A_\mu(x)$, also the fields $\phi_1(x)$ and $\phi_2(x)$. This procedure is equivalent to choosing the gauge $\phi_2'(x) = 0$. After that, the Lagrangian (3.3.6) becomes gauge-non-invariant. Thus, upon spontaneous breaking of the local symmetry and choosing the gauge, only physical particles (in this case, all of them massive) remain in the Lagrangian, while the goldstones disappear.

As can be seen, the effect of spontaneous breaking of global and local symmetries manifests itself in different ways. Initially we had a complex scalar field $\phi(x)$. In the case of spontaneous breaking of the global $U_1$-symmetry, instead of the complex scalar field there appear a real scalar field $\phi_1(x)$ and a goldstone $\phi_2(x)$.

In the case of the local $U_1$-symmetry, in addition to the complex field $\phi(x)$ there is a two-component massless vector gauge field $A_\mu(x)$. In the case of spontaneous symmetry-breaking, the real scalar field $\phi_1(x)$ appears again, but the second degree of freedom is taken by the gauge field that converts into a massive vector boson, while the goldstone $\phi_2(x)$ disappears. In a way, the unphysical Goldstone boson has been exchanged for the physical state of the vector boson with longitudinal polarization. This remarkable property of gauge fields is called the *Higgs mechanism*; the scalar particles described by the fields $\phi_1(x)$ are referred to as the *Higgs bosons.*

**Non-Abelian Group**

Consider a model described by the Lagrangian

$$L=(\partial_\mu\phi^{a*})(\partial_\mu\phi^a)+m^2\phi^{a*}\phi^a-\frac{f}{4}(\phi^{a*}\phi^a)^2. \tag{3.3.7}$$

Here $\phi^a=\begin{pmatrix}\phi^1\\ \phi^2\end{pmatrix}$ is an isotopic doublet of scalar fields. The Lagrangian (3.3.7) is invariant under the global non-Abelian group $SU_2$ acting in the isotopic space. The functions $\phi(x)$ transform according to

$$\delta\phi^a=-\frac{\mathrm{i}}{2}g\varepsilon_k(\tau_k)_{ab}\phi^b,\qquad \delta\phi^{a*}=\frac{\mathrm{i}}{2}g\varepsilon_k\phi^{b*}(\tau_k)_{ba}. \tag{3.3.8}$$

Suppose the model is invariant under the local group $SU_2$. Then, with the formulae of Sect. 2.4, the locally invariant Lagrangian is written as

$$\begin{aligned}\mathscr{L}_1=&-\frac{1}{4}F^k_{\mu\nu}F^k_{\mu\nu}+\left(\partial_\mu\phi^{a*}-\frac{\mathrm{i}}{2}g\phi^{b*}(\tau_k)_{ba}A^k_\mu\right)\\ &\times\left[\partial_\mu\phi^a+\frac{\mathrm{i}}{2}g(\tau_k)_{ab}\phi^bA^k_\mu\right]+m^2\phi^{a*}\phi^a-\frac{f}{4}(\phi^{a*}\phi^a)^2.\end{aligned} \tag{3.3.9}$$

The system has an infinite number of vacuum states. Take the vacuum expectation value of the function $\phi^a(x)$ in the form

$$\phi^a_{\mathrm{v}}(x)=\frac{\sqrt{2}m}{\sqrt{f}}\begin{pmatrix}0\\1\end{pmatrix} \tag{3.3.10}$$

and introduce new scalar fields $\sigma(x)$ and $\theta^k(x)$ (where $k=1, 2, 3$) in such a way that

$$\phi^a = \frac{1}{\sqrt{2}}\left(\frac{2m}{\sqrt{f}} + \sigma + \mathrm{i}\tau_k\theta^k\right)\begin{pmatrix}0\\1\end{pmatrix},$$
$$\phi^{a*} = \frac{1}{\sqrt{2}}(0, 1)\left(\frac{2m}{\sqrt{f}} + \sigma - \mathrm{i}\theta^k\tau_k\right). \tag{3.3.11}$$

Substituting these expressions into (3.3.9) gives a Lagrangian $\mathscr{L}_2$ which contains a triplet of the Goldstone particles $\theta^k(x)$:

$$\mathscr{L}_2 = -\frac{1}{4}(\partial_\mu A_\nu^k - \partial_\nu A_\mu^k)^2 + \frac{g^2m^2}{2f}A_\mu^k A_\mu^k + \frac{1}{2}\partial_\mu\sigma\partial_\mu\sigma$$
$$-m^2\sigma^2 + \frac{1}{2}\partial_\mu\theta^k\partial_\mu\theta^k + \frac{gm}{\sqrt{f}}\partial_\mu\theta^k A_\mu^k + \mathscr{L}_\mathrm{I},$$

where $\mathscr{L}_\mathrm{I}$ is the interaction Lagrangian for the fields $A_\mu^k(x)$, $\sigma(x)$, $\theta^k(x)$.

By fixing the gauge $\theta'^k(x)=0$ for which the Goldstone particles go away we obtain the final expression for the Lagrangian:

$$\mathscr{L}_3 = -\frac{1}{4}(\partial_\mu A_\nu^k - \partial_\nu A_\mu^k)^2 + \frac{g^2m^2}{2f}A_\mu^k A_\mu^k + \frac{1}{2}\partial_\mu\sigma\partial_\mu\sigma - m^2\sigma^2 + \mathscr{L}_\mathrm{I}. \tag{3.3.12}$$

This Lagrangian describes a model containing a triplet of real vector fields $A_\mu^k(x)$ with the same mass $gm/\sqrt{f}$ and a real scalar (neutral) field $\sigma(x)$ with mass $\sqrt{2}m$. As a result, the three states associated with the goldstones are transformed into the three longitudinal components of massive vector fields so that one massive neutral scalar field remains.

It should be noted that $\mathscr{L}_1$, $\mathscr{L}_2$, and $\mathscr{L}_3$ are equivalent Lagrangians describing the same physical system. The Lagrangian $\mathscr{L}_1$ is invariant under an ordinary gauge transformation, $\mathscr{L}_2$ is invariant under somewhat more complicated transformations (e.g. (3.3.4), in the Abelian case), whereas $\mathscr{L}_3$ is non-invariant under gauge transformations and is derived from $\mathscr{L}_2$ provided that in the latter a special gauge is chosen.

## 3.4 Residual Symmetry

The Lagrangian (3.3.7) is invariant under the global group $SU_2$ as well as under the global group $U_1$, that is to say, it is invariant under the global group $SU_2 \times U_1$ which transforms the functions $\phi^a(x)$ as follows

$$\phi^a \to (\phi^a)' = \exp\left(-\frac{\mathrm{i}}{2}g\tau_k\varepsilon_k - \frac{\mathrm{i}}{2}g_1\varepsilon_4\right)\phi^a, \tag{3.4.1}$$

$$\phi^{a*} \to (\phi^{a*})' = \exp\left(\frac{i}{2} g \tau_k \varepsilon_k + \frac{i}{2} g_1 \varepsilon_4\right) \phi^{a*}. \tag{3.4.1}$$

Let us localize the group $SU_2 \times U_1$. The Lagrangian invariants under this group can be obtained by making use of the formulae of Sects. 2.3 and 2.4, yielding

$$\mathscr{L}_1 = -\frac{1}{4} F^k_{\mu\nu} F^k_{\mu\nu} - \frac{1}{4} F_{\mu\nu} F_{\mu\nu} + \left(\partial_\mu \phi^{a*} - \frac{i}{2} g \phi^{b*} (\tau_k)_{ba} A^k_\mu - \frac{i}{2} g_1 \phi^{a*} A_\mu\right)$$
$$\times \left[\partial_\mu \phi^a + \frac{i}{2} g (\tau_k)_{ab} \phi^b A^k_\mu + \frac{i}{2} g_1 \phi^a A_\mu\right] + m^2 \phi^{a*} \phi^a - \frac{f}{4} (\phi^{a*} \phi^a)^2 . \tag{3.4.2}$$

Let us fix the vacuum expectation value of the function $\phi^a(x)$ in the form (3.3.10) and introduce new fields $\sigma(x)$ and $\theta^k(x)$ related to the fields $\phi^a(x)$ through (3.3.11). The substitution of (3.3.11) into (3.4.2) leads to the Lagrangian $\mathscr{L}_2$ containing a triplet of the goldstones $\theta^k(x)$. Choosing now the gauge $\theta'^k(x) = 0$ in which the goldstones disappear we get the Lagrangian

$$\mathscr{L}_3 = -\frac{1}{4} F^k_{\mu\nu} F^k_{\mu\nu} - \frac{1}{4} F_{\mu\nu} F_{\mu\nu} + \frac{g^2 m^2}{2f} A^k_\mu A^k_\mu + \frac{g_1^2 m^2}{2f} A_\mu A_\mu$$
$$- \frac{g g_1 m^2}{f} A_\mu A^3_\mu + \frac{1}{2} \partial_\mu \sigma \partial_\mu \sigma - m^2 \sigma^2 + \mathscr{L}_{\mathrm{I}} . \tag{3.4.3}$$

The free Lagrangian $\mathscr{L}_0$ in (3.4.3) contains a mixed term $A_\mu A^3_\mu$, and thus is non-diagonal in the fields $A_\mu(x)$ and $A^3_\mu(x)$. To diagonalize $\mathscr{L}_0$ in (3.4.3) we introduce the new fields

$$W_\mu = \frac{1}{\sqrt{2}} (A^1_\mu + i A^2_\mu) , \qquad W^*_\mu = \frac{1}{\sqrt{2}} (A^1_\mu - i A^2_\mu) , \tag{3.4.4}$$

$$Z_\mu = A^3_\mu \cos\theta - A_\mu \sin\theta , \qquad B_\mu = A^3_\mu \sin\theta + A_\mu \cos\theta , \tag{3.4.5}$$

where $\theta$ is an unknown parameter. From (3.4.4) and (3.4.5) it follows that

$$A^1_\mu = \frac{1}{\sqrt{2}} (W_\mu + W^*_\mu) ,$$
$$A^2_\mu = \frac{1}{i\sqrt{2}} (W_\mu - W^*_\mu) , \tag{3.4.6}$$
$$A^3_\mu = Z_\mu \cos\theta + B_\mu \sin\theta , \qquad A_\mu = B_\mu \cos\theta - Z_\mu \sin\theta .$$

The terms in (3.4.3) which contain the functions $A_\mu(x)$ and $A^3_\mu(x)$ then assume the form

$$\frac{g^2m^2}{2f}A_\mu^3A_\mu^3+\frac{g_1^2m^2}{2f}A_\mu A_\mu-\frac{gg_1m^2}{f}A_\mu A_\mu^3$$

$$=\frac{m^2}{f}\left\{Z_\mu Z_\mu\left(\frac{1}{2}g^2\cos^2\theta+\frac{1}{2}g_1^2\sin^2\theta+gg_1\cos\theta\sin\theta\right)\right.$$

$$+B_\mu B_\mu\left(\frac{1}{2}g^2\sin^2\theta+\frac{1}{2}g_1^2\cos^2\theta-gg_1\cos\theta\sin\theta\right)$$

$$\left.+B_\mu Z_\mu\left[\frac{1}{2}(g^2-g_1^2)\sin 2\theta-gg_1\cos 2\theta\right]\right\}. \qquad (3.4.7)$$

We can now choose the parameter $\theta$ in such a way that the coefficient of $B_\mu Z_\mu$ vanishes,

$$\frac{1}{2}(g^2-g_1^2)\sin 2\theta-gg_1\cos 2\theta=0 .$$

From this condition it follows that

$$\cos\theta=\frac{g}{\sqrt{g^2+g_1^2}}, \qquad \sin\theta=\frac{g_1}{\sqrt{g^2+g_1^2}} . \qquad (3.4.8)$$

As can be seen, physically meaningful are the combinations (3.4.4, 5) of the fields $A_\mu$ and $A_\mu^k$ rather than these fields themselves. The angle $\theta$ is called the *mixing angle.*

The substitution of (3.4.6 – 8) into (3.4.3) diagonalizes $\mathscr{L}_0$, and the resulting Lagrangian reads

$$\mathscr{L}_3=-\frac{1}{2}(\partial_\mu W_\nu^*-\partial_\nu W_\mu^*)(\partial_\mu W_\nu-\partial_\nu W_\mu)+\frac{g^2m^2}{f}W_\mu^* W_\mu$$

$$-\frac{1}{4}(\partial_\mu Z_\nu-\partial_\nu Z_\mu)(\partial_\mu Z_\nu-\partial_\nu Z_\mu)+\frac{m^2(g^2+g_1^2)}{2f}Z_\mu Z_\mu$$

$$-\frac{1}{4}(\partial_\mu B_\nu-\partial_\nu B_\mu)(\partial_\mu B_\nu-\partial_\nu B_\mu)+\frac{1}{2}\partial_\mu\sigma\partial_\mu\sigma-m^2\sigma\sigma+\mathscr{L}_1 . \qquad (3.4.9)$$

This Lagrangian contains a charged vector field $W_\mu(x)$ with mass $gm/\sqrt{f}$, a real vector field $Z_\mu(x)$ with mass $m\sqrt{(g^2+g_1^2)/f}$, a massless real vector field $B_\mu(x)$, and a real scalar field $\sigma(x)$ with mass $\sqrt{2}m$, as well as the interaction terms for these fields. Consequently, for the chosen way of spontaneous symmetry-breaking, only three out of the four gauge fields acquire a mass in the case of the $SU_2\times U_1$-invariance.

The presence of a massless particle is explained by the fact that the vacuum state (3.3.10) remains invariant under the new group $U_1'$ with the generator $-ig'(1+\tau_3)/2$:

$$\phi_{\mathrm{v}}^{a}(x) \to (\phi_{\mathrm{v}}^{a}(x))' = \phi_{\mathrm{v}}^{a}(x) - \frac{\mathrm{i}g'}{2}(1+\tau_3)\,\phi_{\mathrm{v}}^{a}(x)\,\varepsilon' = \phi_{\mathrm{v}}^{a}(x)\,,$$

where

$$g' = \frac{g g_1}{\sqrt{g^2+g_1^2}}\,.$$

The fields $B_\mu(x)$, $Z_\mu(x)$, $W_\mu(x)$, and $\sigma(x)$ transform under $U_1'$ as follows:

$$\begin{aligned}
&B_\mu(x) \to B_\mu'(x) = B_\mu(x) + \frac{\sqrt{g^2+g_1^2}}{g}\,\partial_\mu \varepsilon'(x)\,,\\
&W_\mu(x) \to W_\mu'(x) = W_\mu(x) + \mathrm{i} g_1 \varepsilon'(x)\, W_\mu(x)\,,\\
&Z_\mu(x) \to Z_\mu'(x) = Z_\mu(x)\,, \qquad \sigma(x) \to \sigma'(x) = \sigma(x)\,.
\end{aligned} \tag{3.4.10}$$

With the aid of these relations the invariance of the Lagrangian under the group $U_1'$ can be proven. Taking into account (3.4.10) and the fact that the Lagrangian for the field $B_\mu(x)$ entering (3.4.9) coincides with the free Lagrangian of the electromagnetic field, we can identify the field $B_\mu(x)$ with the electromagnetic field.

It can be seen that in the model considered, the vacuum state of spontaneously broken symmetry, chosen in the form of (3.3.10), is invariant under the subgroup $U_1'$. The Lagrangian obtained by going over to the new fields (3.3.11) and eliminating the goldstones (i.e., fixing the gauge) is invariant under the same subgroup $U_1'$ as well. Such a remaining invariance of the Lagrangian is called *residual symmetry*.

Residual symmetry can also be exhibited by models invariant under other groups. Generally, residual symmetry is determined by the number of multiplets of Higgs' scalar field and by the transformation properties of these (vector and tensor) multiplets under the initial symmetry group. The number of generators of the residual symmetry group is equal to the number of gauge fields remaining massless.

# Part II

# Quantum Theory of Gauge Fields

# 4. Path Integrals and Transition Amplitudes

There are various formulations of quantum field theory, differing in the form of a basic quantity, viz. the transition amplitude. In the most commonly used *operator approach*, the transition amplitude is expressed as the vacuum expectation value of the product of particle creation and annihilation operators. These operators obey certain commutation relations (cf. Sect. 1.1). Another formulation is based on expressing the transition amplitude in terms of *path integrals* over the fields. In studying the gauge fields, the path-integral formalism has proven to be the most convenient.

In this chapter we shall discuss the path integrals and shall find the expressions for the amplitude of the vacuum-vacuum transition in terms of the path-integral formalism. We shall start by considering unconstrained fields and then go over to fields with constraints. Gauge fields belong to the latter type.

## 4.1 Unconstrained Fields

### 4.1.1 Systems with One Degree of Freedom

Consider a classical system with one degree of freedom. Such a system is described by the canonical coordinate $q$ and the conjugate momentum $p$ and its dynamics is described by the Hamiltonian $H(p, q)$. Going over to quantum mechanics consists in replacing $q$ and $p$ by the corresponding operators $\hat{q}$ and $\hat{p}$ and the function $H(p, q)$ by the operator $\hat{H}(\hat{p}, \hat{q})$. Since the operators $\hat{p}$ and $\hat{q}$ do not commute, an ordering rule should be introduced for them. In a formal derivation of the path integral, a specific choice of the ordering rule is immaterial, and we choose, for the sake of definiteness, that all operators $\hat{q}$ be placed to the left of the operators $\hat{p}$.

The evolution of the system is determined by the Schrödinger equation

$$\mathrm{i}\,\frac{\partial \psi(x, t)}{\partial t} = \hat{H}(\hat{p}, \hat{q})\, \psi(x, t)\,. \tag{4.1.1}$$

Its formal solution can be expressed as

$$\psi(t'') = \hat{U}(t'', t')\, \psi(t')\,, \tag{4.1.2}$$

where $\hat{U}(t'', t')$ is the evolution operator given by

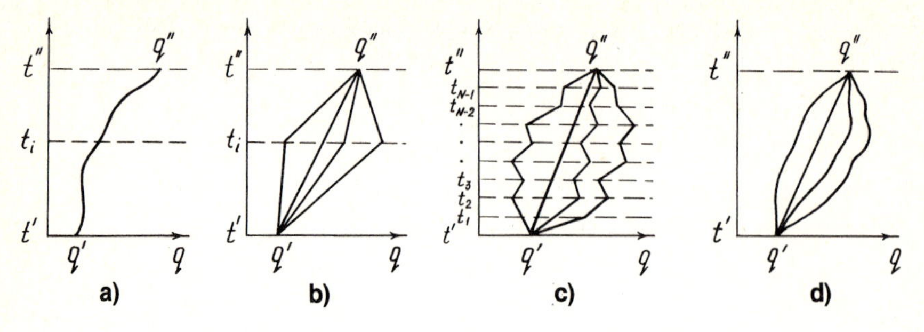


**Fig. 4.1a – d.** Possible trajectories of a system: (**a**) classical; (**b – d**) quantum

$$\hat{U}(t'', t') = \exp[-\mathrm{i}\hat{H}(\hat{p}, \hat{q})(t''-t')] . \tag{4.1.3}$$

The transition amplitude of the system from a state $|q', t'\rangle$ into a state $\langle q'', t''|$ is determined, by virtue of (4.1.3), as follows

$$\langle q'' | \hat{U}(t'', t') | q' \rangle = \langle q'' | \exp[-\mathrm{i}\hat{H}(t''-t')] | q' \rangle \equiv \langle q'', t'' | q', t' \rangle . \tag{4.1.4}$$

Let us demonstrate that this amplitude can be put into another (equivalent) form, namely, into the form of a path-integral.

The basic idea of this approach is as follows. Let the system have the coordinate $q'$ at the moment $t'$ and the coordinate $q''$ at the moment $t''$. If the system were classical, its coordinate $q_i$ at a fixed moment $t_i$ would be strictly determined (Fig. 4.1a). A quantum-mechanical system can be found at the moment $t_i$, with various probabilities, in a point with any coordinate $q_i$ (Fig. 4.1b). Let us make use of the fact that the sum of the transition amplitudes, taken over all intermediate states $q_i$ for a given moment $t_i$, is equal to unity:

$$\sum_{q_i} |q_i, t_i\rangle\langle q_i, t_i| = 1 .$$

This permits rewriting (4.1.4) as

$$\langle q'', t'' | q', t' \rangle = \sum_{q_i} \langle q'', t'' | q_i, t_i \rangle \langle q_i, t_i | q', t' \rangle . \tag{4.1.5}$$

Since the coordinate $q_i$ may assume arbitrary values the summation in (4.1.5) can be replaced by the integration:

$$\langle q'', t'' | q', t' \rangle = \int dq_i \langle q'', t'' | q_i, t_i \rangle \langle q_i, t_i | q', t' \rangle . \tag{4.1.6}$$

Let us divide the time interval $(t'', t')$ into $N$ equal intervals $t', t_1, t_2, \ldots, t_{N-1}, t''$ (Fig. 4.1c):

$$t_{i+1} - t_i = \frac{t''-t'}{N}, \qquad i = 0, 1, 2, \ldots, N-1, \qquad t_0 \equiv t', \qquad t_N \equiv t'' .$$

By analogy with (4.1.6), the amplitude then can be written as a finite-dimensional integral:

$$\langle q'', t'' | q', t' \rangle = \int dq_1 \int dq_2 \ldots \int dq_{N-1} \langle q'', t'' | q_{N-1}, t_{N-1} \rangle$$
$$\ldots \langle q_2, t_2 | q_1, t_1 \rangle \langle q_1, t_1 | q', t' \rangle . \tag{4.1.7}$$

By virtue of (4.1.4), this expression can be rewritten as

$$\langle q'', t'' | q', t' \rangle = \int dq_1 \ldots \int dq_{N-1} \langle q'' | \exp[-\mathrm{i}\hat{H}(t'' - t_{N-1})] | q_{N-1} \rangle$$
$$\ldots \langle q_1 | \exp[-\mathrm{i}\hat{H}(t_1 - t')] | q' \rangle . \tag{4.1.8}$$

In this way we have explicitly taken into account the evolution of the system with time. In order to evaluate the matrix elements in the integrand of (4.1.8) we make use of the smallness of the time intervals $(t_{i+1} - t_i)$. Thus, the exponents can be expanded in powers of $(t_{i+1} - t_i)$:

$$\exp[-\mathrm{i}\hat{H}(t_{i+1} - t_i)] \simeq 1 - \mathrm{i}\hat{H}(t_{i+1} - t_i) + \ldots ,$$

and we obtain, instead of (4.1.8)

$$\langle q'', t'' | q', t' \rangle = \int dq_1 \ldots \int dq_{N-1} \langle q'' | [1 - \mathrm{i}\hat{H}(t'' - t_{N-1})] | q_{N-1} \rangle$$
$$\ldots \langle q_1 | [1 - \mathrm{i}\hat{H}(t_1 - t')] | q' \rangle .$$

Then the matrix element of the operator $\hat{H}(\hat{q}, \hat{p})$ can be found. Indeed, by definition, together with our ordering convention,

$$\langle q_{i+1} | \hat{H} | q_i \rangle = \int dq \, \bar{\psi}(q; q_{i+1}) \hat{H}\left(q, -\mathrm{i}\frac{\partial}{\partial q}\right) \psi(q; q_i) ,$$

where $\psi(q; q_i)$ are the eigenfunctions of the coordinate operator $\hat{q}$. In the coordinate representation these functions have the form

$$\psi(q; q_i) = \delta(q - q_i) \equiv \frac{1}{2\pi} \int dp \exp[-\mathrm{i}(q_i - q)p] ;$$

therefore

$$\langle q_{i+1} | \hat{H} | q_i \rangle = \frac{1}{(2\pi)^2} \int dq \, dp \, d\tilde{p} \exp[-\mathrm{i}(q - q_{i+1})p]$$
$$\times \hat{H}\left(q, -\mathrm{i}\frac{\partial}{\partial q}\right) \exp[-\mathrm{i}(q_i - q)\tilde{p}] .$$

Taking into account that

$$\hat{H}\left(q, -\mathrm{i}\frac{\partial}{\partial q}\right) \exp[-\mathrm{i}(q_i - q)\tilde{p}] = \sum_{n=0}^{\infty} a_n(q) \left(-\mathrm{i}\frac{\partial}{\partial q}\right)^n \exp[-\mathrm{i}(q_i - q)\tilde{p}]$$
$$= H(q, \tilde{p}) \exp[-\mathrm{i}(q_i - q)\tilde{p}] ,$$

we find

$$\langle q_{i+1}|\hat{H}|q_i\rangle = \int \frac{dp_i}{2\pi} \exp[\mathrm{i}(q_{i+1}-q_i)p_i] H(q_i, p_i) \,. \tag{4.1.8'}$$

It can be seen that the matrix element of the operator $\hat{H}(\hat{p}, \hat{q})$ is expressed in terms of the *classical* Hamilton function $H(p, q)$. At this stage we get rid of the operator quantities $\hat{p}_i$, $\hat{q}_i$ and go over to the classical variables $p_i, q_i$. In deriving the expression (4.1.8′) we have taken into account the ordering adopted ($\hat{q}_i$ being placed to the left of $\hat{p}_i$). When another ordering procedure is chosen, then in general a different expression is obtained instead of (4.1.8′), i.e., different ordering procedures lead to different expressions for the transition amplitudes. However, the physical quantities (matrix elements etc.) do not depend on the ordering procedure.

By using (4.1.8′), we arrived at the expression for the amplitude which contains integration over both of the classical variables $p_i$, $q_i$,

$$\begin{aligned}
&\langle q'', t''|q', t'\rangle \\
&= \int dq_1 \ldots \int dq_{N-1} \int \exp[\mathrm{i}(q''-q_{N-1})p_N - \mathrm{i}H(p_N, q_N)(t''-t_{N-1})] \frac{dp_N}{2\pi} \\
&\quad \ldots \int \exp[\mathrm{i}(q_1-q')p_1 - \mathrm{i}H(p_1, q_1)(t_1-t')] \frac{dp_1}{2\pi} \\
&= \int \prod_{i=1}^{N-1} dq_i \prod_{i=1}^{N} \frac{dp_i}{2\pi} \exp\left\{\mathrm{i} \sum_{j=1}^{N} [(q_j-q_{j-1})p_j - H(p_j, q_j)(t_j-t_{j-1})]\right\},
\end{aligned} \tag{4.1.9}$$

where $t_N = t''$, $q_N = q''$, $t_0 = t'$, and $q_0 = q'$.

In the limit of $N \to \infty$, $t_j - t_{j-1} = dt$; the number of integration variables tends to infinity. One can consider the integration as that over the values of the functions $p(t)$ and $q(t)$ at all times $t$ in the interval $(t'', t')$, i.e., the integration is carried out over all possible trajectories (Fig. 4.1 d). The following boundary conditions are placed on the function $q(t)$:

$$q(t') = q', \quad q(t'') = q'' \,. \tag{4.1.10}$$

Hence, in the limit, the transition amplitude from the state $|q', t'\rangle$ into the state $\langle q'', t''|$ is written as the following *infinite-dimensional* integral:

$$\begin{aligned}
&\langle q'', t''|q', t'\rangle \\
&= \lim_{N\to\infty} \int \prod_{i=1}^{N-1} dq_i \prod_{i=1}^{N} \frac{dp_i}{2\pi} \exp\left\{\mathrm{i} \sum_{j=1}^{N} \left[p_j \frac{q_j - q_{j-1}}{t_j - t_{j-1}} - H(p_j, q_j)\right](t_j - t_{j-1})\right\} \\
&\equiv \int \prod_t \frac{\mathscr{D}p(t)\, \mathscr{D}q(t)}{2\pi} \exp\left\{\mathrm{i} \int_{t'}^{t''} dt\, [p\dot{q} - H(p, q)]\right\}.
\end{aligned} \tag{4.1.11}$$

This integral is called the *path integral*, or the *continual integral*, or the *functional integral*. To stress that the integration is carried out over the functions rather than over the variables, we write $\mathscr{D}p(t)$ rather than $dp(t)$.

Remembering (1.1.8, 17) we eventually obtain the desired expression for the transition amplitude (4.1.4) in the path-integral form:

$$\langle q'', t'' | q', t' \rangle = \int \prod_t \frac{\mathscr{D} p(t) \mathscr{D} q(t)}{2\pi} \exp [\mathrm{i} I(t'', t')] . \tag{4.1.12}$$

Here $I(t'', t')$ is the action, i.e., the path integral is a functional integral of the classical action over all possible trajectories in the phase space $(p, q)$, with the boundary conditions given by (4.1.10).

The expression

$$\prod_t \frac{\mathscr{D} p(t) \mathscr{D} q(t)}{2\pi}$$

is called the *integration measure.*

As one can see, we came to the path integral by means of introducing an infinite number of intermediate times $t_i$. Furthermore, all possible values of the coordinates $q_i$ at each moment $t_i$ were taken into account and, on the other hand, the operator quantities were replaced by the classical quantities in the description of the evolution of the system with time. This enabled us to express the quantum-mechanical amplitude in terms of classical quantities alone, but at the price of introducing an infinite-dimensional integral.

The transition probability $P$ for the system is determined by the square of the modulus of the amplitude:

$$P = |\langle q'', t'' | q', t' \rangle|^2 .$$

We would like to make some remarks concerning the path integral.

1) Subsequent transitions between all allowed states are characterized by amplitudes, and not by probabilities themselves. This leads to interference effects characteristic of quantum theory. This fact makes it impossible to reduce the quantum mechanics to any classical statistical mechanics, despite the appearance of the transition amplitude (4.1.11) which is expressed in terms of classical quantities.

2) Let us consider the classical limit (i.e., the limit of the Planck constant $\hbar \to 0$) of the amplitude (4.1.12). Then only classical trajectories are allowed. In the classical limit, $\hbar^{-1} I(t'', t') \gg 1$, i.e., the action $I$ is large as compared to $\hbar$. If a trajectory is not a solution of the classical equation of motion, a small variation of this trajectory will lead to a very large variation of the ratio $I/\hbar$ in the formula (4.1.12) and to fast oscillations of the amplitude. As a result, the contributions from all such trajectories will vanish so that these trajectories do not need to be regarded. However, for the trajectory determined by the classical equation of motion, the action has an extremum, i.e., $\delta I = 0$; small deviations from this trajectory do not change the magnitude of $I$. Therefore, in the amplitude the contributions of the trajectories which are close to the classical ones do not cancel, their phases being about the same (equal to

$\hbar^{-1} I_{\text{class.}}$). Hence, in the classical limit, the main contribution comes from the classical trajectory.

Taking into account small fluctuations around the classical trajectory leads to the quasi-classical approximation with the action expressed as

$$I = I_{\text{class.}} + \Delta I ,$$

where $\Delta I$ stands for the deviations from the classical trajectory.

3) No general methods for calculating path integrals are available. The magnitude of the path integral depends on the ordering convention for the operators in (4.1.1): different ordering procedures lead to different values of the path integrals. It is also not clear which limitations have to be imposed on the class of the functionals and on the functional space to ensure the convergence of the limiting integral.

4) Path integrals with Hamiltonians which depend quadratically on the momentum and the coordinate are referred to as Gaussian. In the Gaussian integrals, which we shall consider, there is no mixing of the variables. Besides, the Gaussian integrals can be calculated exactly. Hence, for the Gaussian integrals, there is no problem of ordering and of the existence of the limit.

In the simplest case of a Hamiltonian quadratic in the momentum,

$$H = \frac{p^2}{2m} + V(q) ,$$

one has the following path integral:

$$\langle q'', t'' | q', t' \rangle = \int \prod_t \frac{\mathscr{D}p(t)\, \mathscr{D}q(t)}{2\pi} \exp\left\{ \mathrm{i} \int_{t'}^{t''} dt \left[ p\dot{q} - \frac{p^2}{2m} - V(q) \right] \right\} .$$

To calculate this integral, we perform a shift

$$p(t) \to p(t) + m\dot{q} .$$

The integrations over $p$ and $q$ are then separated:

$$\langle q'', t'' | q', t' \rangle = \frac{1}{N} \int \prod_t \mathscr{D}q \exp\left\{ \mathrm{i} \int_{t'}^{t''} dt \left[ \frac{m\dot{q}^2}{2} - V(q) \right] \right\} ,$$

where

$$N^{-1} = \int \prod_t \frac{\mathscr{D}p}{2\pi} \exp\left\{ -\mathrm{i} \int_{t'}^{t''} dt \frac{p^2}{2m} \right\} .$$

The normalizing factor $N$ is independent of $q'$ and $q''$. Usually it is included in the definition of the integration measure. In the case of $V(q)$ which depends on $q$ quadratically, the integral over $q$ becomes Gaussian and can be calculated explicitly.

In what follows we shall only deal with the path integrals of the Gaussian type.

### 4.1.2 Systems with a Finite Number of Degrees of Freedom

In a similar way, the following expression for the transition amplitude of a system with a finite number of degrees of freedom can be obtained:

$$\langle q_1'', q_2'', \ldots, q_n''; t'' | q_1', q_2', \ldots, q_n'; t' \rangle$$
$$= \int \prod_t \prod_{k=1}^{n} \frac{\mathcal{D}p_k(t)\, \mathcal{D}q_k(t)}{2\pi} \exp[\mathrm{i} I(t'', t')] \, . \tag{4.1.13}$$

Here $q_i$ and $p_i$ are the canonical coordinates and momenta, respectively, $i = 1, 2, \ldots, n$, where $n$ is the number of the degrees of freedom; $I(t'', t')$ is the action of the system depending on $p_1, \ldots, p_n, q_1, \ldots, q_n$.

### 4.1.3 The Boson Fields

From the viewpoint of Hamiltonian dynamics a field is a system with an infinitely large number of degrees of freedom, for a field is characterized by a generalized coordinate $u(x)$ and a generalized momentum $\pi(x)$ at each point $x$. That is why a system with a finite number of degrees of freedom is considered first, to obtain path integrals for the fields by taking the appropriate limit.

We begin with *boson* fields. Consider, for simplicity, a neutral scalar field described by a single wave function $\phi(x)$. In order to reduce the field to a system with a finite number of degrees of freedom, we take a finite volume $V$ in the three-dimensional space and subdivide it into $m$ small equal cubicles of volume $v_m$. Within each cubicle $v_m$, the function $\phi(x) \equiv \phi(x_0, \boldsymbol{x})$ will be approximated by the function $\phi_m(x_0) \equiv [\int_{v_m} d\boldsymbol{x}\, \phi(x_0, \boldsymbol{x})]/v_m$. We shall consider the functions $\phi_m(x_0)$ and $\dot{\phi}_m(x_0) = \partial \phi_m(x_0)/\partial x_0$ as the generalized coordinates and momenta of the field, respectively. Let $L_m$ be the density of the Lagrangian for the field within the volume $v_m$. The Lagrangian $L(x_0)$ for the field is then written as a sum over all $L_m$,

$$L(x_0) = \sum_m v_m L_m(\dot{\phi}_m(x_0), \phi_m(x_0)) \, . \tag{4.1.14}$$

With the help of this Lagrangian, the generalized momenta $\pi_m$ of the field conjugate to the generalized coordinates can be defined in the usual way:

$$\pi_m(x_0) = \frac{\partial L(\dot{\phi}_m, \phi_m)}{\partial \dot{\phi}_m} \, .$$

Accordingly, the Hamiltonian for the field reads

$$H = \sum_m v_m [\pi_m(x_0)\, \dot{\phi}_m(x_0) - L_m(x_0)] \, .$$

According to (4.1.13), the expression for the amplitude is written, in the case under consideration, as

$$S = \int \prod_{i=1}^{N-1} \prod_m d\phi_m(x_{0i}) \prod_{i=1}^{N} \prod_m \int \frac{d\pi_m(x_{0i})}{2\pi}$$

$$\times \exp\left\{ i \sum_{j=1}^{N} \left[ \sum_m v_m(\pi_m(x_{0j})\,\dot{\phi}_m(x_{0j}) - H_m(\pi_m(x_{0j}),\, \phi_m(x_{0j})))\, \Delta t_j \right] \right\}. \tag{4.1.15}$$

This expression describes the evolution of the system from the state corresponding to the moment $t'$ into the state corresponding to the moment $t''$. However, as already mentioned (cf. Sect. 1.1), the quantum mechanics deals with transitions of a system from the state at $t = -\infty$ into the state at $t = +\infty$. In order to obtain the amplitude for such a transition, one has to take in (4.1.15) the limit $t' \to -\infty$, $t'' \to +\infty$. For this purpose, we act by the operator $\hat{H}$ on the functions of the initial and the final states in (4.1.4), assuming that $t' < -T$ and $t'' > T$, where $T$ is some finite time interval. As a result we find that the time dependence of the amplitude is determined by the terms

$$\exp[-iE_n(t''+T)] \quad \text{and} \quad \exp[iE_n(t'-T)]\,,$$

where $E_n$ is the energy of a stationary state of the system. In the limit $t \to \pm\infty$, only the contribution from the transition from the ground state (with the minimal energy $E_0$) determines the amplitude. In field theory, it is the physical vacuum which plays the role of the ground state. Hence, at $t \to \pm\infty$, the amplitude (4.1.15) describes the vacuum-vacuum transition.

Taking into account that with $v_m \to 0$,

$$\pi_m(x_0) \to \pi(x)\,, \qquad \phi_m(x_0) \to \phi(x)\,, \qquad H_m(\pi_m(x),\, \phi_m(x)) \to H(\pi(x),\, \phi(x))\,,$$

and taking in (4.1.15) the limits $V \to \infty$, $v_m \to 0$, $t'' \to \infty$, and $t' \to -\infty$, we obtain the desired expression for the vacuum-vacuum transition amplitude for the scalar field:

$$S = \int \mathscr{D}\mu(\phi(x),\, \pi(x)) \times \exp\{ i \int dx [\pi(x)\,\dot{\phi}(x) - H(\pi(x),\, \phi(x))] \}\,, \tag{4.1.16}$$

where the integration measure is

$$\mathscr{D}\mu(\phi(x),\, \pi(x)) = \lim_{\substack{v_m \to 0 \\ V \to \infty}} \prod_{x_0} \prod_m \frac{1}{2\pi} d\phi_m(x_0)\, d\pi_m(x_0) = \prod_x \frac{\mathscr{D}\phi(x)\, \mathscr{D}\pi(x)}{2\pi}\,. \tag{4.1.17}$$

In cases where the Hamiltonian is quadratic in the momentum, we arrive at the Gaussian path integral which can be rewritten in a special form, namely, as a *path integral over all the fields*. Let us illustrate this by an example of the Lagrangian

$$L = \frac{1}{2}(\partial_\mu \phi)(\partial_\mu \phi) - \frac{m^2}{2}\phi^2 - \frac{g}{3!}\phi^3 . \tag{4.1.18}$$

According to (1.1.16, 17), to this Lagrangian correspond the generalized momenta

$$\pi(x) = \frac{\partial L}{\partial \dot{\phi}(x)} = \dot{\phi}(x) \tag{4.1.19}$$

and the Hamiltonian

$$H(\pi, \phi) = \pi \dot{\phi} - L = \frac{1}{2}\pi^2(x) + \frac{1}{2}(\nabla \phi(x))^2 + \frac{m^2}{2}\phi^2(x) + \frac{g}{3!}\phi^3(x) . \tag{4.1.20}$$

Substituting (4.1.20) into (4.1.16) yields

$$S = \int \prod_x \frac{\mathscr{D}\phi(x)\mathscr{D}\pi(x)}{2\pi} \times \exp\left\{ \mathrm{i}\int dx \left[ \pi\dot{\phi} - \frac{1}{2}\pi^2 - \frac{1}{2}(\nabla\phi)^2 - \frac{m^2}{2}\phi^2 - \frac{g}{3!}\phi^3 \right]\right\} . \tag{4.1.21}$$

By a shift of the variable

$$\pi(x) \to \pi(x) + \dot{\phi}(x)$$

and by taking into account that the Jacobian of this transformation is equal to unity, we arrive at the desired expression for the vacuum-vacuum transition amplitude in the form of a path integral over all fields

$$S = N \int \prod_x \mathscr{D}\phi(x) \exp\{\mathrm{i}\int dx L[\phi(x), \partial_\mu \phi(x)]\} , \tag{4.1.22}$$

where

$$N = \int \prod_x \frac{\mathscr{D}\pi(x)}{2\pi} \exp\left[ -\frac{\mathrm{i}}{2}\int dx\, \pi^2(x) \right]$$

is a normalization factor usually included in the integration measure, as mentioned above. Equation (4.1.22) is also applicable for an arbitrary Lagrangian for the self-interaction of a field without derivatives.

Later (in Chap. 5) we shall use the described method of deriving the field transition amplitude to find the form of the Green's functions and subsequently that of the elements of the $S$-matrix.

### 4.1.4 The Fermion Fields

In quite the same manner, the expression for the vacuum-vacuum transition amplitude of *fermion* fields can be obtained in terms of a path integral over all the fields. For instance, for the free fermion Dirac field, as described by the Lagrangian (1.1.13), we have

$$S = N \int \prod_x \mathscr{D}\bar{\psi}(x)\, \mathscr{D}\psi(x) \exp\{\mathrm{i}\int dx [\mathrm{i}\,\bar{\psi}(x)\,\gamma_\mu \partial_\mu \psi(x) - M\bar{\psi}(x)\,\psi(x)]\} . \tag{4.1.22'}$$

However, in contrast to the boson fields, the integration now is carried out over the *anticommuting* independent variables, viz. the fermion fields $\bar{\psi}(x)$ and $\psi(x)$. Operations with anticommuting variables have some specific features. Consider, for simplicity, a system characterized by two anticommuting variables $\eta_1$ and $\eta_2$: generalization to the case of a larger number of variables $(\eta_1, \eta_2, \eta_3, \ldots)$ and to the case of mixed sets of variables $(\eta_i, \alpha_i)$ can be obtained in an obvious way. The variables $\eta_1$ and $\eta_2$ obey the following commutation relations

$$[\eta_1, \eta_2]_+ \equiv \eta_1\eta_2 + \eta_2\eta_1 = 0 . \tag{4.1.23}$$

From this it follows that

$$\eta_1^2 = \eta_2^2 = 0 . \tag{4.1.24}$$

For this reason a function of the variables $\eta_1$, $\eta_2$ can only be a finite polynomial, linear in both $\eta_1$ and $\eta_2$:

$$f(\eta_1, \eta_2) = a_0 + a_1\eta_1 + a_2\eta_2 + a_{12}\eta_1\eta_2 . \tag{4.1.25}$$

The derivatives of the function $f(\eta_1, \eta_2)$ with respect to $\eta_1$ and $\eta_2$ are determined by the formula

$$\frac{\partial}{\partial\eta_i}\eta_k = \delta_{ik} . \tag{4.1.26}$$

For instance,

$$\frac{\partial}{\partial\eta_1}f(\eta_1, \eta_2) = a_1 + a_{12}\eta_2 .$$

By definition, the derivatives with respect to $\eta_1$ and $\eta_2$ anticommute

$$\frac{\partial}{\partial\eta_1}\frac{\partial}{\partial\eta_2} = -\frac{\partial}{\partial\eta_2}\frac{\partial}{\partial\eta_1} . \tag{4.1.27}$$

Besides

$$\left[\frac{\partial}{\partial\eta_k}, \eta_m\right]_+ \equiv \frac{\partial}{\partial\eta_k}\eta_m + \eta_m\frac{\partial}{\partial\eta_k} = 0 , \qquad k \neq m .$$

Introduce the differentials of the variables $\eta_1$ and $\eta_2$. These obey the commutation relations

$$[d\eta_k, d\eta_m]_+ \equiv d\eta_k\, d\eta_m + d\eta_m\, d\eta_k = 0 . \tag{4.1.28}$$

The definition of the integral over the variables $\eta_1$ and $\eta_2$, compatible with (4.1.28), has the following, somewhat unusual, form

$$\int d\eta_k = 0 , \tag{4.1.29}$$

$$\int d\eta_m \eta_k = \delta_{km} . \tag{4.1.30}$$

By virtue of this definition we have, for example,

$$\int d\eta_1 f(\eta_1, \eta_2) = \int d\eta_1 (a_0 + a_1 \eta_1 + a_2 \eta_2 + a_{12} \eta_1 \eta_2) = a_1 + a_{12} \eta_2 .$$

Finally, the rule for changing the variables in integrals over anticommuting variables reads

$$\eta_i = \alpha_{ij} \xi_j , \quad \prod_i d\eta_i = (\det \alpha)^{-1} \prod_i d\xi_i , \tag{4.1.31}$$

i.e., the variables and the differentials transform by mutually inverse matrices. For example, the change of variables in the integral

$$\int \eta_1 \eta_2 d\eta_1 \, d\eta_2$$

gives, with the aid of (4.1.31),

$$\int d\eta_1 d\eta_2 \eta_1 \eta_2 = \det(\alpha) [\det(\alpha)]^{-1} \int d\xi_1 \, d\xi_2 \xi_1 \xi_2 = \int d\xi_1 \, d\xi_2 \xi_1 \xi_2 .$$

## 4.2 Fields with Constraints

### 4.2.1 Systems with a Finite Number of Degrees of Freedom

Consider a system with $n$ degrees of freedom described by the canonical coordinates $q_i$ and momenta $p_i$ (where $i = 1, 2, \ldots, n$). Let $m$ *constraints* be imposed on the system, which implies that the variables $p_i$ and $q_i$ have to satisfy $m$ equations

$$\phi_a(p_i, q_i) = 0 , \quad a = 1, 2, \ldots, m . \tag{4.2.1}$$

Our aim is to find an expression for the matrix elements of the evolution operator in form of path integrals under these constraints, i.e., for a generalized Hamiltonian system. Equation (4.1.13) is obviously not applicable for a constrained system, since it contains only independent canonical variables $p_i$ and $q_i$.

Let us investigate in which way (4.1.13) has to be modified in the case of constrained systems. The most straightforward way would be to eliminate the dependent variables. For that, the constraint equations (4.2.1) should be solved. In practice, this may prove to be difficult, and in some cases even impossible. Therefore a formalism has been developed which does not require solving explicitly the constraint equations. The basic idea is as follows. First, an expression for the matrix elements of the evolution operator is obtained in terms of a path integral containing the independent variables alone. This expression is, however, an intermediate result. It is then reduced to the ordinary path integration over all canonical variables, the constraints being included as subsidiary functional factors into the measure of integration. Solving the constraint equations is thus avoided.

### Primary and Secondary Constraints. Subsidiary Conditions

Let us characterize a system by the Hamiltonian $H(p_i, q_i)$. The constraints can be taken into account by means of the Lagrange multipliers $\lambda_a$. The canonical equations of motion are then obtained by means of the variation (with respect to the variables $p_i$ and $q_i$) of the action,

$$\int dt\,[p_i \dot{q}_i - H'(p_i, q_i)], \quad H'(p_i, q_i) = H(p_i, q_i) + \sum_{a=1}^{m} \lambda_a \phi_a(p_i, q_i)\,, \tag{4.2.2}$$

and are written as

$$\dot{q}_i = \frac{\partial H'}{\partial p_i} = \frac{\partial H}{\partial p_i} + \sum_{a=1}^{m} \lambda_a \frac{\partial \phi_a}{\partial p_i}\,, \quad \dot{p}_i = -\frac{\partial H'}{\partial q_i} = -\frac{\partial H}{\partial q_i} - \sum_{a=1}^{m} \lambda_a \frac{\partial \phi_a}{\partial q_i}\,. \tag{4.2.3}$$

For an arbitrary function $f(p_i, q_i)$ of the variables $p_i$ and $q_i$ we have

$$\dot{f}(p_i, q_i) = \frac{\partial f}{\partial q_i}\dot{q}_i + \frac{\partial f}{\partial p_i}\dot{p}_i = \{H, f\} + \sum_{a=1}^{m} \lambda_a \{\phi_a, f\}\,, \tag{4.2.4}$$

where the symbol $\{f, g\}$ denotes the Poisson brackets:

$$\{f, g\} = \sum_{i=1}^{n} \left( \frac{\partial f}{\partial p_i}\frac{\partial g}{\partial q_i} - \frac{\partial f}{\partial q_i}\frac{\partial g}{\partial p_i} \right).$$

The variables $p_i$ and $q_i$ cannot be varied independently because they are connected by the condition (4.2.1). Therefore, certain compatibility conditions for the equations of motion (4.2.3) and the constraint equations (4.2.1) must be satisfied for the problem to be self-consistent. To find these conditions, we start with the time-independence of the constraints (4.2.1):

$$\dot{\phi}_a(p_i, q_i) = 0\,.$$

Putting $f(p_i, q_i) = \phi_a(p_i, q_i)$ in (4.2.4) we arrive at the *consistency conditions*

$$\{H, \phi_a\} + \sum_{b=1}^{m} \lambda_b \{\phi_b, \phi_a\} = 0\,. \tag{4.2.5}$$

(It is understood that henceforth we take $\phi_a(p_i, q_i) = 0$ after calculating Poisson's brackets.)

With regard to the conditions (4.2.5), three cases are possible.

1) Conditions (4.2.5) are certainly fulfilled if each of the Poisson's brackets is zero. Taking into account (4.2.1), the consistency conditions are reduced in this case to

$$\{\phi_a, \phi_b\} = \sum_{c=1}^{m} C_{abc}\phi_c\,, \qquad \{H, \phi_a\} = \sum_{b=1}^{m} C_{ab}\phi_b\,. \tag{4.2.6}$$

2) If the Poisson's brackets in the second term of (4.2.5) are, as before, zero under condition (4.2.1), but are non-zero in the first one, $\{H, \phi_a\} = \Phi_{1a}$, then $\Phi_{1a}$ has to be equated to zero for the conditions (4.2.5) to be fulfilled. Since $\Phi_{1a}$ is independent of $\lambda_a$, the consistency condition in this case is written as

$$\Phi_{1a}(p_i, q_i) = 0\,. \tag{4.2.7}$$

This condition imposes additional constraints on the variables $p_i$ and $q_i$. Dirac called the constraints determined by (4.2.1) the *primary* constraints, and those determined by (4.2.7) the *secondary* constraints. By definition, the secondary constraints cannot be reduced to the primary ones.

The secondary constraints $\Phi_{1a}$ must, in turn, satisfy the consistency condition similar to (4.2.5):

$$\dot{\Phi}_{1a}(p_i, q_i) = \{H, \Phi_{1a}\} + \sum_{b=1}^{m'} \lambda_b\{\Phi_{1b}, \Phi_{1a}\} = 0\,.$$

This condition should be analyzed in the same manner as already done for (4.2.5). If $\{H, \Phi_{1a}\}$ is non-zero, $\{H, \Phi_{1a}\} = \Phi_{2a}$, additional secondary constraints, $\Phi_{2a}(p_i, q_i) = 0$, must be introduced. This may result in several secondary constraints of the type of (4.2.7).

Below, we shall consider the primary and the secondary constraints on equal footing. In both cases the coefficients $\lambda_a$ are arbitrary.

3) If both Poisson's brackets in (4.2.5) are non-zero, then the condition (4.2.5) gives rise to a system of inhomogeneous linear equations with respect to the unknown coefficients $\lambda_a$. The general solution of this system of equations can be written as a sum of solutions for the inhomogeneous system, $\Lambda_a(p, q)$, and of solutions for the homogeneous system, $V_{ba}(p, q)$:

$$\lambda_a = \Lambda_a + \sum_b v_b V_{ba}\,,$$

where $v_b$ are arbitrary coefficients.

If the number of coefficients $v_b$ is less than the number of $\lambda_a$, then part of the coefficients $\lambda_a$ will not be arbitrary. However, in this book we shall not encounter such a case.

Let us consider in more detail the case when both Poisson's brackets in (4.2.5) become zero and all the coefficients $\lambda_a$ are arbitrary. This arbitrariness implies that the dynamical variables $f(p_i, q_i; t)$ at any moment are not uniquely determined by their initial values. To substantiate this, consider an arbitrary dynamical variable $f(p_i, q_i; t)$ at a moment $t$. Its value after a short time interval $\delta t$ will be

$$f(p_i, q_i; t+\delta t) = f(p_i, q_i; t) + \delta t \dot{f}(p_i, q_i; t)$$

or, due to (4.2.4),

$$f(p_i, q_i; t+\delta t) = f(p_i, q_i; t) + \delta t\{H, f(p_i, q_i; t)\} + \delta t \sum_{a=1}^{m} \lambda_a\{\phi_a, f(p_i, q_i; t)\} .$$

Take now some other value $\lambda'_a$ for the arbitrary coefficients $\lambda_a$ and subtract the corresponding expression from the last one. This yields

$$\Delta f(p_i, q_i; t+\delta t) = \delta t \sum_{a=1}^{m} (\lambda_a - \lambda'_a)\{\phi_a, f(p_i, q_i; t)\} .$$

Hence, by varying the coefficients $\lambda_a$, a whole set of values for the dynamical variable $f(p_i, q_i; t)$ at a given moment is found to correspond to a single initial value. All of them characterize the same physical state because a physical state is described by the set of $p_i, q_i$, and consequently should not depend on the choice of $\lambda_a$. The set of values of the dynamical variable $f(p_i, q_i; t)$ for all possible coefficients $\lambda_a$ is called an *orbit*. Running over all times leads to a set of orbits. Since all values of a given dynamical variable on an orbit are physically equivalent, there is no need in considering the whole orbit: it is sufficient to take one point on the orbit to represent it, i.e., to fix one value of $\lambda_a$ on each orbit. To achieve that, we impose on the system the *subsidiary conditions*

$$\chi_b(p_i, q_i) = 0 . \tag{4.2.8}$$

To be *unique* with regard to the choice of $\lambda_a$, the subsidiary conditions (4.2.8) have to satisfy certain requirements. In order to find these requirements we substitute the functions $\chi_b(p_i, q_i)$ into (4.2.4). As a result, we come to the compatibility conditions analogous to (4.2.5):

$$\dot{\chi}_b = \{H, \chi_b\} + \sum_{a=1}^{m} \lambda_a\{\phi_a, \chi_b\} = 0 .$$

These relations enable us to fix the values of $\lambda_a$. To do so, we take, in accordance with the third case, $\{\phi_a, \chi_b\} \neq 0$ which leads to a set of equations for $\lambda_a$. The set will be uniquely solvable with respect to $\lambda_a$ if

i) the number of constraints $\phi_a$ is equal to the number of subsidiary conditions $\chi_b$ (i.e., $a = b$, giving rise to a square matrix with respect to the indices $a$ and $b$);

ii) the relation (i.e., the integrability condition for the system of equations)

$$\det|\{\phi_a, \chi_b\}| \neq 0 , \qquad a = b = 1, 2, \ldots, m, \tag{4.2.9}$$

is satisfied.

Besides, it is convenient to choose the functions $\chi_b(p_i, q_i)$ in such a way that

$$\{\chi_a, \chi_b\} = 0 . \tag{4.2.10}$$

In what follows we shall consider the systems subject to $m$ constraints of the type (4.2.1) and $m$ subsidiary conditions (4.2.8). In this case the variables satisfy $2m$ conditions and are defined in the subspace $\Gamma^*$ [whose dimensionality is $2(n-m)$] of the space $\Gamma$ with dimensionality $2n$.

### Dynamics in the Space $\Gamma^*$

Let us show that in the space $\Gamma^*$, $2(n-m)$ independent canonical variables $p_b^*$ and $q_b^*$ (where $b = m+1, \ldots, n$) can be introduced which satisfy the Hamilton equations with the Hamiltonian $H^*$ obtained from the Hamiltonian $H$ by using (4.2.1) and (4.2.8). First go over from the canonical variables $p_i$ and $q_i$ to the new canonical variables $P_i$ and $Q_i$ in the space $\Gamma$ ($i = 1, 2, \ldots, n$). By virtue of (4.2.10), the functions $\chi_a(p_i, q_i)$ can be chosen as $m$ new momenta, i.e., $\chi_a = P_a$ (where $a = 1, 2, \ldots, m$). Let $Q_a$ be their conjugate coordinates. We denote the remaining canonical coordinates and momenta as

$$P_b = p_b^* \quad \text{and} \quad Q_b = q_b^* \quad (\text{where } b = m+1, \ldots, n) .$$

The variables $P_i$ and $Q_i$ are canonical: by definition they have to obey the relations

$$\{Q_i, Q_k\}_{p,q} = 0 , \quad \{P_i, P_k\}_{p,q} = 0 , \quad \{P_i, Q_k\}_{p,q} = \delta_{ik} . \tag{4.2.11}$$

In terms of the new variables, we have for the Poisson's brackets

$$\{\chi_a, \phi_c\}_{p,q} = \{\chi_a, \phi_c\}_{P,Q} = \{P_a, \phi_c\}_{P,Q} = \sum_i \delta_{ai} \frac{\partial \phi_c}{\partial Q_i} = \frac{\partial \phi_c}{\partial Q_a} .$$

Accordingly, (4.2.9) assumes the form

$$\det \left| \frac{\partial \phi_c(P,Q)}{\partial Q_a} \right| \neq 0 . \tag{4.2.12}$$

This means that the constraint equations (4.2.1) can be solved in terms of the coordinates $Q_a$, i.e., in the form $Q_a = Q_a(p_b^*, q_b^*)$. It should be remembered that in the subspace $\Gamma^*$ the momenta $P_a = 0$.

Let us now obtain the Hamilton equations for the system $\Gamma^*$. The Hamiltonian of the system is

$$H^*(p_b^*, q_b^*) = H(p_i, q_i)\,|_{\phi_a=0,\,\chi_a=0} . \tag{4.2.13}$$

The equations of motion (4.2.3) are rewritten in terms of the new canonical variables $P_a$, $Q_a$, $p_b^*$, and $q_b^*$ as

$$\dot{q}_b^* = \frac{\partial H(P_a, Q_a, p_b^*, q_b^*)}{\partial p_b^*} + \sum_a \lambda_a \frac{\partial \phi_a(P_a, Q_a, p_b^*, q_b^*)}{\partial p_b^*} ,$$

$$\dot{p}_b^* = -\frac{\partial H}{\partial q_b^*} - \sum_a \lambda_a \frac{\partial \phi_a}{\partial q_b^*}, \qquad \dot{Q}_a = \frac{\partial H}{\partial P_a} + \sum_c \lambda_c \frac{\partial \phi_c}{\partial P_a}, \tag{4.2.14}$$

$$\dot{P}_a = \dot{\chi}_a = -\frac{\partial H}{\partial Q_a} - \sum_c \lambda_c \frac{\partial \phi_c}{\partial Q_a} = 0\,.$$

As already mentioned, choosing the subsidiary conditions in the form $\chi_a = 0$ leads to eliminating the ambiguity connected with the functions $\lambda_a$. Due to this circumstance, the last of the equations (4.2.14) makes it possible to express the functions $\lambda_a$ in terms of the canonical variables $p_b^*$ and $q_b^*$.

Furthermore, from (4.2.13) we have

$$\frac{\partial H^*}{\partial q_b^*} = \frac{\partial H}{\partial q_b^*} + \frac{\partial H}{\partial Q_a(p^*,q^*)} \frac{\partial Q_a(p^*,q^*)}{\partial q_b^*},$$

or, with the use of the last of equations (4.2.14) and the conditions $\phi_a = 0$,

$$\frac{\partial H^*}{\partial q_b^*} = \frac{\partial H}{\partial q_b^*} - \sum_c \lambda_c \frac{\partial \phi_c}{\partial Q_a}\frac{\partial Q_a}{\partial q_b^*} = \frac{\partial H}{\partial q_b^*} + \sum_a \lambda_a \frac{\partial \phi_a}{\partial q_b^*}.$$

A comparison of this formula with (4.2.14) yields

$$\dot{p}_b^* = -\,\partial H^*(p^*,q^*)/\partial q_b^*\,.$$

An equation for $q_b^*$ is found in a similar way. The resulting Hamilton equations for the system $\Gamma^*$ with $2(n-m)$ independent variables read

$$\dot{q}_b^* = \frac{\partial H^*(p^*,q^*)}{\partial p_b^*}, \qquad \dot{p}_b^* = -\frac{\partial H^*(p^*,q^*)}{\partial q_b^*}. \tag{4.2.15}$$

Thus, a Hamilton system with $2n$ variables $p_i$ and $q_i$ subject to $m$ constraints (4.2.1) and $m$ subsidiary conditions (4.2.8) can be reduced to a Hamilton system with $2(n-m)$ independent variables $p_b^*, q_b^*$.

## Transition Amplitude

It follows from the above considerations that the transition amplitude for a constrained system can also be expressed in the path-integral form. To do so, we have to solve the constraint equations and the equations of subsidiary conditions with respect to the canonical variables supplementary to $p^*$ and $q^*$. As a result, according to (4.1.13) we have

$$\langle q_1'',\ldots,q_n'';t''|q_1',\ldots,q_n';t'\rangle$$

$$= \int \prod_{t,b} \frac{\mathscr{D}p_b^*(t)\,\mathscr{D}q_b^*(t)}{2\pi} \exp\left\{ \mathrm{i}\int_{t'}^{t''} dt \left[ \sum_{b=m+1}^{n} p_b^* \dot{q}_b^* - H(Q_a(p_b^*,q_b^*), q_b^*, 0, p_b^*) \right] \right\}. \tag{4.2.16}$$

Solving the constraint equations is a tedious problem. It turns out, however, that this can be avoided by passing from the variables $p_b^*, q_b^*$ to the original variables $p_i$ and $q_i$. To that end, we rewrite (4.2.16) in the following equivalent form:

$$\langle q_1'', \ldots, q_n''; t'' | q_1', \ldots, q_n'; t' \rangle$$

$$= \int \prod_{t,b} \frac{\mathscr{D} p_b^*(t) \, \mathscr{D} q_b^*(t)}{2\pi} \prod_{a=1}^{m} \frac{\mathscr{D} P_a(t) \, \mathscr{D} Q_a(t)}{2\pi} \delta(P_a) \, \delta(Q_a - Q_a(p^*, q^*))$$

$$\times \exp\left\{ \mathrm{i} \int_{t'}^{t''} dt \left[ \sum_{a=1}^{m} P_a \dot{Q}_a + \sum_{b=m+1}^{n} p_b^* \dot{q}_b^* - H(P_a, Q_a, p_b^*, q_b^*) \right] \right\}. \quad (4.2.17)$$

Taking into account that

$$\delta(Q_a - Q_a(p^*, q^*)) = \det \left| \frac{\partial \phi_c}{\partial Q_a} \right| \delta(\phi_c)$$

we obtain for (4.2.17)

$$\langle q_1'', \ldots, q_n''; t'' | q_1', \ldots, q_n'; t' \rangle = \int \prod_t \mathscr{D} \mu(P_i(t), Q_i(t))$$

$$\times \exp\left\{ \mathrm{i} \int_{t'}^{t''} dt \left[ \sum_{a=1}^{m} P_a \dot{Q}_a + \sum_{b=m+1}^{n} p_b^* \dot{q}_b^* - H(P_a, Q_a, p_b^*, q_b^*) \right] \right\}, \quad (4.2.18)$$

where

$$\mathscr{D} \mu(P_i(t), Q_i(t)) = \prod_{b=m+1}^{n} \frac{\mathscr{D} p_b^* \, \mathscr{D} q_b^*}{2\pi} \prod_{a,c=1}^{m} \mathscr{D} P_a(t) \, \mathscr{D} Q_a(t) \, \delta(P_a) \, \delta(\phi_c)$$

$$\times \det \left| \frac{\partial \phi_c}{\partial Q_a} \right|$$

is the integration measure.

Let us carry out a canonical transformation from the variables $P_i$, $Q_i$ to the variables $p_i, q_i$. According to Liouville's theorem, the measures $\prod \mathscr{D} P_i \mathscr{D} Q_i$ and $\prod \mathscr{D} p_i \mathscr{D} q_i$ are invariant, and the canonical variables $p_i$ and $q_i$ are related to the canonical variables $P_i$ and $Q_i$ through the generating function $F(p, q)$ for which

$$dF = \sum_i p_i \, dq_i + \sum_i Q_i \, dP_i + (H' - H) \, dt .$$

Here

$$p_i = \frac{\partial F}{\partial q_i}, \qquad Q_i = \frac{\partial F}{\partial P_i}, \qquad H' = H + \frac{\partial F}{\partial t};$$

$H$ and $H'$ denote the Hamiltonian in terms of the variables $P_i, Q_i$ and $p_i, q_i$, respectively. In the case under consideration, the function $F(p, q)$ does not explicitly depend on the time, i.e., $H' = H$. Consequently,

$$(\sum_i p_i\dot{q}_i - H)\,dt = \sum_i (P_i\dot{Q}_i - H)\,dt + d(F - \sum_i P_i Q_i)\,,$$

and after integrating both sides of the equation over $t$

$$\int_{t'}^{t''} dt\left(\sum_i P_i\dot{Q}_i - H\right) = \int_{t'}^{t''} dt\left(\sum_i p_i\dot{q}_i - H\right) - (F - \sum_i P_i Q_i)\Big|_{t'}^{t''}\,.$$

We substitute this formula into (4.2.18). The last term leads to changing the integrand only at the integration limits. This term can thus be singled out from the integrand as a factor; we omit this immaterial factor (which is equivalent to carrying out a unitary transformation). This yields the desired expression for the amplitude in terms of a *path integral* over the variables $p_i$ and $q_i$:

$$\begin{aligned}&\langle q_1'',\ldots,q_n''; t''\,|\,q_1',\ldots,q_n'; t'\rangle \\ &= \int \prod_t \mathscr{D}\mu(q(t),p(t))\,\exp\left\{\mathrm{i}\int_{t'}^{t''} dt\left[\sum_i p_i\dot{q}_i - H(p_i,q_i)\right]\right\}\end{aligned} \tag{4.2.19}$$

with the integration measure

$$\mathscr{D}\mu(q(t),p(t)) = \prod_{a,c} \delta(\chi_a)\,\delta(\phi_c)\,\det|\{\chi_a,\phi_c\}|\prod_{i=1}^{n}\frac{\mathscr{D}p_i(t)\,\mathscr{D}q_i(t)}{(2\pi)^{n-m}}\,. \tag{4.2.20}$$

Thus, in the obtained path-integral expression for the amplitude the integration is carried out over all canonical variables $p_i$ and $q_i$, while the constraints are incorporated in the integration measure. Solving the constraint equations has been avoided in this way.

Finally, the amplitude can be expressed as a path integral of the form

$$\begin{aligned}&\langle q_1'',\ldots,q_n''; t''\,|\,q_1',\ldots,q_n'; t'\rangle \\ &= \int \prod_t \prod_{a,c=1}^{m} \delta(\chi_a)\,\det|\{\chi_a,\phi_c\}|\prod_{i=1}^{n}\frac{\mathscr{D}p_i(t)\,\mathscr{D}q_i(t)}{(2\pi)^{n-m}}\,\frac{\mathscr{D}\lambda_c(t)}{2\pi} \\ &\quad\times\exp\left\{\mathrm{i}\int_{t'}^{t''} dt\left[\sum_i p_i\dot{q}_i - H(p_i,q_i) - \sum_{a=1}^{m}\lambda_a\phi_a\right]\right\}.\end{aligned} \tag{4.2.21}$$

This follows from the fact that the integral over $\lambda_a(t)$ in (4.2.21) coincides with the function $\delta(\phi_a)$ entering (4.2.19).

### 4.2.2 The Electromagnetic Field

Let us now turn to constrained systems with an infinite number of degrees of freedom. Examples we have in mind are the gauge fields, viz. the electromagnetic field or the Yang-Mills field. The transition amplitude for these fields can be obtained in much the same way as for the unconstrained fields. One should proceed from the expression for the amplitude of a system with a finite number of degrees of freedom. This can further be generalized to the case of an infinite number of degrees of freedom. First we consider the free electromagnetic field.

1) As independent variables one can choose either the field $A_\mu(x)$ which leads to the Lagrangian

$$\mathscr{L}_e = -\tfrac{1}{4}(\partial_\mu A_\nu - \partial_\nu A_\mu)^2 , \tag{4.2.22}$$

or the field $A_\mu(x)$ and the tensor $F_{\mu\nu}(x)$, in which case the Lagrangian

$$\mathscr{L}_e = -\tfrac{1}{2}(\partial_\mu A_\nu - \partial_\nu A_\mu - \tfrac{1}{2}F_{\mu\nu})F_{\mu\nu} \tag{4.2.23}$$

is obtained. One speaks of the second order formalism and the first order formalism in the first and the second case, respectively. Of course, both lead to the same physical result. We shall use the first order formalism, i.e., the Lagrangian (4.2.23).

Let us introduce in (4.2.23) the three-dimensional notation, i.e., denote $\mu = 0, k$; $\nu = 0, l$; $k, l = 1,2,3$. Then, instead of (4.2.23), we obtain, omitting the full divergences and not writing the dependence on $A_0$ and $E_0$ explicitly (though this can be done, as will be shown below)

$$\mathscr{L}_e = E_k \dot{A}_k - H(E_k, A_k) , \tag{4.2.24}$$

where

$$\dot{A}_k = \partial_0 A_k , \qquad H(E_k, A_k) = \tfrac{1}{2}(E_k^2 + G_k^2) ,$$

$$E_k = F_{k0} , \qquad G_k = \tfrac{1}{2}\varepsilon_{ijk} F_{ji} .$$

By comparing (4.2.24) and (1.1.17) one can recognize that $H(E_k, A_k)$ is the Hamiltonian, $E_k(x)$ and $A_k(x)$ being the canonically conjugate variables whose Poisson's brackets are

$$\{E_k(x), A_l(y)\} = \delta_{kl}\delta(x-y) . \tag{4.2.25}$$

2) In this case, the role of the generalized coordinate is played by the field $A_\mu(x)$ and of the generalized momentum, according to (1.1.15), by the function

$$B_\mu(x) = F_{\mu 0}(x) = E_\mu(x) .$$

Consequently, $B_0(x) = E_0(x) = 0$. This is the primary constraint. Three other momenta $B_k(x)$ are equal to the components $E_k(x)$ of the electric field. Since

$$\{\textstyle\int d^3x H(E_k, A_k), E_0(x)\} = \partial_k E_k , \tag{4.2.26}$$

the condition of consistency implies that the secondary constraint is $\partial_k E_k = 0$. The secondary constraint satisfies the conditions

$$\{\partial_k E_k(x), \partial_i E_i(y)\} = 0\,, \tag{4.2.27}$$

$$\{\int d^3x H(E_k, A_k), \partial_i E_i(y)\} = 0\,, \tag{4.2.28}$$

i.e., there are no additional secondary constraints in this case. Thus, the electromagnetic field is a constrained system, the condition $E_0(x) = 0$ being its primary and the condition $\partial_k E_k(x) = 0$ its secondary constraint. In this case both Poisson's brackets in (4.2.5) vanish.

3) Fix the subsidiary conditions (4.2.8). In field theory, they are usually called *gauge conditions* or simply *gauges*. The situation with subsidiary conditions in the case of fields is similar to that with the systems having a finite number of degrees of freedom. The finite transformations of the gauge fields $A^k_\mu(x)$ have the form

$$A_\mu(x) \to A^\omega_\mu(x) = \omega(x) A_\mu(x)\, \omega^{-1}(x) + \omega^{-1}(x)\, \partial_\mu \omega(x)\,,$$

where $\omega(x)$ are the elements of the gauge group; for infinitesimal transformations,

$$\omega(x) = 1 + \varepsilon_k(x)\, T^k\,. \tag{4.2.28'}$$

To these transformations at a given point $x$ corresponds a class of fields, or an *orbit*: a set of the fields $A^{k\omega}_\mu(x)$, with $\omega(x)$ running over the entire gauge group. Gauge invariance means that the fields $A^{k\omega}_\mu(x)$ and $A^{k\omega'}_\mu(x)$ describe the same physical states. That is to say, a physical state is described not by one set of the fields $A^k_\mu(x)$ but by a whole class of physically equivalent sets of $A^{k\omega}_\mu(x)$. In practice, when dealing with classes of gauge-equivalent fields, it is sufficient to choose one representative from each class. For that, gauge conditions have to be imposed. In electrodynamics, the following gauges are most commonly used:

$$\partial_k A_k(x) = 0 \quad (k = 1,2,3) \qquad (\textit{Coulomb gauge})\,, \tag{4.2.29}$$

$$\partial_\mu A_\mu(x) = 0 \quad (\mu = 0,1,2,3) \qquad (\textit{Lorentz gauge})\,, \tag{4.2.30}$$

$$A_0(x) = 0 \qquad (\textit{Hamilton gauge})\,. \tag{4.2.31}$$

The gauge condition has to lead to a set of equations for the functions $\omega(x)$ which has a *unique* solution. For that, relations similar to (4.2.9) must be satisfied.

An orbit can be depicted as a line whose points are all physically equivalent and can be converted into each other by means of the gauge transformations. The gauge condition can be represented as a surface which crosses each orbit once (Fig. 4.2).

4) As already mentioned, in this case the number of the gauge conditions has to be equal to the number of constraints. Let us consider first the constraint $\partial_k E_k = 0$. Let us associate with it the Coulomb gauge (4.2.29). For

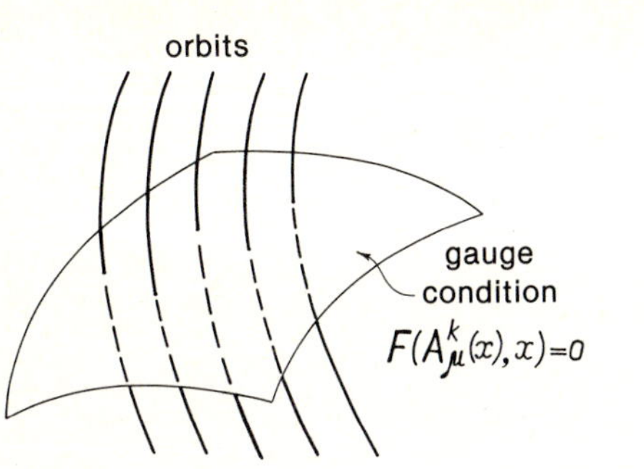


**Fig. 4.2.** Graphical representation of the orbits and of the gauge condition

this gauge, the theory can be formulated in the Hamiltonian form. It is this formulation which enabled us to develop the path-integral formalism. For the Coulomb gauge, the conditions (4.2.10) and (4.2.9) read

$$\{\partial_k A_k(x),\, \partial_i A_i(y)\} = 0\,, \tag{4.2.32}$$

$$\{\partial_k E_k(x),\, \partial_i A_i(y)\} = M_{ec}\,\delta(x-y)\,, \tag{4.2.33}$$

where $M_{ec} = \Delta = \partial_k\partial_k$. In perturbation theory, this operator is reversible and thus has no zero eigenvalues so that $\det|\{\partial_k E_k, \partial_i A_i\}| \neq 0$. That is to say, in perturbation theory, the gauge (4.2.29) obeys the condition of uniqueness. It should be noted that in the cases when perturbation theory is inapplicable, a solution may prove not to be unique (the *Gribov multi-valuedness*, or *ambiguity*).

A similar analysis can also be carried out for the second constraint $E_0 = 0$, if the gauge $A_0 = 0$ is associated with it.

5) If we take into account the constraints and the gauge chosen, then, according to (4.2.19), the transition amplitude for the free electromagnetic field can be expressed as a path integral over the canonical variables $A_k(x)$, $E_k(x)$, $A_0(x)$, $E_0(x)$:

$$\begin{aligned} S = \int \prod_x \prod_k \mathscr{D}A_k(x)\,\mathscr{D}E_k(x)\,\mathscr{D}A_0(x)\,\mathscr{D}E_0(x)\,\delta(\partial_k A_k(x))\,\delta(\partial_k E_k(x))\det|M_{ec}| \\ \times\,\delta(A_0(x))\,\delta(E_0(x))\exp\{\mathrm{i}\int dx[E_k\dot{A}_k + E_0\dot{A}_0 - H(E_k, A_k, E_0, A_0)]\}\,. \end{aligned} \tag{4.2.34}$$

From this we go over to the path integral over all the fields $A_\mu(x)$:

$$S = \int \prod_x \mathscr{D}A_\mu(x)\,\delta(\partial_k A_k(x))\exp\{\mathrm{i}\int dx\,\mathscr{L}_e(x)\}\,. \tag{4.2.35}$$

The Lagrangian $\mathscr{L}_e$ is determined by (4.2.22). The operator $M_{ec}$ is independent of $A_k(x)$ so that its determinant can be included in the normalization factor, which has been omitted here.

It should be emphasized that due to the presence of $\delta(A_0)$ and $\delta(E_0)$ in (4.2.34), the integration over $A_0$ and $E_0$ can be carried out, i.e., as a matter of fact, the expression (4.2.34) does not depend on the variables $A_0$ and $E_0$. This is why we have not written down the explicit dependence on the variables $A_0$ and $E_0$ in the formula (4.2.24).

### 4.2.3 The Yang-Mills Field

Let us now consider the Yang-Mills field in the case of $SU_2$-symmetry as the simplest example.

1) The Lagrangian of the Yang-Mills field, in the first order formalism, reads

$$\mathscr{L}_{\mathrm{YM}} = -\frac{1}{2}\left\{\partial_\mu A_\nu^k - \partial_\nu A_\mu^k - \frac{g}{2}\varepsilon_{lmk}(A_\mu^l A_\nu^m - A_\nu^l A_\mu^m) - \frac{1}{2}F_{\mu\nu}^k\right\}F_{\mu\nu}^k \tag{4.2.36}$$

or, in the three-dimensional form (dropping the full divergences and not writing down the explicit dependence on the variables $A_0$ and $E_0$)

$$\mathscr{L}_{\mathrm{YM}} = E_i^k \dot{A}_i^k - H(E_i^k, A_i^k) . \tag{4.2.37}$$

Here

$$H(E_i^k, A_i^k) = \tfrac{1}{2}[(E_i^k)^2 + (G_i^k)^2] , \quad E_i^k = F_{i0}^k , \quad G_i^k = \tfrac{1}{2}\varepsilon_{lji}F_{jl}^k .$$

Comparing (4.2.37) with (1.1.17) shows that $H(E_i^k, A_i^k)$ is the Hamiltonian and $E_i^k(x)$ and $A_i^k(x)$ are canonically conjugate variables with Poisson's brackets

$$\{E_i^k(x), A_j^l(y)\} = \delta_{kl}\delta_{ij}\delta(x-y) .$$

2) For the gauge field, the primary and the secondary constraints, respectively, are written as

$$E_0^k(x) = 0 ,$$
$$C^k(x) = \partial_i E_i^k(x) - g\varepsilon_{klm}A_i^l E_i^m = 0 .$$

The secondary constraint $C^k(x)$ obeys conditions analogous to (4.2.6):

$$\begin{aligned} &\{C^k(x), C^l(y)\} = g\varepsilon_{klm}C^m(x)\,\delta(x-y) , \\ &\{\textstyle\int d^3x H(E_i^k, G_i^k), C^l(y)\} = 0 , \end{aligned} \tag{4.2.38}$$

i.e., there are no subsidiary secondary constraints on the gauge field. In this case both Poisson's brackets in (4.2.5) vanish.

3) For the Yang-Mills fields, quite a number of gauges have been chosen; the most common ones are

$$\partial_i A_i^k(x) = 0 \quad (\textit{Coulomb gauge}) , \tag{4.2.39}$$

$$\partial_\mu A_\mu^k(x) = 0 \quad (\textit{Lorentz gauge}) , \tag{4.2.40}$$

$$n_\mu A_\mu^k(x) = 0 \quad (\textit{axial gauge}) , \tag{4.2.41}$$

where $n_\mu$ is a unit four-vector.

4) Consider first the constraint $C^k(x)=0$ and put it in correspondence with the Coulomb gauge (4.2.39). Equations (4.2.9, 10) in this case become

$$\begin{aligned}&\{\partial_i A_i^k(x),\, \partial_j A_j^l(y)\}=0\\ &\{C^k(x),\, \partial_i A_i^l(y)\}=M_{\mathrm{YC}}^{kl}(A_i^m)\,\delta(x-y)\,,\end{aligned}\tag{4.2.42}$$

where

$$M_{\mathrm{YC}}^{kl}(A_i^m)=\Delta\,\delta_{kl}+g\,\varepsilon_{klm}A_i^m\partial_i\,. \tag{4.2.43}$$

This operator is reversible (within perturbation theory); therefore $\det|M_{\mathrm{YC}}^{kl}|\neq 0$, i.e., the uniqueness condition is satisfied for this gauge.

If the constraint $E_0^k(x)=0$ is put in correspondence with the gauge $A_0^k(x)=0$, a similar analysis can be carried out.

5) Combining (4.2.38, 42) and (4.2.19) gives the transition amplitude for the Yang-Mills field expressed in terms of the path integral over the canonical variables:

$$\begin{aligned}S=\int\prod_{x,k}\mathscr{D}A_i^k(x)\,\mathscr{D}E_i^k(x)\,\mathscr{D}A_0^k(x)\,\mathscr{D}E_0^k(x)\,\det|M_{\mathrm{YC}}^{kl}(A_i^m)|\\ \times\,\delta(\partial_i A_i^k(x))\,\delta(C^k(x))\,\delta(A_0^k(x))\,\delta(E_0^k(x))\\ \times\exp\{\mathrm{i}\int dx\,[E_i^k\dot{A}_i^k+E_0^k\dot{A}_0^k-H(E_i^k,A_i^k,E_0^k,A_0^k)]\}\end{aligned}\tag{4.2.44}$$

or of the path integral over all the fields

$$S=\int\prod_{x,k}\mathscr{D}A_\mu^k(x)\,\delta(\partial_i A_i^k(x))\,\det|M_{\mathrm{YC}}^{kl}(A_i^m)|\,\exp[\mathrm{i}\int dx\,\mathscr{L}_{\mathrm{YM}}(x)]\,. \tag{4.2.45}$$

6) Because the Coulomb gauge (4.2.39) is relativistically non-invariant, the path integrals (4.2.44, 45) are manifestly relativistically non-invariant as well, this being their drawback. To eliminate this drawback, it is instructive to pass in the integrals (4.2.44, 45) to a manifestly relativistically invariant gauge. The simplest of them is the Lorentz gauge (4.2.40). The transition from the Coulomb gauge to the Lorentz gauge, e.g., in (4.2.45), can be performed with the aid of the following formal procedure (the *Faddeev-Popov trick*).

i) Let us introduce the functional $\Delta_{\mathrm{L}}(A_\mu^k(x))$ defined by

$$\Delta_{\mathrm{L}}(A_\mu^k)\int\prod_{k,x}\mathscr{D}\,\omega(x)\,\delta(\partial_\mu A_\mu^{k\omega}(x))=1\,. \tag{4.2.46}$$

Here $\mathscr{D}\omega(x)$ is the integration measure invariant with respect to the gauge group (i.e., the product of the differentials of the group parameters and the weight determined by the invariance of this measure under the group of gauge transformations): $\mathscr{D}(\omega\omega^\circ)=\mathscr{D}(\omega^\circ\omega)=\mathscr{D}(\omega)$. Since the integration measure is invariant, the functional $\Delta_{\mathrm{L}}(A_\mu^k)$ is also gauge-invariant:

$$\Delta_{\mathrm{L}}(A_\mu^{k\omega})=\Delta_{\mathrm{L}}(A_\mu^k)\,.$$

Substituting (4.2.46) into (4.2.45) yields

$$S = \int \prod_{x,k} \mathscr{D}\omega(x)\, \mathscr{D}A_\mu^k(x) \exp[\mathrm{i}\, dx\, \mathscr{L}_{\mathrm{YM}}(x)]\, \delta(\partial_i A_i^k(x)) \det|M_{\mathrm{YC}}^{kl}(A_i^k)|$$
$$\times \Delta_{\mathrm{L}}(A_\mu^k)\, \delta(\partial_\mu A_\mu^{k\omega}(x)) \,. \tag{4.2.47}$$

ii) Let us introduce another gauge-invariant functional $\Delta_{\mathrm{C}}(A_\mu^k)$ given on the surface $\partial_i A_i^k = 0$ and defined by

$$\Delta_{\mathrm{C}}(A_\mu^k) \int \prod_{x,k} \mathscr{D}\omega(x)\, \delta(\partial_i A_i^{k\omega}(x)) = 1 \,. \tag{4.2.48}$$

The functional $\Delta_{\mathrm{C}}(A_\mu^k)$ coincides with the functional $\det|M_{\mathrm{YC}}^{kl}(A_i^k)|$. Indeed, due to the condition $\partial_i A_i^k = 0$ the dominant contribution to the integral is from the vicinity of the unit element $\omega(x) = 1$. Using a series expansion of (4.2.28′) in this vicinity we obtain

$$\int \prod_{x,k} \mathscr{D}\omega(x)\, \delta(\partial_i A_i^{k\omega}(x)) \sim \int \prod_{x,k} \mathscr{D}\varepsilon_k(x)\, \delta(\partial_i A_i^{k\omega}(x))$$
$$= \int \prod_{x,k} \mathscr{D}\varepsilon_k(x)\, \delta(M_{\mathrm{YC}}^{kl} \varepsilon_l(x)) \,. \tag{4.2.49}$$

Let us calculate this integral. The eigenvalues of the operator $M_{\mathrm{YC}}^{kl}$ are determined from the equation

$$M_{\mathrm{YC}}^{kl} \varepsilon_l(x) = \lambda\, \varepsilon_k(x) \,.$$

Substituting the latter into (4.2.49) and utilizing the formula $\delta(\lambda\varepsilon) = |\lambda|^{-1}\delta(\varepsilon)$, we find

$$\int \prod_{x,k} \mathscr{D}\varepsilon_k(x)\, \delta(\lambda\, \varepsilon_k(x)) = \prod_{x,k} |\lambda|^{-1} \tag{4.2.50}$$

or, using the fact that the product of the eigenvalues of the operator (infinite-dimensional matrix) is equal to the determinant of this matrix,

$$\prod_{x,k} |\lambda|^{-1} = (\det|M_{\mathrm{YC}}^{kl}(A_i^k)|)^{-1} \,. \tag{4.2.51}$$

Making use of (4.2.48 – 51) we arrive at the desired result

$$\Delta_{\mathrm{C}}(A_\mu^k)\,|_{\partial_i A_i^k = 0} = \det|M_{\mathrm{YC}}^{kl}(A_i^k)| . \tag{4.2.52}$$

iii) Let us substitute (4.2.52) into (4.2.47) and subsequently make the change of variables $A_\mu^k \to A_\mu^{k\omega^{-1}}$ whose Jacobian is equal to unity. As a result, we obtain, with the use of the gauge-invariance of the action and of the functionals $\Delta_{\mathrm{L}}$ and $\Delta_{\mathrm{C}}$,

$$S = \int \prod_{x,k} \mathscr{D}\omega(x)\, \mathscr{D}A_\mu^k(x) \exp\{\mathrm{i} \int dx\, \mathscr{L}_{\mathrm{YM}}(x)\}\, \delta(\partial_\mu A_\mu^k)\, \Delta_{\mathrm{L}}(A_\mu^k)$$
$$\times\, \delta(\partial_i A_i^{k\omega^{-1}})\, \Delta_{\mathrm{C}}(A_\mu^k) \,. \tag{4.2.53}$$

The change of variables $A_\mu^{k\omega^{-1}} \to A_\mu^{k\omega}$ in the integral over $\omega$ leads, with the aid of (4.2.48), to the desired expression for the transition amplitude for the gauge field in the Lorentz gauge:

$$S = \int \prod_{x,k} \mathscr{D} A_\mu^k(x) \Delta_L(A_\mu^k(x)) \, \delta(\partial_\mu A_\mu^k(x)) \exp[\mathrm{i} \int dx \, \mathscr{L}_{YM}(x)] \, . \tag{4.2.54}$$

In the case $\partial_\mu A_\mu^k = 0$, a relation holds which is similar to (4.2.52):

$$\Delta_L(A_\mu^k(x)) = \det |\Box \, \delta_{kl} + g \, \varepsilon_{klm} A_\mu^m(x) \, \partial_\mu| . \tag{4.2.55}$$

7) The determinant entering (4.2.55) can be expressed as a path integral over the anti-commuting scalar fields $c^k(x)$ and $\bar{c}^k(x)$ which are commonly referred to as the *Faddeev-Popov ghosts*:

$$\Delta_L(A_\mu^k) = \int \prod_{x,k} \mathscr{D}\bar{c}^k(x) \, \mathscr{D}c^k(x) \times \exp\{\mathrm{i} \int dx [\bar{c}^k \Box \, c^k + g f_{klm} \bar{c}^k A_\mu^m \partial_\mu c^l]\} \, . \tag{4.2.56}$$

To prove this, we introduce the integral

$$I = \int \prod_{x,k} \mathscr{D}\bar{c}^k(x) \, \mathscr{D}c^k(x) \exp\{\mathrm{i} \int dx \, dy \, \bar{c}^k(x) M_{YL}^{kl}(A_\mu^k) c^l(y)\} \tag{4.2.57}$$

and carry out a change of variables

$$\bar{c}^k(x) = \bar{\xi}^k(x) \, , \quad M_{YL}^{kl}(A_\mu^k) c^l(x) = \xi^k(x) \, .$$

Then, making use of the rule (4.1.31) for changing the anti-commuting variables, the integral (4.2.57) can be rewritten as

$$I = \det |M_{YL}^{kl}(A_\mu^k)| \, . \tag{4.2.58}$$

From this and (4.2.55) Eq. (4.2.56) follows.

8) A similar procedure can be used for passing from the Coulomb gauge to the Lorentz gauge in the integrals (4.2.34, 35) for the transition amplitude for the electromagnetic field. However, in this case, the functionals $\Delta_C$ and $\Delta_L$ are independent of the gauge field $A_\mu(x)$. Therefore, $\Delta_C$ and $\Delta_L$ can be taken out of the integrand; the transition amplitude for the free electromagnetic field is then written, in the Lorentz gauge, as

$$S = \int \prod_x \mathscr{D} A_\mu(x) \exp[\mathrm{i} \int dx \, \mathscr{L}_e(x)] \, \delta(\partial_\mu A_\mu(x)) \, . \tag{4.2.59}$$

9) Relativistically-invariant expressions for the amplitude of the gauge field pertaining to a more general gauge can be produced in much the same way. Let us consider, for instance, the generalized Lorentz gauge

$$\partial_\mu A_\mu^k(x) = a^k(x) \, , \tag{4.2.60}$$

where $a^k(x)$ is an arbitrary function. To go over to this gauge, we introduce the functional $\Delta_a(A_\mu^k)$ defined by

$$\Delta_a(A_\mu^k) \int \prod_{x,k} \mathscr{D}\omega(x) \, \delta[\partial_\mu A_\mu^{k\omega}(x) - a^k(x)] = 1 \, . \tag{4.2.61}$$

On the surface $\partial_\mu A_\mu^k(x) = a^k(x)$ holds the relation

$$\Delta_a(A_\mu^k) = \det|M_a|,$$

where $M_a = \Box\,\delta_{kl} + g\varepsilon_{klm}(A_\mu^m \partial_\mu + \partial_\mu A_\mu^m)$. Therefore the expression for the amplitude is written as

$$S = \int \prod_{x,k} \mathscr{D} A_\mu^k(x)\, \delta(\partial_\mu A_\mu^k(x) - a^k(x)) \det|M_a|$$
$$\times \exp(-\mathrm{i}\int dx \tfrac{1}{4} F_{\mu\nu}^k F_{\mu\nu}^k) . \tag{4.2.62}$$

Because $\det|M_a|$ does not depend on $a^k(x)$, the last formula can be integrated over $a^k(x)$, for example, with the weight

$$\exp\left\{-\mathrm{i}\frac{1}{2\alpha}\int [a^k(x)]^2 dx\right\} .$$

This leads to the following expression for the amplitude of the gauge field

$$S = \int \prod_{x,k} \mathscr{D} A_\mu^k(x) \det|M_a|$$
$$\times \exp\left\{\mathrm{i}\int dx\left[-\frac{1}{4}F_{\mu\nu}^k F_{\mu\nu}^k - \frac{1}{2\alpha}(\partial_\mu A_\mu^k)^2\right]\right\} . \tag{4.2.63}$$

The amplitude (4.2.63) is said to have been given in the *α-gauge*.

Notice that in the case of non-Abelian gauge theories, upon the choice of some specific gauges there appears no ghost state. An example of such a gauge is the axial one (4.2.41), for which the determinant analogous to (4.2.55) does not depend on the gauge fields.

# 5. Covariant Perturbation Theory

In the preceding chapter, the expressions for the vacuum-to-vacuum transition amplitude have been obtained. Of practical interest, however, are transitions of a system of initially free particles into a final system of free particles. As shown in Sect. 1.1, such transitions are described by the matrix elements of the $S$-matrix. Therefore, we shall find in this chapter an expression for the *matrix element of the S-matrix*, or for the *transition amplitude* between two states, in terms of the path integral. Unfortunately, effective general methods of analytic calculations of these integrals are lacking as yet and one has to resort to approximate ones. Most developed are the methods of *perturbation theory* representing the transition amplitude as a series in the coupling constant.

It is convenient for practical calculations to use the Green's function. Therefore, we first survey the basic properties of the Green's functions and their generating functionals. Then we find expression for the amplitude in terms of the Green's functions. With the aid of this expression we shall further construct covariant perturbation theories for the model of $\phi^4$-interaction and for a model with non-Abelian gauge fields. Finally, we shall consider one of the modifications of the perturbation theory, in which the expansion is performed with respect to a parameter $1/N$ instead of the coupling constant.

## 5.1 Green's Functions. Generating Functionals

### 5.1.1 Path-Integral Formulation

The *Green's functions* can be expressed in terms of the vacuum-expectation-value of the chronological product of field operators. *Chronological* refers to a product where the operators are ordered in such a way that the time argument increases from right to left.

Consider, for instance, a scalar field. The chronological product of two operators of the scalar field is

$$T(\phi(x)\,\phi(y)) = \begin{cases} \phi(x)\,\phi(y) & \text{for} \quad x_0 > y_0\,, \\ \phi(y)\,\phi(x) & \text{for} \quad y_0 > x_0\,, \end{cases}$$

and their Green's function has, in the Heisenberg representation, the form

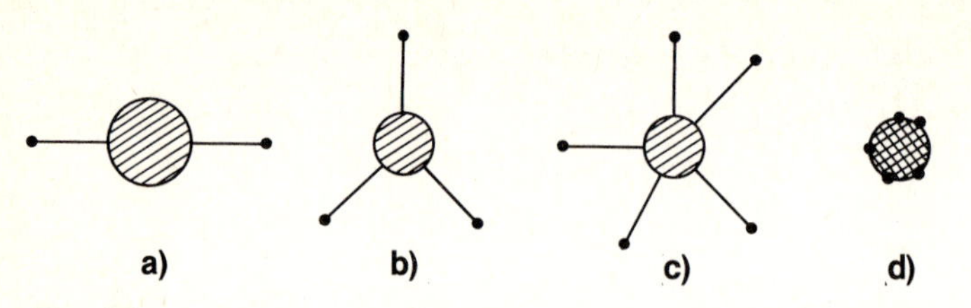


**Fig. 5.1a–d.** Green's functions: **(a)** two-point Green's function; **(b)** three-point Green's function; **(c)** five-point Green's function; **(d)** vertex Green's function

$$\mathrm{i}\langle 0|T(\phi(x)\phi(y))|0\rangle\,. \tag{5.1.1}$$

This function describes the propagation of a particle between the points $x$ and $y$ and is called the *propagation function*, or the *propagator*. Graphically, such a Green's function can be represented by a line with two end-points (Fig. 5.1 a) (two-point Green's function).

The Green's function of three field operators,

$$\mathrm{i}^2\langle 0|T(\phi(x)\phi(y)\phi(z))|0\rangle\,, \tag{5.1.2}$$

describes the interaction of three particles. It is represented by a three-point diagram (cf. Fig. 5.1 b) and can be referred to as the three-point Green's function.

Green's functions of a larger number of field operators, or $n$-point Green's functions, are defined in a similar way (cf. Fig. 5.1 c, for $n = 5$ as an example).

Green's functions can be expressed as path integrals. For this, the same technique can be taken advantage of as was used for deriving the corresponding expressions for the amplitude (cf. Chap. 4). For example, for the Green's function (5.1.1) we find [compare with (4.1.22)]

$$\mathrm{i}\langle 0|T(\phi(y)\phi(z))|0\rangle = \mathrm{i}\int \prod_x \mathscr{D}\phi(x)[\phi(y)\phi(z)\exp\{\mathrm{i}\int dx\,\mathscr{L}(x)\}]\,. \tag{5.1.3}$$

Green's functions of an arbitrary number of the operators $\phi(x)$ can be written down in a similar way. It can be recognized that a Green's function is a weighted average of the product of two or more fields, the weighting factor being $\exp[\mathrm{i}\int dx\,\mathscr{L}(x)]$. Dropping the product $\phi(y)\phi(z)$ in (5.1.3) reduces this expression to (4.1.22).

### 5.1.2 Generating Functional $W(J)$

Green's functions can be derived using the device of the generating functionals. For this purpose, we introduce for each field $u_i(x)$ an auxiliary, unphysical source to correspond to an auxiliary current $J_i(x)$. This gives rise to an additional term $J_i(x)\,u_i(x)$ in the Lagrangian, and the new action assumes the form

$$\int dx\, \mathcal{L}'(x) = \int dx[\mathcal{L}(x) + J_i(x)u_i(x)]\,, \tag{5.1.4}$$

i.e., the action becomes a functional of the currents $J_i(x)$.

Let us introduce the *generating functional* $W(J)$ defined as

$$W(J) = \int \prod_x \mathcal{D}\mu(u_1(x),\ldots,u_n(x)) \\ \times \exp\{\mathrm{i}\int dx[\mathcal{L}(x) + u_1(x)J_1(x) + \ldots + u_n(x)J_n(x)]\}\,, \tag{5.1.5}$$

where $\mathcal{D}\mu(u_i(x))$ is the integration measure. With the help of this generating functional the Green's functions are found by taking variational derivatives with respect to the currents and then setting all these currents equal to zero:

$$\langle 0|T(u_1(x_1)u_2(x_2)\ldots u_n(x_n))|0\rangle \\ = (-\mathrm{i})^n \frac{\delta}{\delta J_1(x_1)} \frac{\delta}{\delta J_2(x_2)} \cdots \frac{\delta}{\delta J_n(x_n)} W(J)\bigg|_{J_1=\ldots=J_n=0}\,. \tag{5.1.6}$$

Consider, as an example, quantum electrodynamics described by the Lagrangian

$$\mathcal{L} = -\tfrac{1}{4}F_{\mu\nu}F_{\mu\nu} + \mathrm{i}\bar{\psi}\gamma_\mu\partial_\mu\psi - M\bar{\psi}\psi - e\bar{\psi}\gamma_\mu\psi A_\mu\,. \tag{5.1.7}$$

The corresponding generating functional $W(J,\eta)$ reads

$$W(J,\eta) = \int \mathcal{D}A_\mu(x)\,\mathcal{D}\bar{\psi}(x)\,\mathcal{D}\psi(x)\,\delta(\partial_\mu A_\mu) \\ \times \exp[\mathrm{i}\int dx(\mathcal{L} + J_\mu A_\mu + \bar{\eta}\psi + \bar{\psi}\eta)]\,. \tag{5.1.8}$$

From this, we have, e.g., for the two-point Green's functions of an electron, of a photon and for the three-point Green's function

$$G(x-y) = -\mathrm{i}\frac{\delta}{\delta\eta(x)}\frac{\delta}{\delta\bar{\eta}(y)}W(J,\eta)\bigg|_{J_\mu=\bar{\eta}=\eta=0}\,, \tag{5.1.9}$$

$$D_{\mu\nu}(x-y) = -\mathrm{i}\frac{\delta}{\delta J_\mu(x)}\frac{\delta}{\delta J_\nu(y)}W(J,\eta)\bigg|_{J_\mu=\eta=\bar{\eta}=0}\,, \tag{5.1.9$'$}$$

$$\Delta_\mu(x,y;z) = (-\mathrm{i})^2\frac{\delta}{\delta\eta(x)}\frac{\delta}{\delta\bar{\eta}(y)}\frac{\delta}{\delta J_\mu(z)}W(J,\eta)\bigg|_{J_\mu=\bar{\eta}=\eta=0}\,, \tag{5.1.10}$$

respectively.

### 5.1.3 Green's Functions in Perturbation Theory

The action in (5.1.8) can be expressed as

$$\int dx\, \mathcal{L}(x) = \int dx\, \mathcal{L}_0(x) + \int dx\, \mathcal{L}_1(x)\,,$$

where $\mathcal{L}_0(x)$ is the Lagrangian of the free fields (i.e., of the spinor and the electromagnetic field) and $\mathcal{L}_1 = -e\bar{\psi}(x)\gamma_\mu\psi(x)A_\mu(x)$ is the interaction Lagrangian between the spinor and the electromagnetic field.

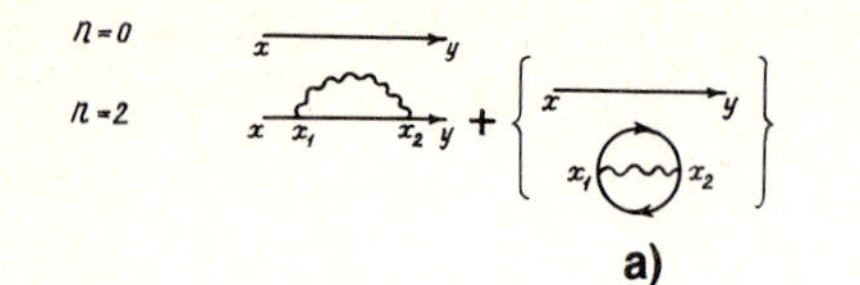


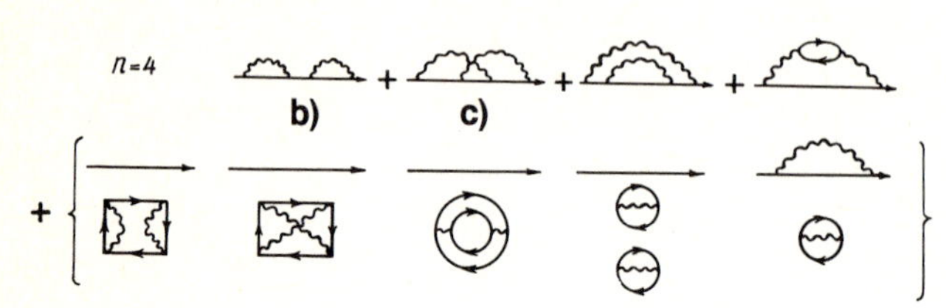


**Fig. 5.2.** Diagrams for the electron Green's function in perturbation theory

*Perturbation theory* is obtained by making a series expansion, with respect to the constant $e$, of $\exp[\mathrm{i}\int dx\, \mathscr{L}_{\mathrm{I}}(x)]$ in the integrand of the path integral,

$$\exp[\mathrm{i}\int dx\, \mathscr{L}_{\mathrm{I}}(x)] = \sum_{n=0}^{\infty} \frac{\mathrm{i}^n}{n!} \int dx_1 \int dx_2 \ldots \int dx_n\, \mathscr{L}_{\mathrm{I}}(x_1)\, \mathscr{L}_{\mathrm{I}}(x_2) \ldots \mathscr{L}_{\mathrm{I}}(x_n)\,, \tag{5.1.11}$$

and by a subsequent term-by-term integration of the resulting series. Substituting (5.1.11), for example, into (5.1.9) we obtain the electron two-point Green's function as given by perturbation theory:

$$G(x-y) = -\mathrm{i} \frac{\delta}{\delta\eta(x)} \frac{\delta}{\delta\bar{\eta}(y)} \Big\{ \int \prod_x \mathscr{D}A_\mu(x)\, \mathscr{D}\bar{\psi}(x)\, \mathscr{D}\psi(x)\, \delta(\partial_\mu A_\mu)$$

$$\times \exp\{\mathrm{i}\int dx [\mathscr{L}_0 + J_\mu A_\mu + \bar{\eta}\psi + \bar{\psi}\eta]\} \Big[ 1 + \mathrm{i}\int \mathscr{L}_{\mathrm{I}}(x)\, dx$$

$$+ \frac{\mathrm{i}^2}{2} \int dx \int dy\, \mathscr{L}_{\mathrm{I}}(x)\, \mathscr{L}_{\mathrm{I}}(y)$$

$$+ \frac{\mathrm{i}^3}{6} \int dx \int dy \int dz\, \mathscr{L}_{\mathrm{I}}(x)\, \mathscr{L}_{\mathrm{I}}(y)\, \mathscr{L}_{\mathrm{I}}(z) + \ldots \Big] \Big\} \Big|_{J_\mu = \bar{\eta} = \eta = 0} .$$

The Feynman diagrams for the electron two-point Green's function which correspond to the zeroth, second, and fourth orders of perturbation theory ($n = 0, 2, 4$) are depicted in Fig. 5.2, where the tadpole diagrams have been omitted.

One can see that the Green's function contains all orders of perturbation theory.

### 5.1.4 Types of Diagrams

In Fig. 5.2, three types of diagrams can be recognized:

i) *disconnected diagrams* containing pieces not connected by lines (e.g., Fig. 5.2a);

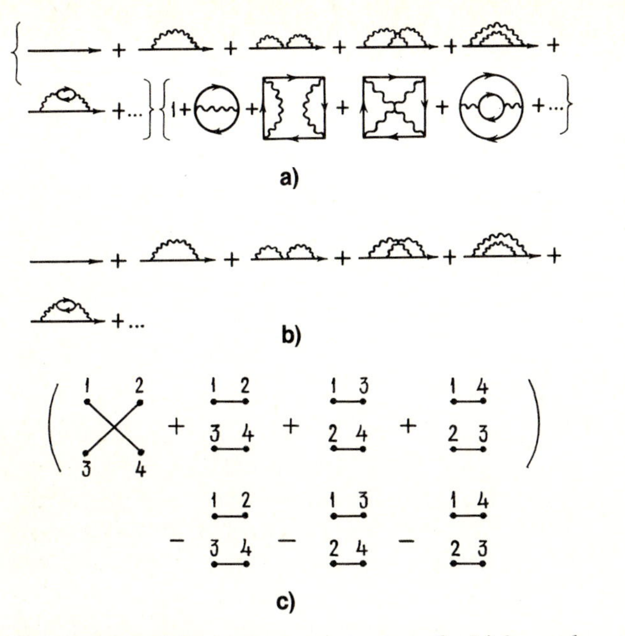


**Fig. 5.3 a – c.** Diagrams for the electron Green's functions corresponding to (**a**) the generating functional $W$; (**b**) the generating functional $W/W_0$; (**c**) the right-hand side of formula (5.1.13)

ii) *connected diagrams* every vertex of which can be reached from any other vertex by moving along the lines of the graph (e.g., Fig. 5.2 b);

iii) *one-particle-irreducible* (OPI in abbreviation) *diagrams* which cannot be converted into disconnected graphs by cutting just one internal line (e.g., Fig. 5.2 c).

These diagrams can be brought into correspondence with *complete, connected* and *one-particle-irreducible Green's functions.*

The diagrams within the curly brackets in Fig. 5.2 contain vacuum-to-vacuum transitions. Since these transitions do not describe scattering, they should be excluded from the consideration. For this purpose, we use the circumstance that, as can be shown, the Green's function diagrams can be expressed, to all orders of the expansion, as the product of two sums: one containing only vacuum-to-vacuum transition diagrams and the other not containing such diagrams (cf. Fig. 5.3 a). The sum of the Green's functions associated with diagrams of vacuum-to-vacuum transitions is described, according to (5.1.5), by the functional $W(0)$. Therefore, by dividing the generating functional $W(J)$ by $W(0)$ we obtain the generating functional $W(J)/W(0)$ not containing vacuum-to-vacuum transitions (cf. Fig. 5.3 b).

### 5.1.5 Generating Functional $Z(J)$

Let us introduce, in addition to the generating functional $W(J)$, the generating functional $Z(J)$ defined by

$$Z(J) = -\mathrm{i} \ln W(J) . \tag{5.1.12}$$

Let us find out for which Green's functions the functional $Z(J)$ is the generating one. To this goal, consider the expression for the four-point Green's function

$$\begin{aligned} \mathrm{i}\frac{\delta^4 Z(J)}{\delta J_1 \delta J_2 \delta J_3 \delta J_4}\bigg|_{J=0} &= \frac{1}{W(0)}\frac{\delta^4 W(J)}{\delta J_1 \delta J_2 \delta J_3 \delta J_4}\bigg|_{J=0} \\ &- \frac{1}{W^2(0)}\frac{\delta^2 W(J)}{\delta J_1 \delta J_2}\bigg|_{J=0} \cdot \frac{\delta^2 W(J)}{\delta J_3 \delta J_4}\bigg|_{J=0} \\ &- \frac{1}{W^2(0)}\frac{\delta^2 W(J)}{\delta J_1 \delta J_3}\bigg|_{J=0} \cdot \frac{\delta^2 W(J)}{\delta J_2 \delta J_4}\bigg|_{J=0} \\ &- \frac{1}{W^2(0)}\frac{\delta^2 W(J)}{\delta J_1 \delta J_4}\bigg|_{J=0} \cdot \frac{\delta^2 W(J)}{\delta J_3 \delta J_2}\bigg|_{J=0} . \end{aligned} \quad (5.1.13)$$

The right-hand side of this formula is graphically represented in Fig. 5.3c. The first term in (5.1.13) contains, as seen, both connected and disconnected four-point Green's functions and, owing to the factor $1/W(0)$, does not contain Green's functions describing vacuum-to-vacuum transitions. The Green's function corresponding to the first term of (5.1.13) is graphically represented by the parenthesised diagrams in Fig. 5.3c. The remaining three terms in (5.1.13) are associated with disconnected Green's functions and are subtracted from the first term. Consequently, the left-hand side of (5.1.13) contains only connected Green's functions.

It can be shown by induction that many-point Green's functions have a similar structure. Thus, the functional $Z(J)$ is the generating functional for connected Green's functions.

### 5.1.6 Generating Functional $\Gamma(\Phi)$

Consider one-particle-irreducible diagrams and the corresponding Green's functions whose external lines are amputated (cf. Fig. 5.1d). Such Green's functions will be referred to as the *vertex Green's functions*. A corresponding functional $\Gamma(\Phi)$ can be attributed to them.

Let us demonstrate that for a given generating functional $Z(J)$, the generating functional $\Gamma(\Phi)$ of the vertex Green's functions is determined by the relation

$$\Gamma(\Phi_i) = Z(J_i) - \int dx\, J_i(x)\, \Phi_i(x) , \quad (5.1.14)$$

where

$$\Phi_i(x) = \frac{\delta Z(J)}{\delta J_i(x)} . \quad (5.1.15)$$

The expression (5.1.14) describes the functional Legendre transformation which introduces a new functional argument, $\Phi_i(x)$, instead of the functional argument $J_i(x)$.

According to (5.1.14), the first derivative of the functional $\Gamma(\Phi)$ with respect to $\Phi_i(x)$ is equal to the current $J_i(x)$:

$$\frac{\delta\Gamma(\Phi)}{\delta\Phi_i(x)} = -J_i(x)\,. \tag{5.1.16}$$

The second derivative of $\Gamma(\Phi)$ with respect to $\Phi_i(x)$ is given by

$$(-\mathrm{i})^2\frac{\delta^2\Gamma(\Phi)}{\delta\Phi_i(x)\,\delta\Phi_j(y)} = X_{ij}(x-y;\Phi)\,. \tag{5.1.17}$$

At $\Phi=0$ this relation reduces to the expression for the inverse propagator. In fact, the second derivative of the functional $Z(J)$ reads as

$$\frac{\delta^2 Z(J)}{\delta J_i(x)\,\delta J_j(y)} = X_{ij}^{-1}(x-y;J)\,. \tag{5.1.18}$$

At $J=0$ the expression for the propagator is recovered. By differentiating (5.1.15, 16) with respect to $J_j(y)$ and $\Phi_j(y)$, respectively, and transforming the expressions obtained, we find:

$$\int dy\frac{\delta^2 Z(J)}{\delta J_i(x)\,\delta J_j(y)}(-\mathrm{i})^2\frac{\delta^2\Gamma(\Phi)}{\delta\Phi_j(y)\,\delta\Phi_k(z)} = \delta_{ik}\delta(x-z)\,. \tag{5.1.19}$$

Since the product of the Green's function and its inverse is a $\delta$-function, (5.1.17) describing the two-point Green's function turns out to coincide with the inverse propagator if both $J$ and $\Phi$ are set equal to zero.

Differentiating further (5.1.19) with respect to $J_k(z)$ and using the derivatives of (5.1.15) with respect to $J_j(y)$ and of (5.1.16) with respect to $\Phi_j(y)$ we arrive at the following expression for the third derivative of the functional $Z(J)$ with respect to $J_i(x)$:

$$\begin{aligned}(-\mathrm{i})^2\frac{\delta^3 Z(J)}{\delta J_i(x)\,\delta J_j(y)\,\delta J_k(z)} &= \int d\zeta\, d\xi\, d\eta\, X_{il}^{-1}(x-\zeta;J)\\ &\times X_{jm}^{-1}(y-\xi;J)X_{kn}^{-1}(z-\eta;J)(-\mathrm{i})^2\frac{\delta^3\Gamma(\Phi)}{\delta\Phi_l(\zeta)\,\delta\Phi_m(\xi)\,\delta\Phi_n(\eta)}\,.\end{aligned} \tag{5.1.20}$$

The integrand contains the product of three propagators and a three-point vertex function (cf. Fig. 5.4a) when $\Phi$ is set equal to zero:

$$\Gamma(\zeta,\eta,\xi) = \left.\frac{(-\mathrm{i})^2\delta^3\Gamma(\Phi)}{\delta\Phi(\zeta)\,\delta\Phi(\xi)\,\delta\Phi(\eta)}\right|_{\Phi=0},$$

which is obtained from the corresponding one-particle-irreducible Green's function by amputating the external propagators (Fig. 5.4b).

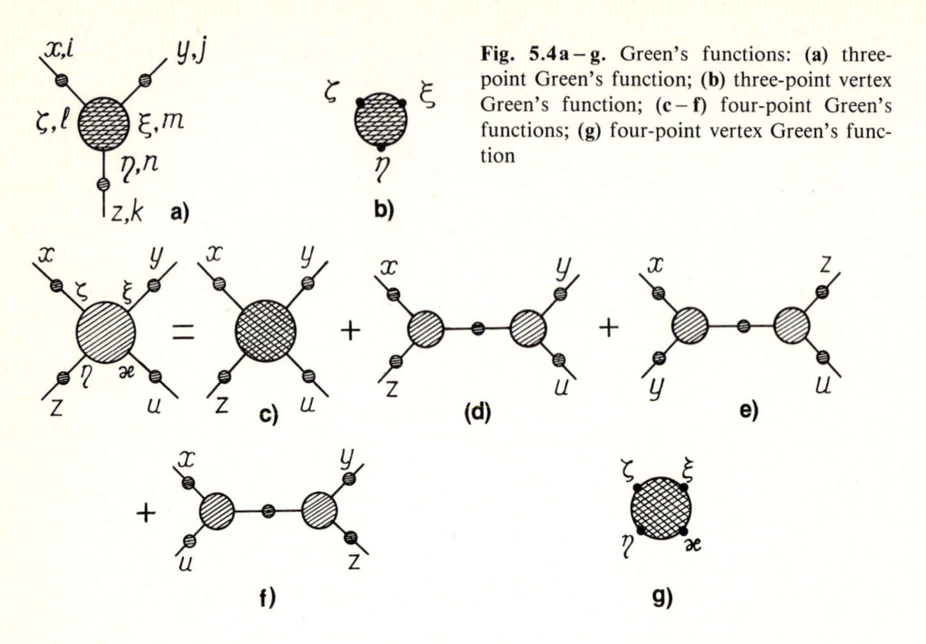


**Fig. 5.4a – g.** Green's functions: **(a)** three-point Green's function; **(b)** three-point vertex Green's function; **(c – f)** four-point Green's functions; **(g)** four-point vertex Green's function

The fourth derivative of the functional $Z(J)$ with respect to $J_i(x)$ is found in a similar way. The corresponding expression is graphically presented in Figs. 5.4c – f; the associated vertex function is depicted in Fig. 5.4g.

It can be proven by induction that the integrand has a similar structure in the case of many-point Green's functions, too. That is to say, $\Gamma(\Phi)$ is the generating functional for the vertex Green's functions.

### 5.1.7 The Tree Approximation for $\Gamma(\Phi)$

Finding an explicit form for the functional $\Gamma(\Phi)$ is generally a rather difficult problem. However, $\Gamma(\Phi)$ has a simple form within the lowest orders of the so-called loopwise expansion of the amplitude, i.e., of the expansion in terms of the number of the independent 4-momenta to be integrated. To Green's functions containing $n$ independent internal momenta corresponds an $n$-loop diagram. In order to find the corresponding Feynman rules, an approximation is required which gives loop-free diagrams. This is referred to as the *tree approximation.* The generating functional $Z(J)$ in the tree approximation can be shown to have the form

$$Z_{\text{tree}}(J) = I(\phi_{0i}) + \int dx\, J_i(x)\, \phi_{0i}(x)\,, \tag{5.1.21}$$

where $I(\phi_{0i})$ is the classical action, $\phi_{0i}$ being the solutions of the classical field equations with external sources $J_i(x)$:

$$\frac{\delta I(\phi_{0i})}{\delta \phi_{0i}(x)} = -J_i(x)\,. \tag{5.1.22}$$

By combining (5.1.15, 21) and using (5.1.22) we obtain, in the tree approximation,

$$\Phi_i(x) = \frac{\delta Z_{\text{tree}}(J)}{\delta J_i(x)} = \int dy \left( \frac{\delta I(\phi_{0i})}{\delta \phi_{0j}(y)} \frac{\delta \phi_{0j}(y)}{\delta J_i(x)} + J_j(y) \frac{\delta \phi_{0j}(y)}{\delta J_i(x)} \right) + \phi_{0i}(x)$$

$$= \phi_{0i}(x) . \tag{5.1.23}$$

Substituting (5.1.21) and (5.1.23) into (5.1.14) gives the desired expression for the generating functional $\Gamma(\Phi)$ in the tree approximation:

$$\Gamma_{\text{tree}}(\Phi) = Z_{\text{tree}}(J) - \int dx\, J_i(x)\, \Phi_i(x) = I(\Phi_i) . \tag{5.1.24}$$

### 5.1.8 Another Expression for $W(J)$

We now transform the generating functional (5.1.5) to a form convenient for calculating the Green's functions.

1) In (5.1.5), we explicitly single out the interaction Lagrangian $\mathscr{L}_\mathrm{I}$:

$$W(J) = \int \prod_x \mathscr{D}\mu(u_k(x))$$
$$\times \exp[\mathrm{i} \int dx\, \mathscr{L}_0(x)] \exp\{\mathrm{i} \int dx [\mathscr{L}_\mathrm{I}(x) + u_k(x) J_k(x)]\} , \tag{5.1.25}$$

where

$$\mathscr{D}\mu(u_k(x)) = \mathscr{D}u_1(x)\, \mathscr{D}u_2(x) \ldots \mathscr{D}u_n(x) .$$

2) Unsing the formula

$$\exp\left( \sum_k a_k u^k + Ju \right) = \exp\left[ \sum_k a_k \left( \frac{\delta}{\delta J} \right)^k \right] \exp(Ju)$$

we can rewrite the last factor in (5.1.25) as

$$\exp\left[ \mathrm{i} \int dx\, \mathscr{L}_\mathrm{I} \left( -\mathrm{i} \frac{\delta}{\delta J_1(x)}, -\mathrm{i} \frac{\delta}{\delta J_2(x)}, \ldots, -\mathrm{i} \frac{\delta}{\delta J_n(x)} \right) \right] \exp(\mathrm{i} \int dx J_k u_k)$$

and take the differential operator entering it out of the integrand of the path integral. Then we get, in place of (5.1.25):

$$W(J) = \exp\left[ \mathrm{i} \int dx\, \mathscr{L}_\mathrm{I} \left( \frac{1}{\mathrm{i}} \frac{\delta}{\delta J_1}, \ldots, \frac{1}{\mathrm{i}} \frac{\delta}{\delta J_n} \right) \right] \int \prod_x \mathscr{D}\mu(u_k(x))$$
$$\times \exp\{\mathrm{i} \int dx [\mathscr{L}_0(x) + u_k(x) J_k(x)]\} . \tag{5.1.26}$$

3) Let us make use of the fact that the integral in (5.1.26) which contains the free Lagrangian $\mathscr{L}_0$ can be expressed as

$$I_0 = \tfrac{1}{2} \int dx\, dy\, u_i(x) K_{ij}(x-y) u_j(y) , \tag{5.1.27}$$

with $K_{ij}(x-y)$ denoting the differential operator which is determined by the form of $\mathscr{L}_0$. The substitution of (5.1.27) into (5.1.26) and a subsequent change of variables

$$u_i(x) \to \tilde{u}_i(x) - \int K_{il}^{-1}(x-y) J_l(y) dy , \tag{5.1.28}$$

eventually yield for the path integral:

$$\begin{aligned} &\int \prod_x \mathscr{D}\mu(u_k(x)) \exp[\tfrac{\mathrm{i}}{2} \int dx\, dy\, u_i(x) K_{ij}(x-y) u_j(y) + \mathrm{i} \int dx\, u_i(x) J_i(x)] \\ &\quad = N \exp[-\tfrac{\mathrm{i}}{2} \int dx\, dy\, J_i(x) K_{ij}^{-1}(x-y) J_j(y)] , \end{aligned} \tag{5.1.29}$$

where

$$N = \int \prod_x \mathscr{D}\mu(\tilde{u}_k(x)) \exp[\tfrac{\mathrm{i}}{2} \int dx\, dy\, \tilde{u}_i(x) K_{ij}(x-y) \tilde{u}_j(y)] .$$

Substituting (5.1.29) into (5.1.26) and dropping the immaterial constant factor $N$, we arrive at the desired expression for $W(J)$:

$$\begin{aligned} W(J) = \exp &\left[ \mathrm{i} \int dx\, \mathscr{L}_\mathrm{I} \left( \frac{1}{\mathrm{i}} \frac{\delta}{\delta J_1(x)}, \ldots, \frac{1}{\mathrm{i}} \frac{\delta}{\delta J_n(x)} \right) \right] \\ &\times \exp \left[ -\frac{\mathrm{i}}{2} \int dx\, dy\, J_i(x) K_{ij}^{-1}(x-y) J_j(y) \right] . \end{aligned} \tag{5.1.30}$$

### 5.1.9 Expression for the Matrix Elements of the $S$-Matrix in Terms of Green's Functions

With the aid of the generating functional (5.1.30), one can construct the matrix elements of the $S$-matrix in the framework of perturbation theory. For this purpose, we introduce a functional containing $W(J)$:

$$\begin{aligned} S(u_k^0) = \exp &\left[ \mathrm{i} \int dx\, \mathscr{L}_\mathrm{I} \left( \frac{1}{\mathrm{i}} \frac{\delta}{\delta J_1(x)}, \ldots, \frac{1}{\mathrm{i}} \frac{\delta}{\delta J_n(x)} \right) \right] \\ &\times \exp \left[ -\mathrm{i} \int dx\, u_k^0(x) J_k(x) \right. \\ &\left. \left. - \frac{\mathrm{i}}{2} \int dx\, dy\, J_i(x) K_{ij}^{-1}(x-y) J_j(y) \right] \right|_{J_1 = \ldots = J_n = 0} . \end{aligned} \tag{5.1.31}$$

Here $u_k^0(x)$ are arbitrary functions. If $u_k^0(x)$ is set equal to zero, the functional (5.1.31) becomes related to (5.1.30).

As it stands, the functional (5.1.31) cannot be calculated exactly, and one is compelled to make use of perturbation theory. The perturbation theory expansion of (5.1.31) leads to diagrams with external free points to which correspond arbitrary functions $u_k^0(x)$. However, in the diagrams for the matrix elements of the $S$-matrix, free external points are contained, associated with the wave functions $u_k(x)$ which are the solutions of free particle equations. (The functions $u_k(x)$ are said to be defined on the mass-shell.) In order to obtain from (5.1.31) the perturbation theory expressions for the matrix elements, the functions $u_k^0(x)$ have to be replaced in (5.1.31) by the functions $u_k(x)$. Consequently, the *matrix element* $S_{m\to n}$ which describes the transition of $m$ free initial particles into a system of $n$ free final particles is obtained according to the rule

$$S_{m\to n} = u_{i_1}(x)\frac{\delta}{\delta u_1^0(x)} \dots u_{i_m}(x)\frac{\delta}{\delta u_m^0(x)} u_{f_{m+1}}(y)\frac{\delta}{\delta u_{m+1}^0(y)} \dots u_{f_{m+n}}(y)\frac{\delta}{\delta u_{m+n}^0(y)} S(u_k^0)\bigg|_{u_1^0=\dots=0} . \tag{5.1.32}$$

Here $u_k^0(x)$ are arbitrary functions, $u_{i_k}(x)$ are the free wave functions of the initial state and $u_{f_j}(x)$ are the free wave functions of the final state. The functions $u_{i_k}(x)$ and $u_{f_j}(x)$ satisfy the equations for the corresponding free particles.

Below we shall apply the outlined method for constructing the covariant perturbation theory to two specific models.

## 5.2 The $\phi^4$-Interaction Model

The model is described by the Lagrangian

$$L = \frac{1}{2}(\partial_\mu\phi)^2 - \frac{m^2}{2}\phi^2 - \frac{g}{4!}\phi^4 . \tag{5.2.1}$$

Making use of the fact that the integration by parts gives

$$\int dx[(\partial_\mu\phi)^2 - m^2\phi^2] = -\int dx\,\phi(x)(\Box + m^2)\phi(x)\,, \quad \Box \equiv \partial_\mu\partial_\mu\,, \tag{5.2.1'}$$

we can write the generating functional $W(J)$, according to (5.1.26), in the form

$$W(J) = \exp\left[-\mathrm{i}\int dy\frac{g}{4!}\left(\frac{1}{\mathrm{i}}\frac{\delta}{\delta J(y)}\right)^4\right]\int\prod_x \mathscr{D}\phi(x)$$
$$\times \exp\left\{\frac{\mathrm{i}}{2}\int dx[-\phi(\Box + m^2)\phi] + \mathrm{i}\int dx\,\phi J\right\} .$$

After some transformations, this reduces to the form of (5.1.30):

$$W(J)=\exp\left[-\mathrm{i}\frac{g}{4!}\int dy\left(\frac{1}{\mathrm{i}}\frac{\delta}{\delta J(y)}\right)^4\right]$$
$$\times\exp\left[-\frac{\mathrm{i}}{2}\int dx\,dy\,J(x)K^{-1}(x-y)J(y)\right]. \qquad (5.2.2)$$

According to (5.1.31), we then have for the generating functional $S(\phi_0)$

$$S(\phi_0)=\exp\left[-\mathrm{i}\frac{g}{4!}\int dy\left(\frac{1}{\mathrm{i}}\frac{\delta}{\delta J(y)}\right)^4\right]$$
$$\times\exp\left[-\mathrm{i}\int dx\,\phi_0(x)\,J(x)-\frac{\mathrm{i}}{2}\int dx\,dy\,J(x)K^{-1}(x-y)J(y)\right]\Bigg|_{J=0}. \qquad (5.2.3)$$

Let us find the operator $K^{-1}(x-y)$. The operator $K(x-y)$ is given by (5.2.1′)

$$K(x-y)=-(\Box+m^2)\,\delta(x-y)\,.$$

By definition, the operator $K(x-y)$ and its inverse, the operator $K^{-1}(x-y)$, are related by

$$\int dy\,K(x-y)K^{-1}(y-z)=\delta(x-z)=\frac{1}{(2\pi)^4}\int dp\,\mathrm{e}^{-\mathrm{i}p(x-z)}\,. \qquad (5.2.4)$$

The substitution of $K(x-y)$ into (5.2.4) gives

$$-(\Box+m^2)_x\frac{1}{(2\pi)^4}\int dp\,K^{-1}(p)\,\mathrm{e}^{-\mathrm{i}p(x-z)}$$
$$=\frac{1}{(2\pi)^4}\int dp\,(p^2-m^2)K^{-1}(p)\,\mathrm{e}^{-\mathrm{i}p(x-z)}\,.$$

From this it follows that the inverse operator $K^{-1}(p)$, in the momentum representation, is

$$K^{-1}(p)=\frac{1}{p^2-m^2}\,. \qquad (5.2.5)$$

In the coordinate representation,

$$K^{-1}(x-y)=\frac{1}{(2\pi)^4}\int dp\,\frac{1}{p^2-m^2}\,\mathrm{e}^{-\mathrm{i}p(x-y)}. \qquad (5.2.6)$$

In order to derive $S(\phi_0)$ in the framework of perturbation theory, we take a series expansion with respect to the coupling constant $g$ of the differential operator entering (5.2.3); this is the perturbation series we looked for:

$$S(\phi_0)=\left[1-\mathrm{i}\frac{g}{4!}\int dx\left(\frac{\delta}{\delta J(x)}\right)^4+\frac{1}{2}\left(\mathrm{i}\frac{g}{4!}\right)^2\int dx\,dy\left(\frac{\delta}{\delta J(x)}\right)^4\left(\frac{\delta}{\delta J(y)}\right)^4+\dots\right]$$
$$\times\exp\left[-\mathrm{i}\int dx\,\phi_0(x)J(x)-\frac{\mathrm{i}}{2}\int dx\,dy\,J(x)K^{-1}(x-y)J(y)\right]\Bigg|_{J=0}. \tag{5.2.7}$$

The expression for $S(\phi_0)$ in different orders of perturbation theory can be found from this expression:

i) In the absence of interaction ($n=0$):

$$S^{(0)}(\phi_0)=1\,;$$

ii) In the first order ($n=1$):

$$S^{(1)}(\phi_0)=-\mathrm{i}\frac{g}{4!}\int dx\left(\frac{\delta}{\delta J(x)}\right)^4\exp\left[-\mathrm{i}\int dx\,J(x)\,\phi_0(x)-\frac{\mathrm{i}}{2}\int dx\,dy\,J(x)K^{-1}(x-y)J(y)\right]\Bigg|_{J=0}=-\mathrm{i}\frac{g}{4!}\int dx\,\phi_0^4(x)$$
$$+\frac{6g}{4!}\int dx\,\phi_0^2(x)K^{-1}(0)+\frac{\mathrm{i}3g}{4!}\int dx[K^{-1}(0)]^2\,;$$

iii) In the second order ($n=2$):

$$S^{(2)}(\phi_0)=-\frac{1}{2}\frac{g^2}{(4!)^2}\int dx\,dy\Big\{\phi_0^4(x)\,\phi_0^4(y)$$
$$+16\,\phi_0^3(x)\,\phi_0^3(y)\,\mathrm{i}K^{-1}(x-y)+72\,\phi_0^2(x)\,\phi_0^2(y)\,[\mathrm{i}K^{-1}(x-y)]^2$$
$$+96\,\phi_0(x)\,\phi_0(y)\,[\mathrm{i}K^{-1}(x-y)]^3+24\,[\mathrm{i}K^{-1}(x-y)]^4$$
$$+12\mathrm{i}\,\phi_0^2(x)K_x^{-1}(0)\,\phi_0^4(y)-6\,[K_x^{-1}(0)]^2\,\phi_0^4(y)$$
$$-96\,\phi_0(x)K_x^{-1}(0)K^{-1}(x-y))\,\phi_0^3(y)$$
$$-36\,\phi_0^2(x)K_x^{-1}(0)K_y^{-1}(0)\,\phi_0^2(y)-144K_x^{-1}(0)\,\mathrm{i}\,[K^{-1}(x-y)]^2\,\phi_0^2(y)$$
$$-36\,[K_x^{-1}(0)]^2\mathrm{i}K_y^{-1}(0)\,\phi_0^2(y)$$
$$-144\,\phi_0(x)K_x^{-1}(0)\,\mathrm{i}K^{-1}(x-y)K_y^{-1}(0)\,\phi_0(y)$$
$$+72K_x^{-1}(0)\,[K^{-1}(x-y)]^2K_y^{-1}(0)\Big\}\,,\quad K_x^{-1}(0)\equiv K^{-1}(x-y)\,|_{y=x}\,. \tag{5.2.8}$$

Let us find, for instance, the amplitude $S^{(2)}_{2\to 2}$ for a two-particle-to-two-particle scattering process in the second order of perturbation theory. By rewriting (5.2.8) and (5.1.32) in the momentum-representation we get

$$S^{(2)}_{2\to 2} = \phi(q_1)\frac{\delta}{\delta\phi_0(q_1)}\cdots\phi(q_4)\frac{\delta}{\delta\phi_0(q_4)}S^{(2)}(\phi_0(q_i))\bigg|_{\phi_0(q_i)=0},$$

where $\phi(q_1)$ and $\phi(q_2)$ are the wave functions of the initial particles with the momenta $q_1$ and $q_2$ and $\phi(q_3)$ and $\phi(q_4)$ are those of the final particles with the momenta $q_3$ and $q_4$.

Carrying out the calculations we eventually find

$$S^{(2)}_{2\to 2} = -(\mathrm{i}g)^2\phi(q_3)\phi(q_4)[\int dp\, K^{-1}(p)K^{-1}(p-q_1-q_2)]$$
$$\times\phi(q_1)\phi(q_2)\delta(q_1+q_2-q_3-q_4), \tag{5.2.8'}$$

where $K^{-1}(p)$ is given by (5.2.5).

In the general case, the expressions for the amplitude in the model under consideration contain various combinations of free-particle functions, propagators, and four-particle interaction vertices.

Henceforth, the vertex functions in the tree approximation will be called the *interaction vertices.*

Let us look for an explicit form of the propagator and of the interaction vertices. The propagator $\mathscr{D}(x-y)$ of a free scalar field $[\mathscr{L}_\mathrm{I}(x)=0]$ is determined by

$$\mathscr{D}(x-y) = \frac{\delta}{\delta J(x)}\frac{\delta}{\delta J(y)}Z_0(J)\bigg|_{J=0},$$

where $Z_0(J)$ is the value of the functional $Z(J)$ at $\mathscr{L}_\mathrm{I}(x)=0$. Substituting (5.1.12) into the above expression and taking into account (5.2.2) and (5.2.6) we find, in the momentum-representation,

$$\mathscr{D}(p) = \frac{-1}{p^2-m^2} \tag{5.2.9}$$

or, in the coordinate representation,

$$\mathscr{D}(x-y) = \frac{-1}{(2\pi)^4}\int dp\frac{1}{p^2-m^2}\mathrm{e}^{-\mathrm{i}p(x-y)}. \tag{5.2.10}$$

As is well known, a unique definition of the Green's function involves a definition of the integration contour. Below we shall assume that such a contour is chosen which leads to the conventional Feynman rule for dealing with the poles of the Green's function, i.e., to the Green's functions in Feynman's formulation.

To find the interaction vertices, we utilize (5.1.24). Then

$$\Gamma_{\text{tree}}(\Phi) = \int dx\, L(\Phi) = \int dx \left( \frac{1}{2} \partial_\mu \Phi \partial_\mu \Phi - \frac{m^2}{2} \Phi^2 - \frac{g}{4!} \Phi^4 \right), \qquad (5.2.11)$$

where $\Phi(x) = \delta Z_{\text{tree}}(J)/\delta J(x)$, $Z_{\text{tree}}(J)$ being given by (5.1.21).

By definition, the four-point interaction vertex is given by

$$\Gamma_0(x,y,z,u) = (-\mathrm{i})^3 \frac{\delta}{\delta \Phi(x)} \frac{\delta}{\delta \Phi(y)} \frac{\delta}{\delta \Phi(z)} \frac{\delta}{\delta \Phi(u)} \Gamma_{\text{tree}}(\Phi) \bigg|_{\Phi=0} .$$

Since

$$\frac{\delta}{\delta \Phi(x)} = \int dk\, \mathrm{e}^{\mathrm{i}kx} \frac{\delta}{\delta \Phi(k)} ,$$

we obtain, by retaining only the $\Phi^4$-term whose contribution is non-zero,

$$\begin{aligned}\Gamma_0(x,y,z,u) &= (-\mathrm{i})^3 \int dk\, dp\, dq\, df\, \mathrm{e}^{\mathrm{i}kx+\mathrm{i}py+\mathrm{i}qz+\mathrm{i}fu} \frac{\delta}{\delta \Phi(k)} \\ &\times \frac{\delta}{\delta \Phi(p)} \frac{\delta}{\delta \Phi(q)} \frac{\delta}{\delta \Phi(f)} \frac{1}{(2\pi)^{16}} \int d\xi\, \mathrm{e}^{-\mathrm{i}\xi r - \mathrm{i}\xi s - \mathrm{i}\xi t - \mathrm{i}\xi v} dr\, ds\, dt\, dv \\ &\times \left( -\frac{g}{4!} \right) \Phi(r)\, \Phi(s)\, \Phi(t)\, \Phi(v) \\ &= \frac{-\mathrm{i}g}{(2\pi)^{12}} \int dk\, dq\, dp\, df\, \delta(k+p+q+f)\, \mathrm{e}^{-\mathrm{i}kx-\mathrm{i}py-\mathrm{i}qz-\mathrm{i}fu} .\end{aligned}$$

Comparing the last expression with that for $\Gamma_0(x,y,z,u)$ in the momentum representation,

$$\Gamma_0(x,y,z,u) = \frac{1}{(2\pi)^{12}} \int dk\, dp\, dq\, df\, \Gamma_0(k,p,q,f)\, \mathrm{e}^{-\mathrm{i}kx-\mathrm{i}py-\mathrm{i}qz-\mathrm{i}fu} ,$$

we obtain the four-point interaction vertex, omitting the $\delta(k+p+q+f)$:

$$\Gamma_0(k,p,q,f) = -\mathrm{i}g . \qquad (5.2.12)$$

We shall not write down explicitly the four-momentum conservation $\delta$-functions in the expressions for the interaction vertices given in this book.

Analytic expressions for the matrix elements can be represented graphically. To that end, the rules of correspondence between the analytic expressions and the graphs have to be established.

The *Feynman correspondence rules* for the amplitude in the $\phi^4$-interaction model are listed, in the momentum representation, in Table 5.1. The scattering amplitude (5.2.8′), for example, is depicted in Fig. 5.5a, the scattering amplitude $S^{(2)}_{3\to 3}$ is presented in Fig. 5.5b. Other amplitudes, not associated with scattering processes, can graphically be represented in a similar way; this will not be done here, however.

**Table 5.1.** Correspondence rules for the $\phi^4$-interaction model

| Physical state | Mathematical expression | Diagram |
|---|---|---|
| Free scalar particle | $\phi(p)$ | $p$ |
| Propagator | $-\dfrac{1}{p^2-m^2}$ | $p$ |
| Interaction vertex | $-\mathrm{i}\,g$ | $p$, $q$, $k$, $f$ |

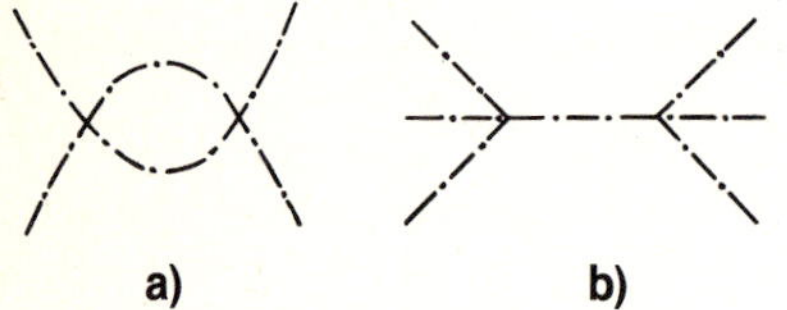


**Fig. 5.5 a, b.** Feynman diagrams for the $\phi^4$-interaction model in the second order of perturbation theory

## 5.3 A Model with Non-Abelian Gauge Fields

Consider a typical model with non-Abelian fields, namely that containing a multiplet of spinor fields $\psi^a(x)$, a multiplet of charged scalar fields $\phi^a(x)$ and a multiplet of gauge fields $A^k_\mu(x)$. The Lagrangian of the model has the form

$$\begin{aligned}\mathscr{L} = &-\frac{1}{4}F^k_{\mu\nu}F^k_{\mu\nu}-\frac{1}{2\alpha}(\partial_\mu A^k_\mu)^2\\ &+\mathrm{i}\bar{\psi}^a\gamma_\mu(\partial_\mu\delta_{ab}+\mathrm{i}g(t_k)_{ab}A^k_\mu)\psi^b-M\bar{\psi}^a\psi^a\\ &+|(\partial_\mu\delta_{ab}+\mathrm{i}g(\theta_k)_{ab}A^k_\mu)\phi^b|^2-m^2\phi^{a\dagger}\phi^a-h_{abc}\bar{\psi}^a\psi^b\phi^c\\ &-h^\dagger_{abc}\bar{\psi}^a\psi^b\phi^{c\dagger}-\tfrac{1}{4}f_{abcd}\phi^{a\dagger}\phi^{b\dagger}\phi^c\phi^d-(\partial_\mu\bar{c}^k)(\partial_\mu c^k)\\ &-gf_{klm}(\partial_\mu\bar{c}^k)A^m_\mu c^l\,, \end{aligned} \tag{5.3.1}$$

where $c^k(x)$ are the Faddeev-Popov ghosts; $t$ and $\theta$ are generators of the representations according to which the spinor and the scalar fields, respectively, transform.

The first two terms describe the gauge field; the third and the fourth one the spinor field and its interaction with the gauge field, the fifth and the sixth one the scalar field and its interaction with the gauge field, the seventh and the eighth one the Yukawa interaction between the fermion and the scalar fields, the ninth one the self-interaction of the scalar field, and the two last terms the ghosts fields and their interaction. The terms corresponding to the ghost fields in the $\alpha$-gauge are obtained in the same way as in (4.2.56).

The Lagrangian of the gauge field and the ghost fields is conveniently expressed as the sum of two Lagrangians, namely,

$$\mathscr{L}_0 = -\frac{1}{2}(\partial_\mu A_\nu^k)^2 - \frac{1}{2\alpha}(\partial_\mu A_\mu^k)^2 + \frac{1}{2}(\partial_\mu A_\nu^k)(\partial_\nu A_\mu^k) - \partial_\mu \bar{c}^k \partial_\mu c^k \tag{5.3.2}$$

and

$$\mathscr{L}_{0\mathrm{I}} = g f_{jpk}(\partial_\mu A_\nu^k) A_\mu^j A_\nu^p - \frac{g^2}{4} f_{jpk} f_{klm} A_\mu^j A_\nu^p A_\mu^l A_\nu^m - g f_{klm}(\partial_\mu \bar{c}^k) A_\mu^m c^l . \tag{5.3.3}$$

The first Lagrangian is associated with the free gauge field and the $c$-fields, while the second one incorporates the self-interaction of the gauge field and its interaction with the $c$-fields.

According to (5.1.26), the generating functional $W$ reads, in the $\alpha$-gauge,

$$\begin{aligned} W(J_\mu^k, J^{a\dagger}, J^a, \bar{\eta}^a, \eta^a, \bar{\chi}^k, \chi^k) &= \exp\left[\mathrm{i}\int dx\, \mathscr{L}_\mathrm{I}\left(-\mathrm{i}\frac{\delta}{\delta J_\mu^k(x)}, \dots\right)\right] \int \prod_{x,\mu,k,a} \mathscr{D}A_\mu^k(x)\, \mathscr{D}\bar{\psi}^a(x)\, \mathscr{D}\psi^a(x) \\ &\times \mathscr{D}\phi^{a\dagger}(x)\, \mathscr{D}\phi^a(x)\, \mathscr{D}\bar{c}^k(x)\, \mathscr{D}c^k(x) \exp\left\{\mathrm{i}\int dx \left[\frac{1}{2} A_\mu^k(x)\right.\right. \\ &\times \delta^{kl}\left(\Box g_{\mu\nu} - \left(1-\frac{1}{\alpha}\right)\partial_\mu\partial_\nu\right) A_\nu^l(x) + \bar{\psi}^a(x)\,\delta_{ab}(\mathrm{i}\gamma_\mu\partial_\mu - M)\,\psi^b(x) \\ &- \phi^{a\dagger}(x)\,\delta_{ab}(\Box + m^2)\,\phi^b(x) + \bar{c}^k(x)\,\delta_{kl}\Box c^l(x) \\ &+ J_\mu^k(x) A_\mu^k(x) + \phi^{a\dagger}(x) J^a(x) + J^{a\dagger}(x)\,\phi^a(x) + \bar{\eta}^a(x)\,\psi^a(x) \\ &\left.\left. + \bar{\psi}^a(x)\,\eta^a(x) + \bar{\chi}^k(x) c^k(x) + \bar{c}^k(x)\,\chi^k(x)\right]\right\}, \end{aligned}$$

where $\mathscr{L}_\mathrm{I}(-\mathrm{i}\delta/\delta J_\mu^k(x), \dots)$ is the interaction Lagrangian in which all the fields are replaced by the corresponding functional derivatives.

Transforming this expression to the form of (5.1.30) yields

$$\begin{aligned} W(J_\mu^k, \dots) &= \exp\left[\mathrm{i}\int dx\, \mathscr{L}_\mathrm{I}\left(-\mathrm{i}\frac{\delta}{\delta J_\mu^k(x)}, \dots\right)\right] \\ &\times \exp\{-\mathrm{i}\int dx\, dy [\tfrac{1}{2} J_\mu^k(x) (K_{\mu\nu}^{kl}(x-y))^{-1} J_\nu^l(y) \\ &+ J^{a\dagger}(x) K_{ab}^{-1}(x-y) J^b(y) + \bar{\eta}^a(x)\, \mathscr{K}_{ab}^{-1}(x-y)\, \eta^b(y) \\ &+ \bar{\chi}^k(x)\, \mathscr{K}_{kl}^{-1}(x-y)\, \chi^l(y)]\} . \end{aligned} \tag{5.3.4}$$

The generating functional $S(A^k_{\mu 0}, \ldots)$ for the matrix elements then follows from (5.1.31):

$$S(A^k_{\mu 0}, \ldots) = \exp\left[\mathrm{i}\int dx\, \mathscr{L}_\mathrm{I}\left(-\mathrm{i}\frac{\delta}{\delta J^k_\mu(x)}, \ldots\right)\right]$$
$$\times \exp\{-\mathrm{i}\int dx(A^k_{\mu 0}J^k_\mu + J^{a\dagger}\phi^a_0 + \phi^{a\dagger}_0 J^a + \bar{\psi}^a_0\eta^a + \bar{\eta}^a\psi^a_0)$$
$$-\mathrm{i}\int dx\,dy[\tfrac{1}{2}J^k_\mu(x)(K^{kl}_{\mu\nu}(x-y))^{-1}J^l_\nu(y)$$
$$+J^{a\dagger}(x)K^{-1}_{ab}(x-y)J^b(y) + \bar{\eta}^a(x)\mathscr{K}^{-1}_{ab}(x-y)\eta^b(y)$$
$$+\bar{\chi}^k(x)\mathscr{K}^{-1}_{kl}(x-y)\chi^l(y)]\}|_{J^k_\mu=\ldots=0}\,. \tag{5.3.5}$$

Here the operators $[K^{kl}_{\mu\nu}(x-y)]^{-1}$, $K^{-1}_{ab}(x-y)$, $\mathscr{K}^{-1}_{ab}(x-y)$ and $\mathscr{K}^{-1}_{kl}(x-y)$ are defined by

$$K^{-1}_{ab}(x-y) = \frac{\delta_{ab}}{(2\pi)^4}\int\frac{dk}{k^2-m^2}\,\mathrm{e}^{-\mathrm{i}k(x-y)}, \tag{5.3.6}$$

$$[K^{kl}_{\mu\nu}(x-y)]^{-1} = -\frac{\delta_{kl}}{(2\pi)^4}\int\frac{dk}{k^2}\left[g_{\mu\nu}-(1-\alpha)\frac{k_\mu k_\nu}{k^2}\right]\mathrm{e}^{-\mathrm{i}k(x-y)}, \tag{5.3.7}$$

$$\mathscr{K}^{-1}_{ab}(x-y) = \frac{\delta_{ab}}{(2\pi)^4}\int\frac{dp}{\not{p}-M}\,\mathrm{e}^{-\mathrm{i}p(x-y)}, \tag{5.3.8}$$

$$\mathscr{K}^{-1}_{kl}(x-y) = -\frac{\delta_{kl}}{(2\pi)^4}\int\frac{dk}{k^2}\,\mathrm{e}^{-\mathrm{i}k(x-y)}. \tag{5.3.9}$$

A series expansion, with respect to the coupling, constants of the differential operator in the above expression (5.3.5) for $S(A^k_{\mu 0}, \ldots)$ gives the perturbation theory form of this formula for the model under consideration:

$$S(A^k_{\mu 0}, \ldots) = \left[1 + \mathrm{i}\int dx\,\mathscr{L}_\mathrm{I}\left(-\mathrm{i}\frac{\delta}{\delta J^k_\mu(x)}, \ldots\right)\right.$$
$$\left.+\frac{\mathrm{i}^2}{2}\int dx\,dy\,\mathscr{L}_\mathrm{I}\left(-\mathrm{i}\frac{\delta}{\delta J^k_\mu(x)}, \ldots\right)\mathscr{L}_\mathrm{I}\left(-\mathrm{i}\frac{\delta}{\delta J^k_\mu(y)}, \ldots\right) + \ldots\right]$$
$$\times \exp\{-\mathrm{i}\int dx(A^k_{\mu 0}J^k_\mu + J^{a\dagger}\phi^a_0 + \phi^{a\dagger}_0 J^a + \bar{\psi}^a_0\eta^a + \bar{\eta}^a\psi^a_0)$$
$$-\mathrm{i}\int dx\,dy[\tfrac{1}{2}J^k_\mu K^{kl^{-1}}_{\mu\nu}J^l_\nu + J^{a\dagger}K^{-1}_{ab}J^b$$
$$+\bar{\eta}^a\mathscr{K}^{-1}_{ab}\eta^b + \bar{\chi}^k\mathscr{K}^{-1}_{kl}\chi^l]\}|_{J^k_\mu=\ldots=0}\,.$$

The expressions for the amplitudes contain various combinations of functions of the free fields, propagators and interaction vertices.

The propagators of the free scalar field and of the fermion field are given by (5.2.9) and (7.1.9, see below), respectively. By definition, for the propagators of the gauge field and of a ghost $c$-field we have

$$\mathscr{D}^{kl}_{\mu\nu}(x-y) = \frac{\delta}{\delta J^k_\mu(x)} \frac{\delta}{\delta J^l_\nu(y)} Z_0(J^k_\mu, \ldots)\bigg|_{J^k_\mu = \ldots = 0} \tag{5.3.10}$$

and

$$\mathscr{D}_{kl}(x-y) = \frac{\delta}{\delta \chi^k(x)} \frac{\delta}{\delta \bar{\chi}^l(y)} Z_0(J^k_\mu, \ldots)\bigg|_{J^k_\mu = \ldots = 0} . \tag{5.3.11}$$

Substituting (5.1.12) into (5.3.10, 11) and taking into account (5.3.4, 7) and (5.3.9) we obtain, in the coordinate representation,

$$\mathscr{D}^{kl}_{\mu\nu}(x-y) = \frac{\delta_{kl}}{(2\pi)^4} \int \frac{dk}{k^2} \left[ g_{\mu\nu} - (1-\alpha)\frac{k_\mu k_\nu}{k^2} \right] \mathrm{e}^{-ik(x-y)} \quad (\alpha\text{-gauge}) , \tag{5.3.12}$$

$$\mathscr{D}_{kl}(x-y) = \frac{\delta_{kl}}{(2\pi)^4} \int \frac{dk}{k^2} \mathrm{e}^{-ik(x-y)} \tag{5.3.13}$$

or, in the momentum representation,

$$\mathscr{D}^{kl}_{\mu\nu}(k) = \frac{\delta_{kl}}{k^2} \left[ g_{\mu\nu} - (1-\alpha)\frac{k_\mu k_\nu}{k^2} \right] \quad (\alpha\text{-gauge}) , \tag{5.3.14}$$

$$\mathscr{D}_{kl}(k) = \frac{\delta_{kl}}{k^2} . \tag{5.3.15}$$

Setting $\alpha = 0$ in (5.3.12) and (5.3.14) we obtain the propagator of the gauge field in the *Landau gauge* and setting $\alpha = 1$ we obtain that in the *Feynman gauge*. In the same way as (7.1.13) is obtained, we can find the expression for the propagator of the gauge field in the *Coulomb gauge*.

Let us find the propagator of the gauge field in the *axial gauge* (4.2.41). In this case

$$K^{kl}_{\mu\nu}(x-y) = \int dk\, \delta_{kl} \left( -k^2 g_{\mu\nu} + k_\mu k_\nu - \frac{1}{\alpha} n_\mu n_\nu \right) \mathrm{e}^{-ik(x-y)} ,$$

and at $\alpha \to 0$ the propagator assumes the form

$$\mathscr{D}^{kl}_{\mu\nu}(x-y) = \frac{\delta_{kl}}{(2\pi)^4} \int \frac{dk}{k^2} \times \left( g_{\mu\nu} + \frac{n^2 k_\mu k_\nu - (nk) k_\mu n_\nu - (nk) n_\mu k_\nu}{(nk)^2} \right) \mathrm{e}^{-ik(x-y)} . \tag{5.3.16}$$

**Table 5.2.** Propagators of the gauge fields in different gauges

| Gauge | Expression for $\mathscr{D}^{kl}_{\mu\nu}(k)$ |
|---|---|
| $\alpha$-gauge | $\frac{\delta_{kl}}{k^2}\left[g_{\mu\nu}-(1-\alpha)\frac{k_\mu k_\nu}{k^2}\right]$ |
| Landau gauge | $\frac{\delta_{kl}}{k^2}\left(g_{\mu\nu}-\frac{k_\mu k_\nu}{k^2}\right)$ |
| Feynman gauge | $\frac{\delta_{kl}}{k^2}g_{\mu\nu}$ |
| Coulomb gauge | $\begin{cases}\mathscr{D}^{kl}_{00}(k)=-\frac{\delta_{kl}}{\boldsymbol{k}^2}; \quad \mathscr{D}^{kl}_{0i}(k)=\mathscr{D}^{kl}_{i0}(k)=0\\ \mathscr{D}^{kl}_{ij}(k)=-\frac{1}{k^2}\delta_{kl}\left(\delta_{ij}-\frac{k_ik_j}{\boldsymbol{k}^2}\right)\end{cases}$ |
| Axial gauge | $\frac{\delta_{kl}}{k^2}\left(g_{\mu\nu}+\frac{n^2k_\mu k_\nu-(nk)k_\mu n_\nu-(nk)n_\mu k_\nu}{(nk)^2}\right)$ |

**Table 5.3.** The Feynman diagrams for the propagators of the considered model

| Propagator of | Analytic expression | Diagram |
|---|---|---|
| Scalar field | $-\frac{\delta_{ab}}{k^2-m^2}$ | $a$ —·—$k$—·→ $b$ |
| Spinor field | $-\frac{\delta_{ab}}{\hat{p}-M}$ | $a$ ——$p$—→ $b$ |
| Ghost field | $\delta_{nl}\frac{1}{k^2}$ | $n$ - - -$k$- - -→ $l$ |
| Gauge field | $\delta_{nl}\left[g_{\mu\nu}-(1-\alpha)\frac{k_\mu k_\nu}{k^2}\right]\frac{1}{k^2}$ | $\mu n$ ~~~$k$~~~ $\nu l$ |

The expressions for the propagators of the gauge fields in various gauges are listed in Table 5.2. The graphical forms of the propagators of the scalar, fermion, gauge, and ghost fields are summarized in Table 5.3, together with their analytic expressions.

To find the interaction vertices for the model, we make use of (5.1.24):

$$\Gamma_{\text{tree}}(\mathscr{A}^k_\mu, \Phi^{a\dagger}, \Phi^a, \bar{\Psi}^a, \Psi^a, \bar{\mathscr{C}}^k, \mathscr{C}^k) = \int dx\, \mathscr{L}(\mathscr{A}^k_\mu(x), \Phi^{a\dagger}(x), \Phi^a(x), \ldots)\,, \tag{5.3.17}$$

where

$$\mathscr{A}^k_\mu(x) = \frac{\delta Z_{\text{tree}}(J^k_\mu, \ldots)}{\delta J^k_\mu(x)}, \qquad \Phi^{a\dagger}(x) = \frac{\delta Z_{\text{tree}}(J^k_\mu, \ldots)}{\delta J^a(x)},$$

$$\Phi^a(x) = \frac{\delta Z_{\text{tree}}(J^k_\mu, \ldots)}{\delta J^{a\dagger}(x)}, \qquad \bar{\Psi}^a(x) = \frac{\delta Z_{\text{tree}}(J^k_\mu, \ldots)}{\delta \eta^a(x)},$$

$$\Psi^a(x) = \frac{\delta Z_{\text{tree}}(J^k_\mu, \ldots)}{\delta \bar{\eta}^a(x)}, \qquad \bar{\mathscr{C}}^k(x) = \frac{\delta Z_{\text{tree}}(J^k_\mu, \ldots)}{\delta \chi^k(x)},$$

$$\mathscr{C}^k(x) = \frac{\delta Z_{\text{tree}}(J^k_\mu, \ldots)}{\delta \bar{\chi}^k(x)}.$$

The model is characterized by nine interaction vertices (cf. Table 5.4).

**Table 5.4.** The Feynman diagrams for the interaction vertices of the model

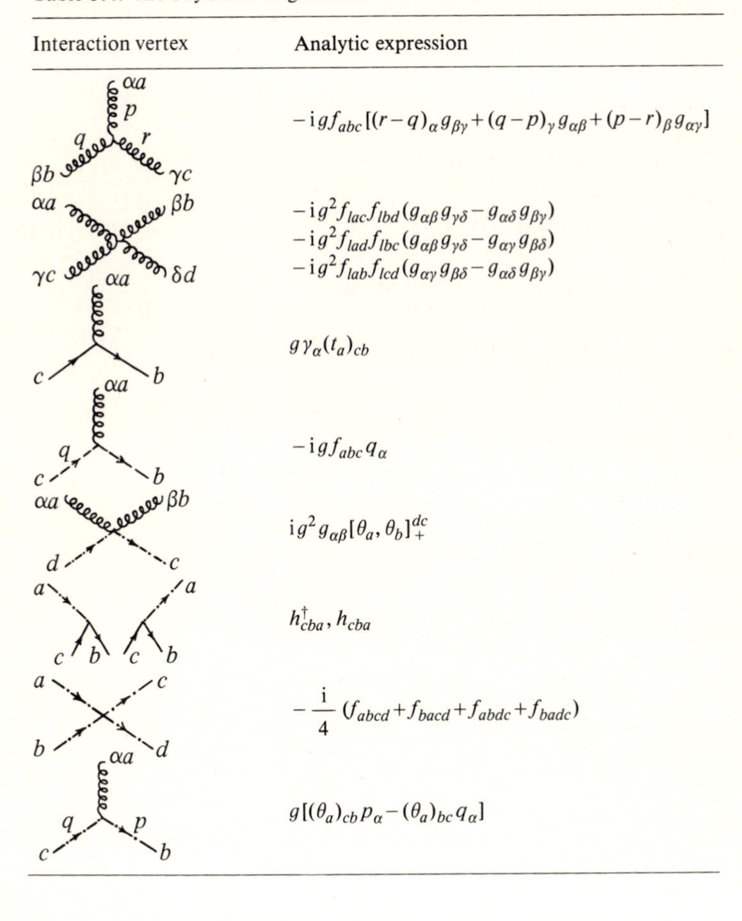

| Interaction vertex | Analytic expression |
|---|---|
| | $-\mathrm{i}gf_{abc}[(r-q)_\alpha g_{\beta\gamma}+(q-p)_\gamma g_{\alpha\beta}+(p-r)_\beta g_{\alpha\gamma}]$ |
| | $-\mathrm{i}g^2 f_{lac}f_{lbd}(g_{\alpha\beta}g_{\gamma\delta}-g_{\alpha\delta}g_{\beta\gamma})$<br>$-\mathrm{i}g^2 f_{lad}f_{lbc}(g_{\alpha\beta}g_{\gamma\delta}-g_{\alpha\gamma}g_{\beta\delta})$<br>$-\mathrm{i}g^2 f_{lab}f_{lcd}(g_{\alpha\gamma}g_{\beta\delta}-g_{\alpha\delta}g_{\beta\gamma})$ |
| | $g\gamma_\alpha(t_a)_{cb}$ |
| | $-\mathrm{i}gf_{abc}q_\alpha$ |
| | $\mathrm{i}g^2 g_{\alpha\beta}[\theta_a, \theta_b]^{dc}_+$ |
| | $h^\dagger_{cba}, h_{cba}$ |
| | $-\frac{\mathrm{i}}{4}(f_{abcd}+f_{bacd}+f_{abdc}+f_{badc})$ |
| | $g[(\theta_a)_{cb}p_\alpha-(\theta_a)_{bc}q_\alpha]$ |

1) The interaction vertex of two spinor fields and the gauge field is defined by

$$\Gamma^{kab}_{0\mu}(x,y,z) = (-\mathrm{i})^2 \frac{\delta}{\delta \mathscr{A}^k_\mu(x)} \frac{\delta}{\delta \Psi^b(z)} \frac{\delta}{\delta \overline{\Psi}^a(y)} \Gamma_{\text{tree}}(\mathscr{A}^k_\mu, \dots)\Big|_{\mathscr{A}=\dots=0} .$$

By retaining in (5.3.17) only the term associated with the interaction between the fermion and the gauge fields, it can be found that

$$\Gamma^{kab}_{0\mu}(p,q,r) = g\gamma_\mu (t_k)_{ab} . \tag{5.3.18}$$

2) The interaction vertex of three-gauge fields is defined by

$$\Gamma^{klm}_{0\mu\nu\varrho}(x,y,z) = (-\mathrm{i})^2 \frac{\delta}{\delta \mathscr{A}^k_\mu(x)} \frac{\delta}{\delta \mathscr{A}^l_\nu(y)} \frac{\delta}{\delta \mathscr{A}^m_\varrho(z)} \Gamma_{\text{tree}}(\mathscr{A}^k_\mu, \dots)\Big|_{\mathscr{A}=\dots=0}$$

Retaining in (5.3.17) only the term associated with the interaction of three gauge fields we get

$$\begin{aligned}\Gamma^{klm}_{0\mu\nu\varrho}(x,y,z) = (-\mathrm{i})^2 \int dp\,dq\,dr\,\mathrm{e}^{\mathrm{i}px+\mathrm{i}qy+\mathrm{i}rz} &\frac{\delta}{\delta \mathscr{A}^k_\mu(p)} \frac{\delta}{\delta \mathscr{A}^l_\nu(q)} \frac{\delta}{\delta \mathscr{A}^m_\varrho(r)} \\ &\times \frac{1}{(2\pi)^{12}} \int ds\,dt\,dv\,d\xi\,\mathrm{e}^{-\mathrm{i}\xi s-\mathrm{i}\xi t-\mathrm{i}\xi v} g f_{jpn}(-\mathrm{i}s_\alpha) \\ &\times \mathscr{A}^n_\beta(s)\mathscr{A}^j_\alpha(t)\mathscr{A}^p_\beta(v) .\end{aligned} \tag{5.3.19}$$

Differentiation, e.g., with respect to $\mathscr{A}^m_\varrho(r)$ yields

$$\begin{aligned}g f_{jpm}(-\mathrm{i}s_\alpha)\mathscr{A}^j_\alpha(t)\mathscr{A}^p_\varrho(v)\delta(s-r) + g f_{mpn}(-\mathrm{i}s_\varrho)\mathscr{A}^n_\beta(s)\mathscr{A}^p_\beta(v)\delta(t-r) \\ + g f_{jmn}(-\mathrm{i}s_\alpha)\mathscr{A}^n_\varrho(s)\mathscr{A}^j_\alpha(t)\delta(v-r) .\end{aligned}$$

Calculating the remaining derivatives in the same manner and substituting the resulting expressions into (5.3.19) we find the interaction vertex of three gauge fields:

$$\Gamma^{klm}_{0\mu\nu\varrho}(p,q,r) = -\mathrm{i} g f_{klm}[g_{\mu\nu}(q-p)_\varrho + g_{\mu\varrho}(p-r)_\nu + g_{\nu\varrho}(r-q)_\mu] . \tag{5.3.20}$$

3) The interaction vertex of four gauge fields is defined by

$$\Gamma^{abcd}_{0\alpha\beta\gamma\delta}(x,y,z,u) = (-\mathrm{i})^3 \frac{\delta}{\delta \mathscr{A}^a_\alpha(x)} \frac{\delta}{\delta \mathscr{A}^b_\beta(y)} \frac{\delta}{\delta \mathscr{A}^c_\gamma(z)} \frac{\delta}{\delta \mathscr{A}^d_\delta(u)} \Gamma_{\text{tree}}(\mathscr{A}^k_\mu, \dots)\Big|_{\mathscr{A}=\dots=0} .$$

Successive differentiation similar to that in the preceding case leads to the following expression for the interaction vertex of four gauge fields:

$$\Gamma^{abcd}_{0\alpha\beta\gamma\delta}(p,q,r,s) = -\mathrm{i}g^2 f_{kac} f_{kbd}(g_{\alpha\beta}g_{\gamma\delta} - g_{\alpha\delta}g_{\beta\gamma})$$
$$-\mathrm{i}g^2 f_{kad} f_{kbc}(g_{\alpha\beta}g_{\gamma\delta} - g_{\alpha\gamma}g_{\beta\delta})$$
$$-\mathrm{i}g^2 f_{kab} f_{kcd}(g_{\alpha\gamma}g_{\beta\delta} - g_{\alpha\delta}g_{\beta\gamma})\,. \tag{5.3.21}$$

The expressions for the rest of the interaction vertices of the model (cf. Table 5.4) are found in the same way. It should be emphasized that the analytic expressions for the interaction vertices given in Table 5.4 and in Table 12.1 (see below) imply that the four-vectors of the particles enter into the vertex, while the arrows characterize the initial and the final states of the particles.

## 5.4 The 1/*N*-Expansion

For some processes, it is instructive to use a modified perturbation theory in order to find the amplitude. This modification, which consists in employing a series expansion with respect to the parameter $1/N$ (where $N \gg 1$) instead of the usual coupling constant, is referred to as the *1/N-expansion.*

We shall illustrate the basics of the $1/N$-expansion technique by considering, as an example, the four-dimensional $\phi^4$-interaction model which is invariant under the $SU_N$-group. Let the scalar field $\phi^a(x)$ be transforming according to the *fundamental* representation of the $SU_N$-group so that the scalar field forms a multiplet $\phi^a$ with $N$ components ($a = 1, 2, \ldots, N$).

1) The Lagrangian of such a model, similar to (5.2.1), is written as

$$L' = \frac{1}{2}\partial_\mu\phi^a\partial_\mu\phi^a - \frac{m^2}{2}\phi^a\phi^a - \frac{1}{8}\lambda(\phi^a\phi^a)^2\,. \tag{5.4.1}$$

2) As will be seen, it is convenient to pass to another Lagrangian,

$$L = L' + \frac{1}{2}\frac{N}{g}\left(\sigma - \frac{1}{2}\frac{g}{N}\phi^a\phi^a\right)^2, \tag{5.4.2}$$

where $g = \lambda N$ and $\sigma(x)$ is an auxiliary *one-component* field.

The additional term in (5.4.2) does not change the dynamics of the system. In fact, the change of variables $\sigma \to \sigma + (g/2N)\phi^a\phi^a$, $\phi^a \to \phi^a$ in the path integral corresponding to the Lagrangian (5.4.2) yields

$$\int\prod_{x,a}\mathscr{D}\sigma(x)\,\mathscr{D}\phi^a(x)\,\mathrm{e}^{\mathrm{i}I} = \int\prod_{x}\mathscr{D}\sigma(x)\,\mathrm{e}^{\mathrm{i}\int dx\,\sigma^2(x)}\int\prod_{x,a}\mathscr{D}\phi^a(x)\,\mathrm{e}^{\mathrm{i}I'}\,, \tag{5.4.3}$$

where $I'(x)$ is the action corresponding to the Lagrangian (5.4.1). The integration over $\sigma(x)$ gives a constant and, consequently, the Lagrangians $L'$ and $L$ describe the systems with the same dynamics.

3) Opening up the second term in (5.4.2) and taking into account that the terms $(\phi^a\phi^a)^2$ cancel out we find

$$L=\frac{1}{2}\phi^a K(\sigma)\phi^a+\frac{1}{2}\frac{N}{g}\sigma^2, \tag{5.4.4}$$

where $K(\sigma)=-(\partial_\mu\partial_\mu+m^2+\sigma)$. The interaction Lagrangian $L_{\rm I}$ of the fields $\phi^a(x)$ and $\sigma(x)$ has the form

$$L_{\rm I}=-\tfrac{1}{2}\phi^a\phi^a\sigma. \tag{5.4.5}$$

4) With the use of (5.4.4), the expression for the generating functional $W(J)$ for the Green's functions of the fields $\phi^a(x)$ takes the form (cf. Sect. 5.1)

$$W(J)=\int\prod_x \mathcal{D}\sigma(x)\prod_{x,a}\mathcal{D}\phi^a(x)$$
$$\times\exp\left\{\mathrm{i}\int dx\left[\frac{1}{2}\phi^a(x)K(\sigma)\phi^a(x)+\frac{1}{2}\frac{N}{g}\sigma^2(x)+J^a(x)\phi^a(x)\right]\right\}, \tag{5.4.6}$$

where $J(x)$ are the auxiliary currents associated with the field $\phi^a(x)$.

The integral over $\phi^a(x)$ in (5.4.6) can be calculated exactly. Making use of the formulae

$$\int\prod_x\mathcal{D}\phi(x)\exp\left\{\frac{\mathrm{i}}{2}\int dx\,dy\,\phi(x)K(x,y)\phi(y)+\mathrm{i}\int dx\,B(x)\phi(x)\right\}$$
$$=\frac{1}{\sqrt{\det K}}\exp\left\{-\frac{\mathrm{i}}{2}\int dx\,dy\,B(x)K^{-1}(x,y)B(y)\right\};$$
$$\det K=\exp[\mathrm{Tr}\ln K],$$

we obtain, in place of (5.4.6), the following expression for the generating functional

$$W(J)=\int\prod_x\mathcal{D}\sigma(x)\exp[\mathrm{i}NI(\sigma(x))],$$

where

$$I(\sigma)=\frac{1}{2}\left[\mathrm{i}\,\mathrm{Tr}\ln K(\sigma)+\frac{1}{g}\int dx\,\sigma^2(x)-\frac{1}{N}\int dx\,dy\,J^a(x)K^{-1}(x,y)J^a(y)\right]. \tag{5.4.7}$$

It is seen that the integral over the multiplet $\phi^a(x)$ is reduced to the integral over a scalar function $\sigma(x)$, which was the actual goal while introducing the fields $\sigma(x)$.

5) In the case under consideration, there appears, in front of the action, a *common factor* $N$. Let $N$ be a large quantity ($N\gg 1$). Then an asymptotic

value for the integral can be found by means of the *stationary phase method.* This method makes use of the fact that the main contribution to the integral is determined by the stationary point $\sigma_0(J)$ which corresponds to an extremum of the action $I(\sigma)$, i.e., is defined through the condition $\delta I/\delta\sigma = 0$, while the higher terms of the series expansion of the action with respect to $\sigma$ in the vicinity of the stationary point give small corrections. The series expansion of the action $I(\sigma)$ in the vicinity of the stationary point $\sigma_0(J)$ has the form

$$I(\sigma) = I(\sigma_0) + \frac{1}{2!} I''(\sigma_0)(\sigma - \sigma_0)^2 + \sum_{n=3}^{\infty} \frac{1}{n!} I^{(n)}(\sigma_0)(\sigma - \sigma_0)^n . \tag{5.4.8}$$

Let us substitute this expression into (5.4.7) and take out of the path integral the last term in (5.4.8). Let us, furthermore, take into account that

$$\int \prod_x \mathscr{D}\sigma(x) \exp\left\{ \mathrm{i}N\left[\frac{1}{2} I''(\sigma_0)\sigma^2(x) + N^{-1} j(x)\sigma(x)\right]\right\}$$
$$= [\det I''(\sigma_0)]^{-1/2} \exp\left\{-\frac{\mathrm{i}}{2} N j(x)[I''(\sigma_0)]^{-1} j(x)\right\},$$

where $j(x)$ is the auxiliary current corresponding to the field $\sigma(x)$. As a result, we find

$$W(J,j) = [\det I''(\sigma_0)]^{-1/2} \exp[\mathrm{i}NI(\sigma_0(J))] \exp\left[\mathrm{i}N \sum_{n=3}^{\infty} \frac{1}{n!} I^{(n)}(\sigma_0)\mathrm{i}^{-n} \frac{\delta^n}{\delta j^n(x)}\right]$$
$$\times \exp\left\{\frac{-\mathrm{i}}{2} N^{-1} j(x)[I''(\sigma_0)]^{-1} j(x)\right\} \tag{5.4.9}$$

or, upon the series expansion of the exponential,

$$W(J,j) = [\det I''(\sigma_0)]^{-1/2} \exp\{NI(\sigma_0)\} \sum_{k=0}^{\infty} \frac{(\mathrm{i}N)^k}{k!} \left[\sum_{n=1}^{\infty} \frac{1}{n!} I^{(n)}(\sigma_0)\mathrm{i}^{-n} \frac{\delta^n}{\delta j^n(x)}\right]^k$$
$$\times \exp\left\{-\frac{\mathrm{i}}{2} N^{-1} j(x)[I''(\sigma_0)]^{-1} j(x)\right\}. \tag{5.4.10}$$

This functional depends on two currents: $J(x)$ and $j(x)$. Functional differentiation of $W(J,j)$ with respect to these currents gives the corresponding Green's functions.

In particular, according to (5.4.10), for the propagator $D(x,y)$ of the field $\sigma(x)$ in the first order in $1/N$ we have:

$$D(x,y) = \left.\frac{\delta W(J,j)}{\delta j(x)\,\delta j(y)}\right|_{J=j=0} = -\mathrm{i}N^{-1}[I''(\sigma_0)]^{-1} .$$

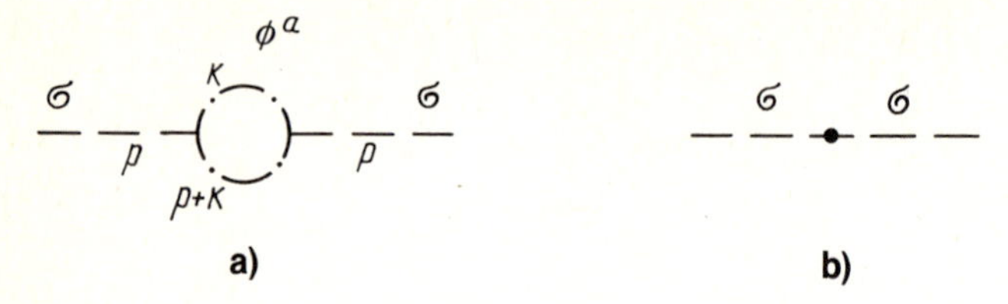


**Fig. 5.6a, b.** Diagrams contributing to the propagator of the $\sigma$-field in the first order in $1/N$

Using (5.4.7) we obtain

$$I''(\sigma)=\left.\frac{\delta I(\sigma)}{\delta\sigma(x)\,\delta\sigma(y)}\right|_{\sigma=\sigma_0}=g^{-1}\delta(x-y)-\frac{\mathrm{i}}{2}G(x,y;\sigma_0)\,G(y,x;\sigma_0)\,,$$

where $G$ is the Green's function of the operator $K$. Choosing $\sigma_0=0$ as the ground state and passing to the momentum representation we arrive at the expression for the propagator of the $\sigma$-field in the first order in $1/N$:

$$D(p)=\left[N\left(\frac{1}{2}\int\frac{dk}{(2\pi)^4}\,\frac{1}{k^2-m^2}\,\frac{1}{(p+k)^2-m^2}+\mathrm{i}g^{-1}\right)\right]^{-1}. \qquad (5.4.11)$$

The diagrams corresponding to the two terms in (5.4.11) are presented in Fig. 5.6.

The formula (5.4.10) describes a series expansion of the generating functional, and thus of the amplitude of the process, in the parameter $1/N$. This expansion leads to the expressions analogous to (5.2.7) and (5.2.8); each term in these expressions can be put into correspondence with a Feynman diagram.

6) In constructing the amplitude of a process in a given order with respect to the parameter $1/N$, a different procedure can be employed: using (5.4.7), it can be first found what power of $N$ corresponds to an arbitrary Feynman diagram; the diagrams contributing to a given order in $1/N$ can be then selected.

Let us consider an arbitrary connected diagram which contains $E$ external lines, $I$ internal lines and $V$ vertices corresponding to the field $\sigma$. According to (5.4.7), each vertex of the diagram involves the parameter $N$. Since the propagator is a quantity reciprocal to the quadratic part of the Lagrangian, to each internal or external line there corresponds a factor of $1/N$. The diagram is thus characterized by the quantity $N^{V-I-E}$. The number of internal lines is equal to the number of the momenta over which the integration is to be carried out. These momenta are, however, not independent because the momenta meeting at each of the vertices $V$ are interrelated through a conservation law; besides, one of the conservation laws (pertaining to the process as a whole) involves the external momenta so that the number of independent internal momenta is $L=I-(V-1)$.

Taking this into account we find for the power of $N$:

$$N^{V-I-E} = N^{-E-L+1} . \tag{5.4.12}$$

In particular, the diagrams containing two external lines ($E = 2$) and no loops ($L = 0$, the tree approximation) contribute to the leading order in $1/N$.

7) Let us consider as an example the elastic scattering of particles which are described by the field $\phi^a$ in the first order in $1/N$: $\phi^a\phi^a \to \phi^a\phi^a$. Within the same order, the propagator of the $\sigma$-field is determined by (5.4.11).

The obtained expression (5.4.11) for the propagator of the $\sigma$-field in the first order in $1/N$ can be represented as the sum of an infinite series of *perturbation theory* with respect to the coupling constant $g$. To such a series corresponds an infinite sum of the one-loop, two-loop, etc. diagrams of perturbation theory with respect to the field $\phi^a(x)$. These diagrams are depicted in Fig. 5.7, with the hatching indicating that the corresponding diagrams are taken into account in all orders in $g$ [assuming the normal form of the Lagrangian (5.4.1)].

The diagrams contributing, according to (5.4.5), to the process $\phi^a\phi^a \to \phi^a\phi^a$ in the first order in $1/N$ are presented in Fig. 5.8.

8) It should be emphasized that the $1/N$-expansion technique is based on the possibility of reducing the integration over the field $\phi^a(x)$ in the generating functional to the integration over the field $\sigma(x)$. This allows to express the

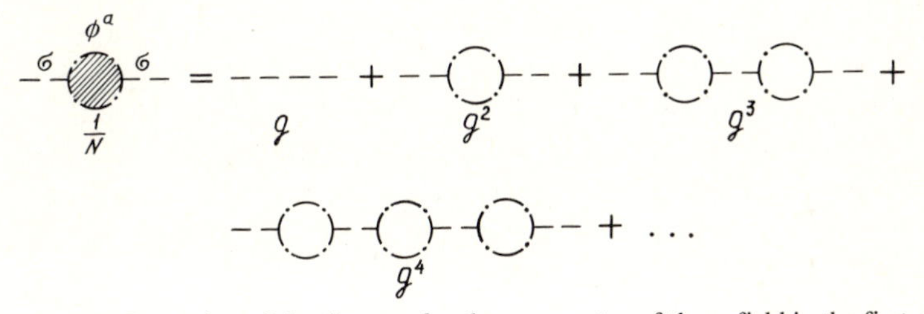


**Fig. 5.7.** Expansion of the diagram for the propagator of the $\sigma$-field in the first order in $1/N$ into an infinite series corresponding to the diagrams of the different orders of perturbation theory in the coupling constant $g$

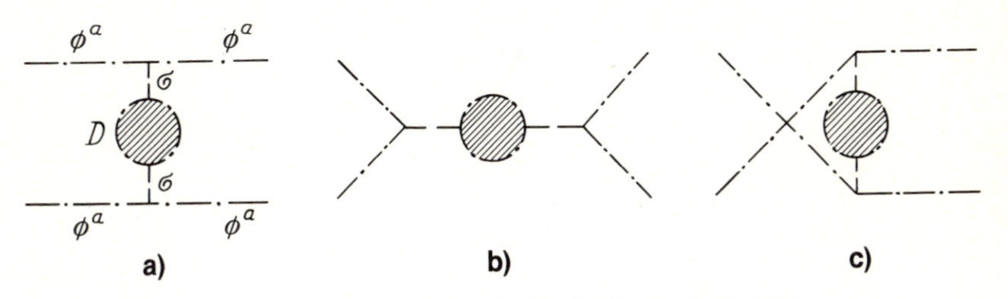


**Fig. 5.8.** Diagrams for the process $\phi^a\phi^a \to \phi^a\phi^a$ in the first order in $1/N$

generating functional in the form (5.4.7) which contains $N$ in front of the action as a common factor. Unfortunately, this possibility can be only realized if the Lagrangian involves the fields which transform according to the fundamental representation of the group. The method of $1/N$ expansion, as it has been outlined here, therefore cannot be applied to the gauge fields of the $SU_N$-group, since these transform according to the adjoint rather than the fundamental representation of the $SU_N$-group (the number of the gauge fields of the group $SU_N$ being equal to $N^2-1$ and not to $N$).

Other models containing the fields which transform according to the fundamental representation of a given group have been analyzed along the same lines.

# Part III

## Gauge Theory of Electroweak Interactions

Gauge fields provide new possibilities of constructing a *unified theory* of elementary-particle *interactions*. In this chapter we consider gauge models which unify electromagnetic and weak interactions. The gauge theory of strong interaction as well as theories unifying strong, electromagnetic and weak interactions will be considered later in Part IV.

The major difficulty in unifying the electromagnetic and the weak interactions originates from the basic differences in their character. Firstly, in contrast to the long-range electromagnetic interaction, the weak interaction has a finite interaction radius. The mediators of the weak interaction therefore must be massive (intermediate vector bosons), as distinct from the massless mediators of the electromagnetic interaction (photons). Secondly, unlike the electromagnetic interaction which conserves spatial parity, the weak interaction is not parity conserving.

This difficulty can be overcome by assuming the gauge fields to be the mediators of both interactions and by using spontaneous symmetry-breaking. A suitable choice of spontaneous symmetry-breaking makes one gauge field remain massless (photon) and to interact with the parity-conserving current, while the rest of gauge fields acquire a mass (intermediate bosons), their interaction not being parity conserving.

The choice of a specific realization of the model is quite arbitrary, and a large number of particular unified models for the electroweak *lepton* interactions have been analyzed in the literature. The most successful is the model suggested by Glashow, Salam and Weinberg which is referred to as the *standard* one. This model was further generalized to the case of the electroweak interaction of *quarks* by Glashow, Iliopoulos, and Maiani.

In this part, we shall obtain the Lagrangians of the standard model for electroweak interactions of both leptons and quarks. Furthermore, in the framework of the standard model, we shall present the theory of electromagnetic and weak interactions of particles. Finally, we shall dwell on the renormalizability of gauge theories and discuss unified models others than the standard one.

# 6. Lagrangians of the Electroweak Interactions

## 6.1 The Standard Model for the Electroweak Interactions of Leptons

We proceed with the models of electromagnetic and weak interactions. To construct the Lagrangian for a unified model, the following steps are required.

1) Choosing the gauge group which determines the interaction-mediating fields; the number of the gauge fields is equal to the dimension of the adjoint representation of this group;
2) Choosing primary fermions to underlie the model;
3) Choosing the representations of the gauge group in which the fermions are placed; the lowest representations are usually chosen.
4) Introducing an appropriate number of multiplets of scalar mesons as well as of interaction terms of these multiplets with the fermions (the Yukawa terms) to obtain massive particles;
5) Specifying the final composition of the model;
6) Writing the globally invariant Lagrangian for the model;
7) Writing the corresponding locally invariant Lagrangian;
8) Using the spontaneous symmetry-breaking mechanism to obtain the expression for the Lagrangian and then to diagonalize its free part.

It should be emphasized that the choice of a specific form of the model is very ambiguous. A large variety of models have been considered for this reason.

### 6.1.1 The Standard Model

Consider the simplest unified model of electromagnetic and weak interactions put forward by Glashow, Salam, and Weinberg and referred to as the *standard model.*

1) To provide the weak interactions between the leptons, at least three gauge fields (corresponding to three intermediate vector bosons) have to be introduced. The minimal unitary group possessing the adjoint representation of dimension three is the group $SU_2$. To induce the electromagnetic interaction between the leptons, one gauge field (iso-singlet) is sufficient. The group possessing an adjoint representation whose dimension is unity is the

group $U_1$. Let us take the gauge group of the model in the form of a direct product of the groups $SU_2$ and $U_1$: $SU_2 \times U_1$.

2) As the primary particles, we choose the electron, the $\mu^-$-lepton, the $\tau^-$-lepton and their neutrinos.

3) Suppose the particles chosen are placed in the lowest representations of the group $SU_2$, as the following $SU_2$-doublets:

$$\begin{pmatrix} \nu_e \\ e^- \end{pmatrix}, \begin{pmatrix} \nu_\mu \\ \mu^- \end{pmatrix}, \quad \text{and} \quad \begin{pmatrix} \nu_\tau \\ \tau^- \end{pmatrix}.$$

Let us single out in each doublet the right-handed and the left-handed particles:

$$\begin{aligned} L &= \frac{1}{2}(1+\gamma_5)\begin{pmatrix} \nu_e \\ e^- \end{pmatrix} = \begin{pmatrix} \nu_e \\ e^- \end{pmatrix}_L \equiv L^a, \\ R &= \frac{1}{2}(1-\gamma_5)\begin{pmatrix} \nu_e \\ e^- \end{pmatrix} = \begin{pmatrix} \nu_e \\ e^- \end{pmatrix}_R \equiv R^a, \quad a = 1, 2, \end{aligned} \tag{6.1.1}$$

where $\gamma_5 = \mathrm{i}\gamma_0\gamma_1\gamma_2\gamma_3$ (cf. page 11).

Experimental evidence shows that the right-handed neutrinos and the doublets of right-handed leptons are not observed within the accessible range of energies. Therefore we shall confine our consideration to the left-handed lepton $SU_2$-doublets and right-handed lepton $SU_2$-singlets

$$\begin{pmatrix} \nu_e \\ e^- \end{pmatrix}_L, \begin{pmatrix} \nu_\mu \\ \mu^- \end{pmatrix}_L, \begin{pmatrix} \nu_\tau \\ \tau^- \end{pmatrix}_L, \ e_R^-, \mu_R^-, \text{ and } \tau_R^- . \tag{6.1.2}$$

4) For the massless vector gauge fields to become massive intermediate vector bosons, we introduce the scalar Higgs fields. As we have seen in Chap. 3, each vector gauge field which acquires mass is brought into correspondence with a scalar field. The latter becomes unphysical (goldstone) and is removed; in addition, at least one physical neutral scalar field appears. We have to provide the three vector gauge fields with mass, while the fourth field (the photon) remains massless. Hence, at least four scalar fields must be introduced. We combine these in a $SU_2$-doublet of complex scalar fields.

5) Thus, it has been established that the model comprises three left-handed lepton doublets, three right-handed lepton singlets, and one doublet of complex scalar mesons:

$$\begin{aligned} L^a(x) &= \begin{pmatrix} \nu_e(x) \\ e^-(x) \end{pmatrix}_L, \begin{pmatrix} \nu_\mu(x) \\ \mu^-(x) \end{pmatrix}_L, \begin{pmatrix} \nu_\tau(x) \\ \tau^-(x) \end{pmatrix}_L; \\ R(x) &= e_R^-(x), \mu_R^-(x), \tau_R^-(x); \ \phi^a(x) . \end{aligned} \tag{6.1.3}$$

The electron sector, the muon sector, and the $\tau^-$-lepton sector are absolutely alike. Therefore we shall confine ourselves to considering only one of them, e.g., the electron sector.

6) The Lagrangian of the electron sector globally invariant under the group $SU_2 \times U_1$ is written in the form

$$L = \mathrm{i}\bar{L}^a \gamma_\mu \partial_\mu L^a + \mathrm{i}\bar{R}\gamma_\mu \partial_\mu R + \partial_\mu \phi^{a*} \partial_\mu \phi^a + m^2 \phi^{a*} \phi^a - h\bar{L}^a \phi^a R - h\bar{R}\phi^{a*} L^a - \tfrac{1}{4} f(\phi^{a*}\phi^a)^2 . \tag{6.1.4}$$

This Lagrangian comprises the free Lagrangians of the massless fermion fields and of the scalar Higgs fields as well as the interaction Lagrangians describing the coupling between the scalar Higgs fields and the fermions (the Yukawa terms, $h$ denoting the coupling constant) and the self-interaction of the scalar Higgs fields ($f$ denoting the self-interaction coupling constant). The fermion mass term $\bar{\psi}\psi = \bar{L}R + \bar{R}L$ does not appear in (6.1.4) since it is not gauge invariant due to the different transformation properties of $L^a(x)$ and $R(x)$. The fermions acquire a mass by virtue of the Yukawa terms. The potential of the Higgs fields,

$$V(\phi) = -m^2 \phi^{a*} \phi^a + \tfrac{1}{4} f(\phi^{a*}\phi^a)^2$$

gives rise to a non-zero vacuum expectation value of the fields.

These fields transform under the group $SU_2$ as

$$L^a(x) \to L'^a(x) = \exp\left(-\frac{\mathrm{i}}{2} g \tau_k \varepsilon_k\right)_{ab} L^b(x) , \qquad R(x) \to R'(x) = R(x) ,$$
$$\phi^a(x) \to \phi'^a(x) = \exp\left(-\frac{\mathrm{i}}{2} g \tau_k \varepsilon_k\right)_{ab} \phi^b(x) , \tag{6.1.5}$$

and under the group $U_1$ as

$$L^a(x) \to L'^a(x) = \exp\left(-\frac{\mathrm{i}}{2} g_1 Y_{\mathrm{L}} \varepsilon_4\right)_{ab} L^b(x) ,$$
$$R(x) \to R'(x) = \exp\left(-\frac{\mathrm{i}}{2} g_1 Y_{\mathrm{R}} \varepsilon_4\right) R(x) , \tag{6.1.6}$$
$$\phi^a(x) \to \phi'^a(x) = \exp\left(-\frac{\mathrm{i}}{2} g_1 \varepsilon_4\right)_{ab} \phi^b(x) .$$

Here $Y$ is the hypercharge operator defined by

$$\tfrac{1}{2} Y = Q - I_3 , \tag{6.1.6$'$}$$

where $Q$ is the charge operator and $I_3 = \frac{1}{2}\tau_3$ is the iso-spin operator.

Substituting into (6.1.6$'$) the eigenvalues of the operators $Q$ and $I_3$ corresponding to the lepton sector, we find

$$\tfrac{1}{2} Y_{\mathrm{L}} = \begin{pmatrix} 0 & 0 \\ 0 & -1 \end{pmatrix} - \begin{pmatrix} \frac{1}{2} & 0 \\ 0 & -\frac{1}{2} \end{pmatrix} = \begin{pmatrix} -\frac{1}{2} & 0 \\ 0 & -\frac{1}{2} \end{pmatrix} , \qquad \tfrac{1}{2} Y_{\mathrm{R}} = -1 .$$

It should be stressed that the iso-spin and the hypercharge thus introduced have nothing to do with the iso-spin and the hypercharge of hadrons.

7) Let us localize the transformations (6.1.5, 6). Using the formulae of Sect. 2.3, 4, and 3.4 we then find for the locally invariant Lagrangian

$$\mathscr{L} = -\frac{1}{4}F^k_{\mu\nu}F^k_{\mu\nu} - \frac{1}{4}F_{\mu\nu}F_{\mu\nu} + \left[\partial_\mu\phi^{a*} - \frac{\mathrm{i}}{2}g\phi^{b*}(\tau_k)^{ba}A^k_\mu - \frac{\mathrm{i}}{2}g_1\phi^{a*}A_\mu\right]$$
$$\times\left[\partial_\mu\phi^a + \frac{\mathrm{i}}{2}g(\tau_k)^{ab}\phi^b A^k_\mu + \frac{\mathrm{i}}{2}g_1\phi^a A_\mu\right]$$
$$+ \mathrm{i}\bar{L}^a\gamma_\mu\left[\partial_\mu L^a + \frac{\mathrm{i}}{2}g(\tau_k)^{ab}L^b A^k_\mu + \frac{\mathrm{i}}{2}g_1(Y_L)_{ab}L^b A_\mu\right]$$
$$+ \mathrm{i}\bar{R}\gamma_\mu\left(\partial_\mu R + \frac{\mathrm{i}}{2}g_1 Y_R R A_\mu\right) - h\bar{L}^a\phi^a R - h\bar{R}\phi^{a*}L^a$$
$$+ m^2\phi^{a*}\phi^a - \frac{f}{4}(\phi^{a*}\phi^a)^2 . \tag{6.1.7}$$

This Lagrangian includes (i) the massless gauge fields $A_\mu(x)$ and $A^k_\mu(x)$; (ii) two coupling constants, $g$ and $g_1$, corresponding to the groups $SU_2$ and $U_1$, respectively.

8) Let us make use of the mechanism of spontaneous symmetry-breaking (cf. Chap. 3). As seen from (6.1.4), the symmetry of $SU_2\times U_1$ is spontaneously broken. Choose the vacuum expectation value of the function $\phi^a(x)$ in the form (3.3.10), introduce the new real scalar fields $\sigma(x)$ and $\theta^k(x)$ (where $k=1,2,3$), and carry out a shift of the field according to (3.3.11) thus fixing the vacuum. Substituting further (3.3.11) into (6.1.7) and fixing the gauge $\theta'^k(x)=0$, which eliminates the Goldstone particles, we arrive at the following expression for the Lagrangian:

$$\mathscr{L} = \mathscr{L}_0 + \mathscr{L}_\mathrm{I}\,, \qquad \text{where}$$

$$\mathscr{L}_0 = -\frac{1}{4}(\partial_\mu A^k_\nu - \partial_\nu A^k_\mu)(\partial_\mu A^k_\nu - \partial_\nu A^k_\mu)$$
$$-\frac{1}{4}(\partial_\mu A_\nu - \partial_\nu A_\mu)(\partial_\mu A_\nu - \partial_\nu A_\mu) + \frac{g^2m^2}{2f}A^k_\mu A^k_\mu$$
$$+\frac{g_1^2m^2}{2f}A_\mu A_\mu - \frac{gg_1m^2}{f}A^3_\mu A_\mu + \frac{1}{2}\partial_\mu\sigma\,\partial_\mu\sigma - m^2\sigma^2 + \mathrm{i}\bar{R}\gamma_\mu\partial_\mu R$$
$$+\mathrm{i}\bar{L}^1\gamma_\mu\partial_\mu L^1 + \mathrm{i}\bar{L}^2\gamma_\mu\partial_\mu L^2 - \frac{\sqrt{2}hm}{\sqrt{f}}\bar{L}^2R - \frac{\sqrt{2}hm}{\sqrt{f}}\bar{R}L^2;$$

$$\mathscr{L}_\mathrm{I} = \frac{g}{4}\varepsilon_{lmk}(A^l_\mu A^m_\nu - A^l_\nu A^m_\mu)(\partial_\mu A^k_\nu - \partial_\nu A^k_\mu)$$
$$-\frac{g^2}{8}(A^l_\mu A^m_\nu - A^l_\nu A^m_\mu)(A^l_\mu A^m_\nu - A^l_\nu A^m_\mu) + \frac{g_1^2 m}{2\sqrt{f}}\sigma A_\mu A_\mu$$

$$+\frac{g_1^2}{8}\sigma^2 A_\mu A_\mu + \frac{g^2 m}{2\sqrt{f}}\sigma A_\mu^k A_\mu^k + \frac{g^2}{8}\sigma^2 A_\mu^k A_\mu^k$$

$$-\frac{g g_1 m}{\sqrt{f}}\sigma A_\mu^3 A_\mu - \frac{g g_1}{4}\sigma^2 A_\mu^3 A_\mu + g_1 \bar{R}\gamma_\mu R A_\mu + \frac{g_1}{2}\bar{L}^1\gamma_\mu L^1 A_\mu$$

$$+\frac{g_1}{2}\bar{L}^2\gamma_\mu L^2 A_\mu - \frac{g}{2}\bar{L}^1\gamma_\mu L^1 A_\mu^3 + \frac{g}{2}\bar{L}^2\gamma_\mu L^2 A_\mu^3$$

$$-\frac{g}{2}\bar{L}^2\gamma_\mu L^1(A_\mu^1 + \mathrm{i}A_\mu^2) - \frac{g}{2}\bar{L}^1\gamma_\mu L^2(A_\mu^1 - \mathrm{i}A_\mu^2)$$

$$-\frac{\sqrt{2}}{2}h\sigma\bar{L}^2 R - \frac{\sqrt{2}}{2}h\sigma\bar{R}L^2 - \frac{1}{2}m\sqrt{f}\sigma^3 - \frac{1}{16}f\sigma^4 . \qquad (6.1.8)$$

By introducing the new fields given by (3.4.4, 5) we diagonalize the free Lagrangian. Subsequent substitution of the explicit expressions for $L^a$ and $R$ given by (6.1.3) into (6.1.8) eventually yields the *Lagrangian of the standard model* expressed in terms of the fields

$$\mathscr{L} = \mathscr{L}_0 + \mathscr{L}_\mathrm{I} , \quad \text{where}$$

$$\mathscr{L}_0 = -\frac{1}{2}(\partial_\mu W_\nu^* - \partial_\nu W_\mu^*)(\partial_\mu W_\nu - \partial_\nu W_\mu) + \frac{g^2 m^2}{f} W_\mu^* W_\mu$$

$$-\frac{1}{4}(\partial_\mu Z_\nu - \partial_\nu Z_\mu)(\partial_\mu Z_\nu - \partial_\nu Z_\mu) + \frac{m^2(g^2+g_1^2)}{2f} Z_\mu Z_\mu$$

$$-\frac{1}{4}(\partial_\mu B_\nu - \partial_\nu B_\mu)^2 + \frac{1}{2}\partial_\mu\sigma\,\partial_\mu\sigma - m^2\sigma^2 + \mathrm{i}\bar{e}\gamma_\mu\partial_\mu e$$

$$-\frac{\sqrt{2}hm}{\sqrt{f}}\bar{e}e + \mathrm{i}\bar{\nu}_\mathrm{e}\gamma_\mu\frac{1+\gamma_5}{2}\partial_\mu\nu_\mathrm{e} \qquad (6.1.9)$$

and

$$\mathscr{L}_\mathrm{I} = -\frac{\mathrm{i}g^2}{\sqrt{g^2+g_1^2}} W_\mu^* W_\nu(\partial_\mu Z_\nu - \partial_\nu Z_\mu)$$

$$-\frac{\mathrm{i}gg_1}{\sqrt{g^2+g_1^2}} W_\mu^* W_\nu(\partial_\mu B_\nu - \partial_\nu B_\mu)$$

$$+\frac{\mathrm{i}g^2}{\sqrt{g^2+g_1^2}}(W_\nu^*\partial_\nu W_\mu - W_\nu\partial_\nu W_\mu^* + W_\nu\partial_\mu W_\nu^* - W_\nu^*\partial_\mu W_\nu) Z_\mu$$

$$+\frac{\mathrm{i}gg_1}{\sqrt{g^2+g_1^2}}(W_\nu^*\partial_\nu W_\mu - W_\nu\partial_\nu W_\mu^* + W_\nu\partial_\mu W_\nu^* - W_\nu^*\partial_\mu W_\nu) B_\mu$$

$$
-\frac{g^2}{2} W_\mu^* W_\mu W_\nu^* W_\nu + \frac{g^2}{2} W_\mu^* W_\mu^* W_\nu W_\nu - \frac{g^4}{g^2+g_1^2} W_\mu^* W_\mu Z_\nu Z_\nu
$$

$$
+\frac{g^4}{g^2+g_1^2} W_\mu^* W_\nu Z_\mu Z_\nu - \frac{g^2 g_1^2}{g^2+g_1^2} W_\mu^* W_\mu B_\nu B_\nu
$$

$$
+\frac{g^2 g_1^2}{g^2+g_1^2} W_\mu^* W_\nu B_\mu B_\nu - \frac{2g^3 g_1}{g^2+g_1^2} W_\mu^* W_\mu Z_\nu B_\nu + \frac{g^3 g_1}{g^2+g_1^2} W_\mu^* W_\nu Z_\mu B_\nu
$$

$$
+\frac{g^3 g_1}{g^2+g_1^2} W_\mu^* W_\nu Z_\nu B_\mu + \frac{g^2 m}{\sqrt{f}} \sigma W_\mu^* W_\mu + \frac{m(g^2+g_1^2)}{2\sqrt{f}} \sigma Z_\mu Z_\mu
$$

$$
+\frac{1}{4} g^2 \sigma^2 W_\mu^* W_\mu + \frac{g^2+g_1^2}{8} \sigma^2 Z_\mu Z_\mu + \frac{g g_1}{\sqrt{g^2+g_1^2}} \bar{e}\gamma_\mu e B_\mu
$$

$$
-\frac{g}{2\sqrt{2}} \bar{\nu}_e \gamma_\mu (1+\gamma_5) e W_\mu^* - \frac{g}{2\sqrt{2}} \bar{e}\gamma_\mu (1+\gamma_5) \nu_e W_\mu - \frac{\sqrt{g^2+g_1^2}}{4}
$$

$$
\times \left\{ \bar{\nu}_e \gamma_\mu (1+\gamma_5) \nu_e - \bar{e}\gamma_\mu \left[ \gamma_5 + \frac{g^2 - 3g_1^2}{g^2+g_1^2} \right] e \right\} Z_\mu - \frac{1}{\sqrt{2}} h\sigma\bar{e}e
$$

$$
-\frac{1}{2} m\sqrt{f}\sigma^3 - \frac{1}{16} f\sigma^4 . \tag{6.1.10}
$$

The Lagrangian (6.1.9) comprises the fields which can be identified with the charged intermediate vector boson field $W_\mu(x)$ with mass $gm/\sqrt{f}$; with the neutral intermediate vector boson field $Z_\mu(x)$ with mass $\bar{g}m/\sqrt{f}$; with the electromagnetic field $B_\mu(x)$; with the real scalar field $\sigma(x)$ with mass $m\sqrt{2}$; with the electron field with mass $mh\sqrt{2/f}$; and with the left-handed neutrino field $\nu_{eL}(x)$; $\bar{g} = \sqrt{g^2+g_1^2}$.

In addition, the Lagrangian (6.1.10) contains the terms describing the interactions of these fields. The following Lagrangians are involved in these terms (the electron, the muon, and the $\tau$-lepton sectors are considered).

i) The Lagrangian for the electromagnetic interaction of leptons

$$
\mathscr{L}_\mathrm{I} = -eJ_\mu^{e,l}(x) B_\mu(x), \; e = -\frac{g g_1}{\bar{g}}, \tag{6.1.11}
$$

where

$$
J_\mu^{e,l}(x) = \bar{e}(x)\gamma_\mu e(x) + \bar{\mu}(x)\gamma_\mu \mu(x) + \bar{\tau}(x)\gamma_\mu \tau(x) \tag{6.1.12}
$$

is the *electromagnetic lepton current*.

ii) The interaction Lagrangian for the *charged* weak lepton currents

$$
\mathscr{L}_\mathrm{I} = -\frac{g}{2\sqrt{2}} J_\mu^{(+),l}(x) W_\mu^*(x) + \text{Hermitian conjugate}, \tag{6.1.13}
$$

where $W_\mu$ and $W_\mu^*$ are the $W^+$ and $W^-$ bosons fields, respectively and

$$J_\mu^{(+),l}(x) = \bar{\nu}_e(x)\gamma_{\mu L}e(x) + \bar{\nu}_\mu(x)\gamma_{\mu L}\mu(x) + \bar{\nu}_\tau(x)\gamma_{\mu L}\tau(x) \tag{6.1.14}$$

is the *charged weak lepton current.*

iii) The interaction Lagrangian for the *neutral* weak lepton current

$$\mathscr{L}_I = -\tfrac{1}{2}\bar{g}[J_\mu^{n,\nu}(x) + J_\mu^{n,l}(x)]Z_\mu(x), \tag{6.1.15}$$

where

$$J_\mu^{n,\nu}(x) = \tfrac{1}{2}[\bar{\nu}_e(x)\gamma_{\mu L}\nu_e(x) + \bar{\nu}_\mu(x)\gamma_{\mu L}\nu_\mu(x) + \bar{\nu}_\tau(x)\gamma_{\mu L}\nu_\tau(x)] \tag{6.1.16}$$

is the *neutral weak neutrino current* and

$$\begin{aligned} J_\mu^{n,l}(x) = (-\tfrac{1}{2}+\xi)[\bar{e}(x)\gamma_{\mu L}e(x) + \bar{\mu}(x)\gamma_{\mu L}\mu(x) + \bar{\tau}(x)\gamma_{\mu L}\tau(x)] \\ +\xi[\bar{e}(x)\gamma_{\mu R}e(x) + \bar{\mu}(x)\gamma_{\mu R}\mu(x) + \bar{\tau}(x)\gamma_{\mu R}\tau(x)] \end{aligned} \tag{6.1.17}$$

is the *neutral weak current of charged leptons.*

Here

$$\gamma_{\mu L} = \gamma_\mu(1+\gamma_5), \qquad \gamma_{\mu R} = \gamma_\mu(1-\gamma_5), \qquad \xi = \sin^2\theta_W = \frac{g_1^2}{\bar{g}^2}; \tag{6.1.17'}$$

with respect to a spinor $\psi$ satisfying the Dirac equation, the following relations hold:

$$\psi_L = \tfrac{1}{2}(1+\gamma_5)\psi, \qquad \psi_R = \tfrac{1}{2}(1-\gamma_5)\psi,$$
$$\bar{\psi}_L = \bar{\psi}\tfrac{1}{2}(1-\gamma_5), \qquad \bar{\psi}_R = \bar{\psi}\tfrac{1}{2}(1+\gamma_5),$$

where $\psi_L$ and $\psi_R$ are the left-handed and the right-handed spinors, respectively. Therefore, by using the relation

$$\gamma_\mu(1+\gamma_5) = \tfrac{1}{2}(1-\gamma_5)\gamma_\mu(1+\gamma_5),$$

we obtain

$$\bar{\psi}\gamma_{\mu L}\psi = 2\bar{\psi}_L\gamma_\mu\psi_L, \qquad \bar{\psi}\gamma_{\mu R}\psi = 2\bar{\psi}_R\gamma_\mu\psi_R,$$

which implies that the expressions for the currents can be presented in two equivalent forms containing either unpolarized or polarized spinors.

For the charged current $J_\mu^{(+)}$ and its Hermitian conjugate we have

$$(J_\mu^{(+)}(x))^\dagger = J_\mu^{(-)}(x).$$

It is seen that the standard model leads to the Lagrangians of both electromagnetic and weak interactions, and in this sense it is a unified model of these interactions, indeed.

The most interesting property of the standard model is the occurrence of *neutral weak currents.* A number of consequences of this prediction have been

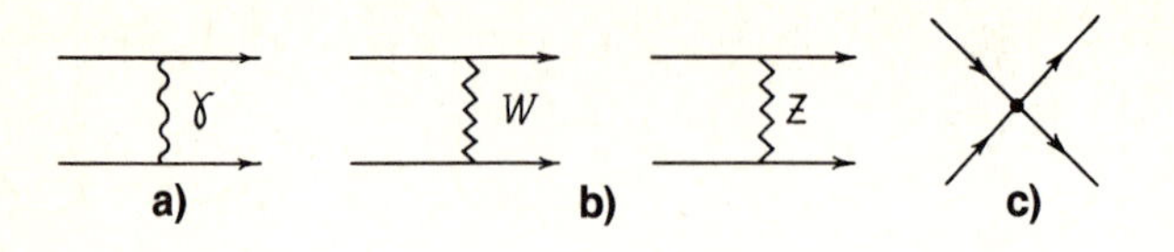


**Fig. 6.1a–c.** Interactions: (**a**) electromagnetic interaction; (**b**) weak interactions via W- or Z-boson exchange; (**c**) point weak interactions

subject to experimental checks. The theoretical results were verified in these experiments. (In Sect. 8.1 this will be discussed in more detail.)

The Lagrangian (6.1.11) of the standard model is analogous to the Lagrangian of the electromagnetic interaction in quantum electrodynamics,

$$\mathscr{L}_{\mathrm{I}} = -e\bar{\psi}\gamma_{\mu}\psi A_{\mu}, \quad \text{with} \tag{6.1.18}$$

$$e = -\frac{g g_1}{\bar{g}}. \tag{6.1.19}$$

That is to say, the electromagnetic interaction between leptons involves the exchange of photons (cf. Fig. 6.1 a). The photons being massless particles, the electromagnetic interaction is long range.

In analogy with the electromagnetic interaction, the weak interaction between leptons in the standard model originates from the exchange of *charged* intermediate *$W^{\pm}$-bosons* (charged currents) and of *neutral Z-bosons* (neutral currents) (cf. Fig. 6.1 b). Intermediate bosons being massive particles, the weak interaction is short range.

In the matrix element $M$ of the processes caused by weak charged currents there appears a factor (propagator)

$$M \sim -\frac{(g/(2\sqrt{2}))^2}{q^2 - m_W^2}, \tag{6.1.20}$$

where $q$ is the momentum and $m_W$ is the mass of the intermediate boson. In the limit of small energies ($|q^2| \ll m_W^2$) (6.1.20) reduces to

$$M \sim \frac{g^2}{8m_W^2}, \tag{6.1.21}$$

which implies that the fact of realization of the weak interaction via the exchange of intermediate bosons turns out to be immaterial. At small energies, the Lagrangian (6.1.13) reduces to an *effective Lagrangian* describing the interaction of four fermions at one point (the so called *point interaction*) (cf. Fig. 6.1 c). Such a four-fermion interaction is characterized by the coupling constant $G$ and by the matrix element proportional to this constant:

$$M \sim \frac{1}{\sqrt{2}} G . \tag{6.1.22}$$

Comparing (6.1.21) with (6.1.22) for any concrete process gives

$$\frac{G}{\sqrt{2}} = \frac{g^2}{8 m_W^2} . \tag{6.1.23}$$

There exist a number of important relations between the constants and the parameters involved in the standard model. By combining (6.1.19) with (6.1.17′) and (3.4.8) we find

$$e = -g \sin\theta_W , \qquad e = -g_1 \cos\theta_W , \qquad e = -\bar{g} \sin\theta_W \cos\theta_W . \tag{6.1.24}$$

The free parameter $\theta_W$ is called the *Weinberg angle*. From (6.1.24) follows

$$\frac{1}{e^2} = \frac{1}{g^2} + \frac{1}{g_1^2} . \tag{6.1.25}$$

Hence, the interaction of all the gauge fields is determined in the standard model by the electric charge and the free parameter $\theta_W$. The appearance of the free parameter is due to the symmetry group of the standard model which is the direct product of two simple groups, $SU_2$ and $U_1$.

From (6.1.19, 23) and (6.1.24) it is found that the mass of the $W$-boson is

$$m_W = \left( \frac{\pi\alpha}{\sqrt{2} G} \right)^{1/2} \frac{1}{\sin\theta_W} = \frac{37.3}{\sin\theta_W} \,\text{GeV} . \tag{6.1.26}$$

Then, by virtue of the relations

$$m_W = g m/\sqrt{f} , \qquad m_Z = \bar{g} m/\sqrt{f} , \qquad m_W/m_Z = \cos\theta_W \tag{6.1.27}$$

we obtain for the mass of the $Z$-boson

$$m_Z = \frac{m_W}{\cos\theta_W} = \frac{37.3}{\sin\theta_W \cos\theta_W} \,\text{GeV} = \frac{74.6}{\sin 2\theta_W} \,\text{GeV} . \tag{6.1.28}$$

One can see that the lower values of the masses of $W$-bosons and $Z$-bosons are quite large. Recently the intermediate vector bosons $W^{\pm}$ and $Z$ have been experimentally discovered.

Not less important is detecting the Higgs bosons. Unfortunately, unlike the mass of $W$-bosons and $Z$-bosons, in the standard model the mass of the Higgs boson is not fixed which makes the search for it a difficult task.

## 6.2 Quark Models of Hadrons

### 6.2.1 Hadrons

Like in the case of leptons, the electromagnetic and the weak interactions also exist between the hadrons. However, as distinct from the leptons, the hadrons additionally exhibit strong interactions. This circumstance appreciably complicates the situation.

In this section we shall consider the quark model of hadrons. Gauge theories of electromagnetic and weak interactions of quarks will be treated in the next section and those of strong interactions in Part IV.

Experimental evidence shows that hadrons combine into *multiplets.* First the *iso*-multiplets of hadrons have been detected, followed by the *unitary* multiplets and – relatively recently – by the *charmed* multiplets. These types of multiplets are attributed to the $SU_2$, $SU_3$, and $SU_4$ symmetries, respectively.

### 6.2.2 $SU_3$-Symmetry. Three Quarks

The octets and the decuplets of baryons (cf. Table 6.1) and the meson octets (cf. Table 6.2) are unitary multiplets. These particles are placed in the corresponding representations of the group $SU_3$.

To account for the existence of the multiplets, Gell-Mann and Zweig proposed the *quark model* of hadrons. The basic assumptions underlying this model are as follows.

1) There exists a fundamental $SU_3$-triplet of strongly interacting particles. These particles are called *quarks* and are denoted $u$, $d$, and $s$. The quarks $u$ and $d$ form an $SU_2$-doublet (of strong iso-spin $I = 1/2$) with zero strangeness; the quark $s$ is an $SU_2$-singlet (of zero iso-spin) with strangeness $S = -1$.

**Table 6.1.** Baryon multiplets

| $Y = B + S$ | | 1 | | 0 | | $-1$ | $-2$ |
|---|---|---|---|---|---|---|---|
| $I$ | | $\frac{1}{2}$ | $\frac{3}{2}$ | 0 | 1 | $\frac{1}{2}$ | 0 |
| | $\frac{1}{2}^-$ | $N^*_{1535}$ | | $\Lambda^*_{1670}$ | | | |
| | $\frac{1}{2}^+$ | $N$ | | $\Lambda$ | $\Sigma$ | $\Xi$ | |
| $s^P$ | $\frac{3}{2}^+$ | | $N^*_{1232}$ | | $\Sigma^*_{1385}$ | $\Xi^*_{1530}$ | $\Omega$ |
| | $\frac{3}{2}^-$ | $N^*_{1520}$ | | $\Lambda^*_{1690}$ | $\Sigma^*_{1670}$ | | |

**Table 6.2.** Meson multiplets

| $Y = S$ | 0 | | $\pm 1$ |
|---|---|---|---|
| $s^P$ | $I = 1$ | $I = 0$ | $I = \frac{1}{2}$ |
| $0^-$ | $\pi$ | $\eta$, $X^0$ | $K$ |
| $0^+$ | $\delta_{980}$ | $S^*_{980}$ | $\varkappa_{1400}$ |
| $1^-$ | $\varrho$, $\varrho'$ | $\omega$, $\phi$ | $K^*_{892}$ |
| $1^+$ | $A_1$ | $D$, $E$ | $Q_1$ |
| $1^+$ | $B$ | | $Q_2$ |
| $2^+$ | $A_2$ | $f_{1270}$, $f'_{1515}$ | $K^*_{1430}$ |

**Table 6.3.** Quarks and their characteristics

| Quark | Electric charge $Q$ | Charm $C$ | Stran-geness $S$ | Baryon number $B$ | Hyper-charge $Y$ | Spatial spin $s$ | Iso-spin projection $I_3$ |
|---|---|---|---|---|---|---|---|
| $u$ | 2/3 | 0 | 0 | 1/3 | 1/3 | 1/2 | 1/2 |
| $d$ | −1/3 | 0 | 0 | 1/3 | 1/3 | 1/2 | −1/2 |
| $s$ | −1/3 | 0 | −1 | 1/3 | −2/3 | 1/2 | 0 |
| $c$ | 2/3 | 1 | 0 | 1/3 | −2/3 | 1/2 | 0 |

2) The observed hadrons are bound states of either quarks, or antiquarks, or both quarks and antiquarks.

The quantum numbers of the quarks $u$, $d$, and $s$ are listed in the first three lines of Table 6.3 where $B$ denotes the baryon number, $Y$ the hypercharge, $I$ the iso-spin, $I_3$ its projection, $Q$ the electric charge, $S = Y - B$ strangeness, and $C$ charm (to be discussed below).

It should be emphasized that the spatial spin of the quarks has to be half-integer, in order to account for the observed values of the spin of baryons and mesons.

The mesons are made up of one quark and one antiquark: for instance, for the pseudoscalar meson octet one has

$$\begin{array}{ccccc} & \bar{K}^0(s\bar{d}) & & K^-(s\bar{u}) & \\ \pi^+(u\bar{d}) & & \pi^0, \eta^0(u\bar{u}, d\bar{d}, s\bar{s}) & & \pi^-(d\bar{u}) \,. \\ & K^+(u\bar{s}) & & K^0(d\bar{s}) & \end{array}$$

The baryons and the baryon resonances are made up of three quarks: for example, a baryon octet and a decuplet of baryon resonances have the form

$$\begin{array}{ccccc} & \Xi^0(uss) & & \Xi^-(dss) & \\ \Sigma^+(uus) & & \Sigma^0, \Lambda^0(dus) & & \Sigma^-(dds)\,, \\ & p(uud) & & n(udd) & \end{array}$$

$$\begin{array}{ccccccc} & & & \Omega^-(sss) & & & \\ & & \Xi^{*0}(uss) & & \Xi^{*-}(dss) & & \\ & \Sigma^{*+}(uus) & & \Sigma^{*0}(uds) & & \Sigma^{*-}(dds) & \\ \Delta^{++}(uuu) & & \Delta^+(uud) & & \Delta^0(udd) & & \Delta^-(ddd)\,. \end{array} \tag{6.2.1}$$

The electric charge $Q$ of a quark (hadron) is determined, in terms of the quantum numbers of the quark (hadron), by the expression

$$Q = I_3 + \frac{Y}{2} = I_3 + \frac{B+S}{2}\,. \tag{6.2.2}$$

### 6.2.3 $SU_4$-Symmetry. Charm

For constructing charmed particles, it is insufficient to have three quarks: an additional quark – the charmed one – must be introduced. Its characteristics are summarized in the last line of Table 6.3.

Again the mesons are made up of quarks and antiquarks. In the $SU_4$-symmetry, the mesons form a 15-plet comprising an octet of non-charmed $SU_3$-mesons, a new meson with hidden charm (consisting of the quark $c$ and antiquark $\bar{c}$), and 6 charmed mesons. The 15-plets of pseudoscalar and vector mesons are presented in Tables 6.4 and 6.5, respectively. All these particles have been experimentally observed. Conventionally, a 15-plet is graphically

**Table 6.4.** Pseudoscalar mesons (including the charmed mesons)

| | $\bar{u}$ | $\bar{d}$ | $\bar{s}$ | $\bar{c}$ |
|---|---|---|---|---|
| $u$ | $\frac{\eta'}{\sqrt{3}} + \frac{\eta}{\sqrt{6}} + \frac{\pi^0}{\sqrt{2}}$ | $\pi^+$ | $K^+$ | $\bar{D}^0$ |
| $d$ | $\pi^-$ | $\frac{\eta'}{\sqrt{3}} + \frac{\eta}{\sqrt{6}} - \frac{\pi^0}{\sqrt{2}}$ | $K^0$ | $D^-$ |
| $s$ | $K^-$ | $\bar{K}^0$ | $\frac{\eta'}{\sqrt{3}} + \frac{2\eta}{\sqrt{6}}$ | $F^-$ |
| $c$ | $D^0$ | $D^+$ | $F^+$ | $\eta_c$ |

**Table 6.5.** Vector mesons (including the charmed mesons)

| | $\bar{u}$ | $\bar{d}$ | $\bar{s}$ | $\bar{c}$ |
|---|---|---|---|---|
| $u$ | $\frac{1}{\sqrt{2}}(\omega^0+\varrho^0)$ | $\varrho^+$ | $K^{*+}$ | $\bar{D}^{*0}$ |
| $d$ | $\varrho^-$ | $\frac{1}{\sqrt{2}}(\omega^0-\varrho^0)$ | $K^{*0}$ | $D^{*-}$ |
| $s$ | $K^{*-}$ | $\bar{K}^{*0}$ | $\phi^0$ | $F^{*-}$ |
| $c$ | $D^{*0}$ | $D^{*+}$ | $F^{*+}$ | $\psi^0$ |

presented as an octahedron, one cut plane of which contains the non-charmed mesons of the $SU_3$-octet (cf. Fig. 6.2).

The baryons are made up of three quarks (both ordinary and charmed). In the $SU_4$-symmetry, three 20-plets can occur; one of them is depicted in Fig. 6.3. Some of the charmed baryons have been experimentally detected.

In the four-quark model the charge $Q$ of a quark (hadron) is determined in terms of the quantum numbers of the quark (hadron) as follows:

$$Q = I_3 + \frac{B+S+C}{2}. \tag{6.2.3}$$

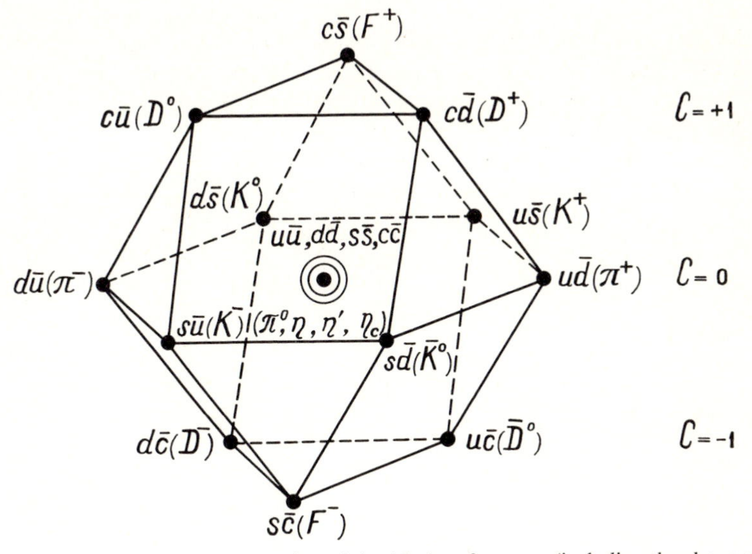


**Fig. 6.2.** Graphical representation of the 15-plet of mesons (including the charmed mesons)

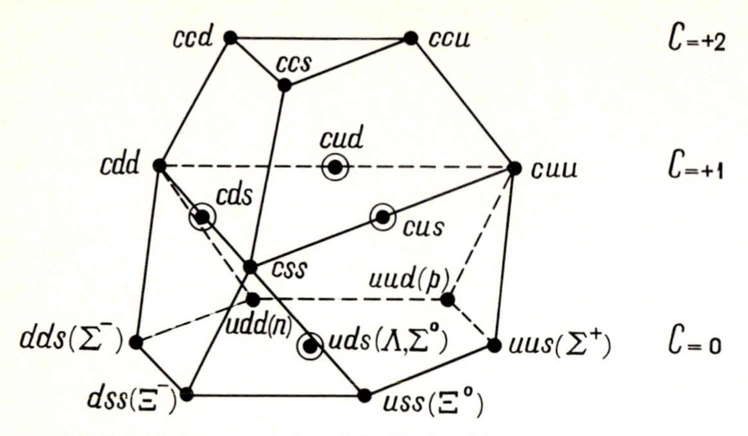


**Fig. 6.3.** Graphical representation of the 20-plet of baryons (including the charmed baryons)

### 6.2.4 Coloured Quarks

As we already know, the quarks must have half-integer spin. Such particles obey the Pauli principle stating that the system cannot include two or more fermions with the same quantum numbers. For particles made up of three quarks this principle is, however, not always satisfied. For example, the $\Omega$-particle contained in the decouplet (6.2.1) is constructed from three identical quarks. To avoid this inconsistency, the quarks have been provided with one more quantum number called *colour*. Then the quarks can also be distinguished by colour, thus circumventing the contradiction with Pauli's principle.

Increasing the number of quarks requires extending the symmetry. The most natural extension of the $SU_4$-symmetry is the $SU_4 \times SU_3'$-symmetry, where $SU_3'$ is the colour symmetry group. Let the quantum number specifying the type of the quarks $u$, $d$, $s$, $c$ be called *flavour* and denote it by a Latin index $a = 1, 2, 3, 4$; let a Greek index $\alpha = 1, 2, 3$ be introduced to denote colour. The twelve quarks can then be described by a matrix function $q_a^\alpha$ which can be presented in the form

$$q_a^\alpha = \begin{pmatrix} u_1 & u_2 & u_3 \\ d_1 & d_2 & d_3 \\ s_1 & s_2 & s_3 \\ c_1 & c_2 & c_3 \end{pmatrix} . \tag{6.2.4}$$

Generators of the ordinary $SU_4$-symmetry act on the Latin indices $a$, those of the colour symmetry on the Greek ones, $\alpha$. That is to say, generators of the ordinary flavour group $SU_4$ will interchange the rows of the matrix (6.2.4), while those of the colour group $SU_3'$ will interchange its columns.

Because coloured hadrons are not observed in experiment one more postulate is added to the quark model of hadrons, namely that all hadrons must form singlets of the colour group. This implies that the baryons consist of three quarks differing in colour and that the mesons consist of quarks and antiquarks of all three colours occurring with equal weight.

### 6.2.5 Quark-Lepton Symmetry

It has been known for a long time that there is a correspondence between the leptons and the quarks:

$$\begin{array}{ccccc} \nu_e & e^- & \nu_\mu & \mu^- & \ldots \\ u & d & c & s & \ldots \end{array}$$

implying that they enter the theory in a symmetric way. This fact is referred to as the *quark-lepton symmetry.* It cannot be ruled out, however, that both the lepton series and the quark series will be extended in the future and it is not clear as yet whether the notion of quark-lepton symmetry will survive.

Also of interest is the question of the maximum possible number of flavours of the quarks and leptons. Various estimates give for quarks or leptons a figure not exceeding $7-8$, i.e., the number of leptons and quarks may prove not to be large.

In the framework of the quark-lepton symmetry, one can attribute colour (lilac one) also to leptons to combine them with the quarks thus extending the symmetry to $SU_4 \times SU_4'$. The sixteen particles are then described by a matrix function $q_a^\alpha(x)$ (where $a, \alpha = 1, 2, 3, 4$) which is expressed as follows:

$$q_a^\alpha(x) = \begin{pmatrix} u_1(x) & u_2(x) & u_3(x) & \nu_e(x) \\ d_1(x) & d_2(x) & d_3(x) & e^-(x) \\ s_1(x) & s_2(x) & s_3(x) & \mu^-(x) \\ c_1(x) & c_2(x) & c_3(x) & \nu_\mu(x) \end{pmatrix} . \qquad (6.2.4')$$

### 6.2.6 Two Types of Models with Coloured Quarks

It is usually assumed that the colour symmetry is non-broken with respect to strong interactions. With regard to electromagnetic and weak interactions two cases are possible: (i) the colour symmetry is non-broken with respect to these interactions; and (ii) these interactions generate colour symmetry-breaking.

In case (i) the electric charge of the quarks does not depend on their colour and must be *fractional.* In case (ii) the quarks can have both fractional and *integer* charge. To prove the latter statement, we consider as an example a model whose flavour group and colour group are $SU_4$ and $SU_4'$, respectively. The group $SU_4$ acts on the flavour index $a$ of the function $q_a^\alpha$ and transforms this function according to the fundamental representation. The group $SU_4'$

acts on the colour index $\alpha$ and transforms the function $q_a^\alpha$ according to the conjugate fundamental representation. Consequently, the function $q_a^\alpha$ transforms according to the representation $(4,4^*)$. The generators $\Lambda_k$ of the group $SU_4$ are analogous to the matrices $\lambda_i$ defined by (1.2.13). The generators $\Lambda_k^c$ of the conjugate representation are equal to the transposed generators $\Lambda_k$ taken with the opposite sign: $\Lambda_k^c = -\Lambda_k^T$. The group $SU_4$ has three linearly-independent diagonal generators. Let us define with their help the matrices of iso-spin, hypercharge, and charm:

$$I_3 = \begin{pmatrix} \frac{1}{2} & 0 & 0 & 0 \\ 0 & -\frac{1}{2} & 0 & 0 \\ 0 & 0 & 0 & 0 \\ 0 & 0 & 0 & 0 \end{pmatrix}, \quad Y = \begin{pmatrix} \frac{1}{3} & 0 & 0 & 0 \\ 0 & \frac{1}{3} & 0 & 0 \\ 0 & 0 & -\frac{2}{3} & 0 \\ 0 & 0 & 0 & 0 \end{pmatrix},$$

$$C = \begin{pmatrix} \frac{1}{4} & 0 & 0 & 0 \\ 0 & \frac{1}{4} & 0 & 0 \\ 0 & 0 & \frac{1}{4} & 0 \\ 0 & 0 & 0 & -\frac{3}{4} \end{pmatrix}. \tag{6.2.5}$$

These matrices act on the flavours. The colours are acted on by the matrices $I_3' = -I_3$, $Y' = -Y$, and $C' = -C$.

Taking into account (6.2.3) we determine the electric charge of the coloured quarks as

$$(Q+Q')_{ab,\alpha\beta} = (I_3 + \tfrac{1}{2}Y - \tfrac{2}{3}C)_{ab}\delta_{\alpha\beta} + (I_3' + \tfrac{1}{2}Y' - \tfrac{2}{3}C')_{\alpha\beta}\delta_{ab}\,. \tag{6.2.6}$$

Using (6.2.5, 6) we find

$$(Q+Q')_{ab,\alpha\beta}q_b^\beta = \begin{pmatrix} \frac{1}{2} & 0 & 0 & 0 \\ 0 & -\frac{1}{2} & 0 & 0 \\ 0 & 0 & -\frac{1}{2} & 0 \\ 0 & 0 & 0 & \frac{1}{2} \end{pmatrix}_{ab} q_b^\alpha + \begin{pmatrix} -\frac{1}{2} & 0 & 0 & 0 \\ 0 & \frac{1}{2} & 0 & 0 \\ 0 & 0 & \frac{1}{2} & 0 \\ 0 & 0 & 0 & -\frac{1}{2} \end{pmatrix}_{\alpha\beta} q_a^\beta\,. \tag{6.2.7}$$

Acting on the function $q_a^\alpha(x)$ by the matrix (6.2.7) we obtain the values of the electric charges of quarks for the case of the group $SU_4 \times SU_4'$:

$$\begin{pmatrix} u_1 & u_2 & u_3 & \nu_e \\ d_1 & d_2 & d_3 & e^- \\ s_1 & s_2 & s_3 & \mu^- \\ c_1 & c_2 & c_3 & \nu_\mu \end{pmatrix} \rightarrow \begin{pmatrix} 0 & 1 & 1 & 0 \\ -1 & 0 & 0 & -1 \\ -1 & 0 & 0 & -1 \\ 0 & 1 & 1 & 0 \end{pmatrix}. \tag{6.2.8}$$

One can see that all quark charges are integers. Nowadays one considers mostly models with fractional charges.

### 6.2.7 Heavy Quarks

Recently, new particles with a mass of about 10 GeV have been discovered which are named $\Upsilon$-mesons. It has been supposed that this meson is a manifestation of some new, *heavier quark b* possessing a new flavour, namely *beauty* (or *bottom*). If this is the case, then the $\Upsilon$-meson consists of the quark $b$ and antiquark $\bar{b}$ and is quite analogous to the $\psi$-meson, the first discovered representative of the family of charmed particles.

By analogy with other quarks, the $b$-quark may have a counterpart referred to as the *t-quark* (or *top quark*), not detected as yet.

## 6.3 The Standard Model of Electroweak Interactions of Quarks

Let us go over to discussing the electromagnetic and weak interactions of quarks.

The simplest way is extending the range of applicability of the standard model to include the quarks.

1) Choose the standard group $SU_2 \times U_1$ as the gauge group. Gauge fields of this group are the mediators of the weak and electromagnetic interactions between the quarks.

2) The initial particles underlying the model are the 12 quarks listed in the matrix (6.2.4) together with the quarks $b$ and $t$.

3) Experiment shows that no right-handed quark doublets exist in the presently accessible energy range. Therefore, by analogy with the leptons, we combine the left-handed quarks into $SU_2$-doublets

$$L_1^\alpha = \frac{1}{2}(1+\gamma_5)\begin{pmatrix} u \\ d' \end{pmatrix}^\alpha = \begin{pmatrix} u \\ d' \end{pmatrix}_L^\alpha, \quad L_2^\alpha = \frac{1}{2}(1+\gamma_5)\begin{pmatrix} c \\ s' \end{pmatrix}^\alpha = \begin{pmatrix} c \\ s' \end{pmatrix}_L^\alpha,$$

$$L_3^\alpha = \frac{1}{2}(1+\gamma_5)\begin{pmatrix} t \\ b' \end{pmatrix}^\alpha = \begin{pmatrix} t \\ b' \end{pmatrix}_L^\alpha, \tag{6.3.1}$$

while the right-handed quarks are put into $SU_2$-singlets

$$\begin{aligned} R_1^\alpha &= \tfrac{1}{2}(1-\gamma_5)u^\alpha = u_R^\alpha, & R_2^\alpha &= \tfrac{1}{2}(1-\gamma_5)c^\alpha = c_R^\alpha, \\ R_3^\alpha &= \tfrac{1}{2}(1-\gamma_5)t^\alpha = t_R^\alpha, & R_4^\alpha &= \tfrac{1}{2}(1-\gamma_5)d'^\alpha = d_R'^\alpha, \\ R_5^\alpha &= \tfrac{1}{2}(1-\gamma_5)s'^\alpha = s_R'^\alpha, & R_6^\alpha &= \tfrac{1}{2}(1-\gamma_5)b'^\alpha = b_R'^\alpha . \end{aligned} \tag{6.3.2}$$

Here $\alpha$ is the colour index assuming three values: $\alpha = 1, 2, 3$. The quarks $d'$, $s'$, and $b'$ are a linear combination (mixture) of the quarks $d$, $s$, and $b$. The necessity of mixing the quarks is dictated by experiment; this will be discussed in some detail below, in Chap. 8.

4) For the Higgs scalar fields we again introduce the $SU_2$-doublet of complex scalar fields $\phi^a(x)$.

5) Hence, the model includes 9 left-handed quark doublets (6.3.1), 18 right-handed quark singlets (6.3.2) and one doublet of complex scalar fields $\phi^a(x)$. The three coloured quark sectors are fully alike.

6) The globally invariant Lagrangian for the model reads

$$L = \mathrm{i}\bar{L}^\alpha_{na}\gamma_\mu\partial_\mu L^\alpha_{na} + \mathrm{i}\bar{R}^\alpha_m\gamma_\mu\partial_\mu R^\alpha_m + \partial_\mu\phi^{a*}\partial_\mu\phi^a + m^2\phi^{a*}\phi^a$$
$$- \tfrac{1}{4}f(\phi^{a*}\phi^a)^2 - h_n\bar{L}^\alpha_{na}\phi^a R^\alpha_{n+3} - h_n\bar{R}^\alpha_{n+3}\phi^{a*}L^\alpha_{na} - \tilde{h}_n\bar{L}^\alpha_{na}\tilde{\phi}^a R^\alpha_n$$
$$- \tilde{h}_n\bar{R}^\alpha_n\tilde{\phi}^{a*}L^\alpha_{na},$$

where

$$a = 1,2; \quad n = 1,2,3; \quad m = 1,\ldots,6; \quad \alpha = 1,2,3 \quad \text{and} \quad \tilde{\phi} = i\tau_2\phi .$$

The transformations of the functions, involved in this expression, under the groups $SU_2$ and $U_1$ are determined by (6.1.5, 6). For the quark sector we have, according to (6.1.6′):

$$\frac{1}{2}Y_\mathrm{L} = \begin{pmatrix} \frac{2}{3} & 0 \\ 0 & -\frac{1}{3} \end{pmatrix} - \begin{pmatrix} \frac{1}{2} & 0 \\ 0 & -\frac{1}{2} \end{pmatrix} = \begin{pmatrix} \frac{1}{6} & 0 \\ 0 & \frac{1}{6} \end{pmatrix}, \quad \frac{1}{2}Y_\mathrm{R} = \begin{pmatrix} \frac{2}{3} & 0 \\ 0 & -\frac{1}{3} \end{pmatrix}.$$

Furthermore, in complete analogy with the Glashow-Salam-Weinberg model, we find the locally invariant Lagrangian for the model and, with the aid of the mechanism of spontaneous symmetry-breaking, obtain an expression for the Lagrangian analogous to (6.1.9, 10).

Among other terms, this Lagrangian includes the following ones:

i) The Lagrangian for electromagnetic interaction of the quarks

$$\mathscr{L}_\mathrm{I} = -eJ^{e,q}_\mu(x)B_\mu(x), \tag{6.3.3}$$

where

$$J^{e,q}_\mu(x) = \tfrac{1}{3}[\bar{d}'(x)\gamma_\mu d'(x) + \bar{s}'(x)\gamma_\mu s'(x) + \bar{b}'(x)\gamma_\mu b'(x)]$$
$$- \tfrac{2}{3}[\bar{u}(x)\gamma_\mu u(x) + \bar{c}(x)\gamma_\mu c(x) + \bar{t}(x)\gamma_\mu t(x)] \tag{6.3.4}$$

is the *electromagnetic quark current*;

ii) The interaction Lagrangian for the charged weak quark currents

$$\mathscr{L}_\mathrm{I} = -\frac{g}{2\sqrt{2}}J^{(+),q}_\mu W^-_\mu + \text{Herm. conj.}, \tag{6.3.5}$$

where

$$J^{(+),q}_\mu(x) = \bar{u}(x)\gamma_{\mu\mathrm{L}}d'(x) + \bar{c}(x)\gamma_{\mu\mathrm{L}}s'(x) + \bar{t}(x)\gamma_{\mu\mathrm{L}}b'(x) \tag{6.3.6}$$

is the *charged weak quark current* and $W^-_\mu(x)$ is the intermediate $W^-$-boson field;

iii) The interaction Lagrangian for the neutral weak quark currents

$$\mathcal{L}_{\mathrm{I}} = -\tfrac{1}{2}\bar{g} J_\mu^{n,q}(x) Z_\mu(x) \tag{6.3.7}$$

where $J_\mu^{n,q}$ is the *neutral weak quark current* ($\xi = \sin^2\theta_{\mathrm{W}}$):

$$\begin{aligned} J_\mu^{n,q}(x) = {} & (\tfrac{1}{2} - \tfrac{2}{3}\xi)[\bar{u}(x)\gamma_{\mu\mathrm{L}}u(x) + \bar{c}(x)\gamma_{\mu\mathrm{L}}c(x) + \bar{t}(x)\gamma_{\mu\mathrm{L}}t(x)] \\ & - \tfrac{2}{3}\xi[\bar{u}(x)\gamma_{\mu\mathrm{R}}u(x) + \bar{c}(x)\gamma_{\mu\mathrm{R}}c(x) + \bar{t}(x)\gamma_{\mu\mathrm{R}}t(x)] \\ & + (-\tfrac{1}{2} + \tfrac{1}{3}\xi)[\bar{d}'(x)\gamma_{\mu\mathrm{L}}d'(x) + \bar{s}'(x)\gamma_{\mu\mathrm{L}}s'(x) + \bar{b}'(x)\gamma_{\mu\mathrm{L}}b'(x)] \\ & + \tfrac{1}{3}\xi[\bar{d}'(x)\gamma_{\mu\mathrm{R}}d'(x) + \bar{s}'(x)\gamma_{\mu\mathrm{R}}s'(x) + \bar{b}'(x)\gamma_{\mu\mathrm{R}}b'(x)]\,. \end{aligned} \tag{6.3.8}$$

The most peculiar feature of the model under consideration is the form of (6.3.7) for the interaction Lagrangian of neutral weak quark currents. This Lagrangian does not involve flavour-changing currents. In particular, processes caused by neutral strangeness-changing currents are forbidden in this model. This conclusion is in good agreement with experiment: with very high accuracy, the processes

$$K_L^0 \to \mu^+\mu^-\,, \qquad K^+ \to \pi^+ e^- e^+$$

appear to be forbidden, indeed.

It should be noted that the processes mentioned are not forbidden in the three-quark model, which had been an argument for introducing a fourth, charmed, quark a long time before the discovery of charmed particles.

Expressions (6.1.14, 6.3.6, 6.1.16, 17, and 6.3.8) for the charged and neutral weak currents of leptons and quarks can be rewritten in the following compact form ($k = 1, 2, 3$):

$$J_\mu^{(+)} = 2\sum_k \bar{L}_k \gamma_\mu \tfrac{1}{2}(\tau_1 + \mathrm{i}\tau_2) L_k\,, \quad J_\mu^{(+)} = J_\mu^{(+),l} + J_\mu^{(+),q}\,, \tag{6.3.9}$$

$$J_\mu^n = 2\sum_k \bar{L}_k \gamma_\mu \tfrac{1}{2}\tau_3 L_k - 2\sin^2\theta_{\mathrm{W}} J_\mu^e\,, \quad J_\mu^n = J_\mu^{n,l} + J_\mu^{n,\nu} + J_\mu^{n,q}. \tag{6.3.10}$$

Here $L_k$ denotes the left-handed $SU_2$-doublets

$$\begin{pmatrix}\nu_e\\ e^-\end{pmatrix}_{\mathrm{L}}, \begin{pmatrix}\nu_\mu\\ \mu^-\end{pmatrix}_{\mathrm{L}}, \begin{pmatrix}\nu_\tau\\ \tau^-\end{pmatrix}_{\mathrm{L}}, \begin{pmatrix}u\\ d'\end{pmatrix}_{\mathrm{L}}, \begin{pmatrix}c\\ s'\end{pmatrix}_{\mathrm{L}}, \begin{pmatrix}t\\ b'\end{pmatrix}_{\mathrm{L}},$$

and $\tau$ the Pauli matrices; the expressions for $J_\mu^e$ are given by (6.1.12) and (6.3.4) for the lepton sector and the quark sector, respectively.

With the use of (6.3.9) and (6.3.10), the Lagrangian of weak interaction is written in the form

$$\mathcal{L} = -\frac{g}{2\sqrt{2}} J_\mu^{(+)}(x) W_\mu^-(x) + \text{Herm. conj.} - \frac{g}{2\cos\theta_{\mathrm{W}}} J_\mu^n(x) Z_\mu(x)\,. \tag{6.3.11}$$

From this it follows that for the *effective Lagrangian* (at $|q^2| \ll m_{Z,W}^2$) we have

$$\mathscr{L}_{\text{eff}} = \frac{G}{\sqrt{2}} \left( J_\mu^{(+)} J_\mu^{(-)} + \frac{m_W^2}{m_Z^2 \cos^2\theta_W} J_\mu^n J_\mu^n \right). \tag{6.3.12}$$

According to (6.1.27),

$$\frac{m_W^2}{m_Z^2 \cos^2\theta_W} = 1 ; \tag{6.3.13}$$

therefore (6.3.12) can be rewritten for the standard model as

$$\mathscr{L}_{\text{eff}} = \frac{G}{\sqrt{2}} (J_\mu^{(+)} J_\mu^{(-)} + J_\mu^n J_\mu^n) . \tag{6.3.14}$$

The Lagrangians and the currents in the standard model have a number of important properties. The major ones are summarized below.

1) The electromagnetic current is a vector. Consequently, the Lagrangian of electromagnetic interaction is invariant under the inversion of space.

2) Charged weak currents have the following space-time structure:

$$\bar{q}\gamma_\mu(1+\gamma_5)q = \bar{q}(\gamma_\mu - \gamma_5\gamma_\mu)q . \tag{6.3.15}$$

This quantity is the difference of a vector $V$ and an axial vector $A$, i.e., the interaction of charged weak currents is a $V-A$-interaction. The form of the charged weak current (6.3.9) also implies that it includes only left-handed components of spinors.

3) It follows from the $V-A$-structure of charged weak currents that the weak interaction is non-invariant under the inversion of space. Indeed, upon the inversion of space, the vector $V$ changes its sign, in contrast to the *axial vector A*. Consequently, in the effective Lagrangian

$$(V-A)(V-A)^\dagger = VV^\dagger + AA^\dagger - VA^\dagger - AV^\dagger \tag{6.3.16}$$

the two first terms remain unchanged under the inversion of space, while the last two terms change sign. The invariance-breaking of the Lagrangian for the weak interactions under the inversion of space has been experimentally confirmed.

4) The weak interaction is also non-invariant with respect to charge conjugation. This follows from the form of the Lagrangian (6.3.16) where the $VV^\dagger$ and $AA^\dagger$ terms do not change sign under charge conjugation, while the $VA^\dagger$ and $AV^\dagger$ terms do.

5) Compared with charged weak currents, the neutral ones exhibit two peculiar features: (i) they are diagonal, i.e., convert the particles into themselves and (ii) they include both left-handed and right-handed components of spinors. The absence of non-diagonal terms and the presense of both left-

handed and right-handed spinor components is confirmed by experiment (Sect. 8.1).

6) The expressions for the charged weak quark currents must contain a linear combination of the quarks rather than the quarks themselves (*quark mixing*).

It should also be mentioned that the coefficients in the expression for the neutral current (6.1.17, 18), (6.3.8), which contain the parameter $\xi$, have been determined by analyzing experimental data without using the standard model. This resulted in a single set of the values of the coefficients which coincide with those obtained for the standard model with $\xi \sim 1/4$.

## 6.4 Non-Standard Models

Besides the standard model, a great number of other models have been analyzed. Let us call them *non-standard.*

The interest in non-standard models has two main reasons. The first one is the state of the art of experiment. At present, only a limited number of predictions of the standard model have been experimentally verified (and, furthermore, the accuracy of experiments is not yet high enough); a lot of important consequences of the standard model have not been checked experimentally as yet. The second reason lies in the problem of uniqueness of the standard model: even if all major predictions of the standard model will be confirmed by experiment, there will remain an open question of possible existence of other models which account for the experimental data equally well.

### 6.4.1 Ambiguity in the Choice of the Model

As already seen, to construct a unified model one has to choose (a) the symmetry group, (b) the representations of the symmetry group which are associated with the fermions, (c) a set of initial particles, (d) the way the particles are placed in the representations chosen, (e) the representations of the symmetry group associated with the Higgs fields, (f) the interaction Lagrangian for fermions and Higgs fields (the Yukawa term), and (g) the Lagrangian for the Higgs fields.

Since a unique choice cannot be made, a great variety of non-standard models can be constructed. The simplest are those based on the same group $SU_2 \times U_1$ as the standard model but using the doublet rather than the scalar representations for right-handed fermions (Table 6.6). For example, the following types of models have been analyzed:

i) *Vector model* in which all right-handed fermions are combined into doublets; one of the generations of this model is

$$\begin{pmatrix} N_e \\ e^- \end{pmatrix}_R, \begin{pmatrix} u \\ b \end{pmatrix}_R, \begin{pmatrix} t \\ d \end{pmatrix}_R;$$

**Table 6.6.** Some gauge models

| | Standard model | Vector model | Asymmetric model |
|---|---|---|---|
| Gauge group | $SU_2 \times U_1$ | $SU_2 \times U_1$ | $SU_2 \times U_1$ |
| Intermediate fields | $W^{\pm}, Z, \gamma$ | $W^{\pm}, Z, \gamma$ | $W^{\pm}, Z, \gamma$ |
| Left-handed fermions | $\begin{pmatrix} \nu_e \\ e^- \end{pmatrix}_L \begin{pmatrix} u \\ d \end{pmatrix}_L$ | $\begin{pmatrix} \nu_e \\ e^- \end{pmatrix}_L \begin{pmatrix} u \\ d \end{pmatrix}_L$ | $\begin{pmatrix} \nu_e \\ e^- \end{pmatrix}_L \begin{pmatrix} u \\ d \end{pmatrix}_L$ |
| Right-handed fermions | $(e^-)_R, (u)_R, (d)_R$ | $\begin{pmatrix} N_e \\ e^- \end{pmatrix}_R \begin{pmatrix} u \\ b \end{pmatrix}_R \begin{pmatrix} t \\ d \end{pmatrix}_R$ | $\begin{pmatrix} N_e \\ e^- \end{pmatrix}_R (u)_R (d)_R$ |
| Minimum number of the Higgs scalar fields | one doublet | one doublet for $L$ one triplet for $R$ | one doublet for $L$ one triplet for $R$ |
| Relations between the masses of mesons ($\xi = \sin^2 \theta_W$) | $m_Z^2(1-\xi) = m_W^2$ | $m_Z^2(1-\xi) \leqslant m_W^2$ | $m_Z^2(1-\xi) \leqslant m_W^2$ |

ii) *Asymmetric model* in which the right-handed fermions $N_e$ and $e^-$ form a doublet, while the fermions $u$ and $d$ occur as singlets:

$$\begin{pmatrix} N_e \\ e^- \end{pmatrix}_R , u_R, d_R .$$

Here $N_e$ is the heavy neutral lepton not yet observed experimentally.

Besides, a number of models were studied based on simplest groups other than $SU_2 \times U_1$, such as

$$SU_3 \times U_1, SU_4 \times U_1, SO_4 \times U_1, O_4 \times U_1, Sp_4 \times U_1, SU_2 \times SU_2, U_{2L} \times U_{2R},$$

$$SU_2 \times U_1 \times U_1, SU_{2L} \times SU_{2R} \times U_1, SU_{4L} \times SU_{4R} \times U_1, SU_3, SU_4, \quad \text{etc.}$$

A non-standard model is constructed in the same way as the standard one. The specifics of the group for a given non-standard model will show up in each stage of the model-building by a few characteristic properties.

### 6.4.2 The $SU_3 \times U_1$-Model

Let us illustrate this, as an example, for a version of a non-standard model, namely that based on the group $SU_3 \times U_1$.

1) For this model, the group $SU_3 \times U_1$ is the gauge group. The model includes a total of 9 gauge fields.

2) Place the fermions in the lowest representations of the group $SU_3$, i.e., in singlets and triplets.

3) Choose the following multiplets of right-handed and left-handed fermions as the initial ones:

i) the lepton triplets

$$\begin{pmatrix} e^- \\ \nu_e \\ E_1^0 \end{pmatrix}_L, \begin{pmatrix} \mu^- \\ \nu_\mu \\ M_1^0 \end{pmatrix}_L, \begin{pmatrix} \tau^- \\ \nu_\tau \\ T_1^0 \end{pmatrix}_L, \begin{pmatrix} e^- \\ E_1^0 \\ E_2^0 \end{pmatrix}_R, \begin{pmatrix} \mu^- \\ M_1^0 \\ M_2^0 \end{pmatrix}_R, \begin{pmatrix} T^- \\ T_1^0 \\ T_2^0 \end{pmatrix}_R; \tag{6.4.1}$$

ii) the lepton singlets

$$\tau_R^-, T_L^-, E_{2L}^0, M_{2L}^0, T_{2L}^0; \tag{6.4.2}$$

iii) the quark triplets

$$\begin{pmatrix} u \\ d \\ b \end{pmatrix}_L, \begin{pmatrix} c \\ s \\ h \end{pmatrix}_L, \begin{pmatrix} u \\ b \\ d \end{pmatrix}_R, \begin{pmatrix} g \\ h \\ s \end{pmatrix}_R; \tag{6.4.3}$$

iv) the quark singlets

$$c_R, g_L. \tag{6.4.4}$$

Here $E_i^0$, $M_i^0$, and $T_i^0$ are new heavy neutral leptons which have not been observed experimentally up to now.

4) Introduce two triplets $\phi_i$ and $\phi_i'$ and two octets $\phi_{ij}$ and $\phi_{ij}'$ of Higgs fields.

5) Consequently, the model includes the multiplets of leptons and quarks, $\Psi$, and the Higgs fields, $\Phi$, listed above.

6) The globally invariant Lagrangian reads

$$L = \mathrm{i}\bar{\Psi}\gamma_\mu \partial_\mu \Psi - \tfrac{1}{2}\partial_\mu \Phi \partial_\mu \Phi + \bar{\Psi}\Gamma\Psi\Phi - P(\Phi), \tag{6.4.5}$$

where $P(\Phi)$ is the interaction Lagrangian for the Higgs fields.

7) Localize the Lagrangian (6.4.5) bearing in mind the expression (2.1.12) for generalized derivatives of the fields.

8) Make use of the mechanism of spontaneous symmetry-breaking. Choose the interaction Lagrangian for the Higgs fields $P(\Phi)$ in a form leading to the following vacuum expectation values:

$$(\phi_{ij})_v = \frac{v_1}{\sqrt{2}} \begin{pmatrix} 0 & 0 & 0 \\ 0 & 0 & 1 \\ 0 & 1 & 0 \end{pmatrix}, \quad (\phi_{ij}')_v = \frac{v_2}{\sqrt{2}} \begin{pmatrix} 0 & 0 & 0 \\ 0 & 0 & -\mathrm{i} \\ 0 & \mathrm{i} & 0 \end{pmatrix},$$

$$(\phi_i)_v = w_1 \begin{pmatrix} 0 \\ 0 \\ 1 \end{pmatrix}, \qquad (\phi_i')_v = w_2 \begin{pmatrix} 1 \\ 0 \\ 0 \end{pmatrix},$$

where $v_1$, $v_2$, $w_1$ and $w_2$ are parameters determining the mass of the particles.

Diagonalize the free Lagrangian; it turns out that two free parameters have to be introduced for that (one of them analogous to the Weinberg angle).

As a result, the Lagrangian includes the fields which can be identified with the fields of three charged and two neutral intermediate bosons, with the electromagnetic field, and with the fields of right- and left-handed leptons and quarks. Each of these particles possesses a definite mass.

In addition, the Lagrangian contains terms describing the interaction among the fields mentioned, particularly the electromagnetic interaction of leptons and quarks as well as the interaction of charged and neutral weak currents of leptons and quarks.

One recognizes that the model considered differs from the standard one. The Lagrangians for other non-standard models can be constructed along the same lines.

At this point we abandon the Lagrangians for the electroweak interactions of particles and pass over to analyzing the electromagnetic and the weak interactions in the framework of the standard model.

# 7. Quantum Electrodynamics

## 7.1 Covariant Perturbation Theory for Quantum Electrodynamics

Quantum electrodynamics describing the electromagnetic interaction between leptons and photons is characterized by the Lagrangian (5.1.7).

Using the result of the integration by parts

$$-\int dx(\partial_\mu A_\nu \partial_\mu A_\nu - \partial_\mu A_\nu \partial_\nu A_\mu) = \int dx\, A_\mu(\Box g_{\mu\nu} - \partial_\mu \partial_\nu) A_\nu\,, \tag{7.1.1}$$

one can write the generating functional $W$ in the $\alpha$-gauge, according to (5.1.26), as

$$\begin{aligned}W(J_\mu, \eta, \bar{\eta}) = R \int \prod_x \mathscr{D}A_\mu(x)\, \mathscr{D}\bar{\psi}(x)\, \mathscr{D}\psi(x) \\ \times \exp\{\mathrm{i}\int dx[\tfrac{1}{2}A_\mu(\Box g_{\mu\nu} - (1-\tfrac{1}{\alpha})\partial_\mu\partial_\nu)A_\nu + \bar{\psi}(i\gamma_\mu\partial_\mu - M)\psi \\ + J_\mu A_\mu + \bar{\eta}\psi + \bar{\psi}\eta]\}\,,\end{aligned}$$

$$R = \exp\left[-\mathrm{i}e\int dy\left(\frac{1}{\mathrm{i}}\frac{\delta}{\delta J_\mu(y)}\right)\left(\frac{1}{\mathrm{i}}\frac{\delta}{\delta \eta(y)}\right)\gamma_\mu\left(\frac{1}{\mathrm{i}}\frac{\delta}{\delta \bar{\eta}(y)}\right)\right]$$

or, after some transformations, in the form of (5.1.30):

$$\begin{aligned}W(J_\mu, \eta, \bar{\eta}) = R \exp\left[-\frac{\mathrm{i}}{2}\int dx\, dy\, J_\mu(x) K^{-1}_{\mu\nu}(x-y) J_\nu(y)\right. \\ \left. - \mathrm{i}\int dx\, dy\, \bar{\eta}(x)\, \mathscr{K}^{-1}(x-y)\, \eta(y)\right].\end{aligned} \tag{7.1.2}$$

According to (5.1.31), it follows for the generating functional $S(A^0_\mu, \bar{\psi}_0, \psi_0) \equiv S_e$:

$$\begin{aligned}S_\mathrm{e} = R \exp\left\{-\mathrm{i}\int dx[A^0_\mu(x) J_\mu(x) + \bar{\psi}_0(x)\eta(x) + \bar{\eta}(x)\psi_0(x)]\right. \\ - \frac{\mathrm{i}}{2}\int dx\, dy\, J_\mu(x) K^{-1}_{\mu\nu}(x-y) J_\nu(y) \\ \left. - \mathrm{i}\int dx\, dy\, \bar{\eta}(x)\, \mathscr{K}^{-1}(x-y)\, \eta(y)\right\}\Bigg|_{J=\eta=\bar{\eta}=0}.\end{aligned} \tag{7.1.3}$$

The operator $\mathscr{K}^{-1}(x-y)$ is determined by the relation

$$(\mathrm{i}\gamma_\mu \partial_\mu - M)\frac{1}{(2\pi)^4}\int dp\, \mathscr{K}^{-1}(p)\,\mathrm{e}^{-\mathrm{i}p(x-y)} = \delta(x-y)\,,$$

and is equal to

$$\mathscr{K}^{-1}(p) = \frac{1}{\not{p} - M} \tag{7.1.4}$$

in the momentum representation and to

$$\mathscr{K}^{-1}(x-y) = \frac{1}{(2\pi)^4}\int dp \frac{1}{\not{p}-M}\,\mathrm{e}^{-\mathrm{i}p(x-y)} \tag{7.1.5}$$

in the coordinate representation.

The inverse operator $K^{-1}_{\mu\nu}(x-y)$ is defined by the relation

$$\left(\Box g_{\mu\nu} - \left(1-\frac{1}{\alpha}\right)\partial_\mu\partial_\nu\right)\frac{1}{(2\pi)^4}\int dk\, K^{-1}_{\nu\varrho}(k)\,\mathrm{e}^{-\mathrm{i}k(x-y)} = \delta(x-y)\,g_{\mu\varrho}$$

and is equal to

$$K^{-1}_{\mu\nu}(k) = -\frac{1}{k^2}\left[g_{\mu\nu} - (1-\alpha)\frac{k_\mu k_\nu}{k^2}\right] \tag{7.1.6}$$

in the momentum representation and to

$$K^{-1}_{\mu\nu}(x-y) = \frac{-1}{(2\pi)^4}\int\frac{dk}{k^2}\left[g_{\mu\nu} - (1-\alpha)\frac{k_\mu k_\nu}{k^2}\right]\mathrm{e}^{-\mathrm{i}k(x-y)} \tag{7.1.7}$$

in the coordinate representation.

A series expansion of the differential operator in (7.1.3) with respect to the coupling constant $e$ yields the perturbation theory form of $S_\mathrm{e}$:

$$\begin{aligned} S_\mathrm{e} = \Bigg[ & 1 + e\int dx \frac{\delta}{\delta\eta(x)}\gamma_\mu\frac{\delta}{\delta\bar{\eta}(x)}\frac{\delta}{\delta J_\mu(x)} \\ & + \frac{e^2}{2}\int dx\,dy \frac{\delta}{\delta\eta(x)}\gamma_\mu\frac{\delta}{\delta\bar{\eta}(x)}\frac{\delta}{\delta J_\mu(x)}\frac{\delta}{\delta\eta(y)}\gamma_\nu\frac{\delta}{\delta\bar{\eta}(y)}\frac{\delta}{\delta J_\nu(y)} + \dots\Bigg] \\ & \times \exp\Bigg[-\mathrm{i}\int dx(A^0_\mu J_\mu + \bar{\psi}_0\eta + \bar{\eta}\psi_0) \\ & - \frac{\mathrm{i}}{2}\int dx\,dy\, J_\mu K^{-1}_{\mu\nu}J_\nu - \mathrm{i}\int dx\,dy\,\bar{\eta}\,\mathscr{K}^{-1}\eta\Bigg]\Bigg|_{J=\eta=\bar{\eta}=0}\,. \end{aligned} \tag{7.1.8}$$

Hence, for example, in the second order of perturbation theory one has for $S_\mathrm{e}$:

$$
\begin{aligned}
S_e^{(2)} = & -\tfrac{1}{2}e^2\int dx\,dy\,A_\mu^0(x)\,\bar\psi_0(x)\,\gamma_\mu\psi_0(x)\,A_\nu^0(y)\,\bar\psi_0(y)\,\gamma_\nu\psi_0(y)\\
& -\mathrm{i}e^2\int dx\,dy\,\bar\psi_0(y)\,\gamma_\nu A_\nu^0(y)\,\mathscr{K}^{-1}(y-x)\,A_\mu^0(x)\,\gamma_\mu\psi_0(x)\\
& -\frac{\mathrm{i}}{2}e^2\int dx\,dy\,\bar\psi_0(x)\,\gamma_\mu\psi_0(x)\,K_{\mu\nu}^{-1}(x-y)\,\bar\psi_0(y)\,\gamma_\nu\psi_0(y)\\
& +\tfrac{1}{2}e^2\int dx\,dy\,A_\mu^0(x)\,\gamma_\mu\mathscr{K}^{-1}(x-y)\,\gamma_\nu\mathscr{K}^{-1}(y-x)\,A_\nu^0(y)\\
& +e^2\int dx\,dy\,\bar\psi_0(y)\,\gamma_\nu K_{\nu\mu}^{-1}(y-x)\,\mathscr{K}^{-1}(y-x)\,\gamma_\mu\psi_0(x)\\
& +\frac{\mathrm{i}}{2}e^2\int dx\,dy\,\mathscr{K}^{-1}(x-y)\,\gamma_\nu K_{\nu\mu}^{-1}(y-x)\,\mathscr{K}^{-1}(y-x)\,\gamma_\mu\\
& +\tfrac{1}{2}e^2\int dx\,dy\,A_\mu^0(x)\,\gamma_\mu\mathscr{K}_x^{-1}(0)\,\mathscr{K}_y^{-1}(0)\,\gamma_\nu A_\nu^0(y)\\
& +\frac{\mathrm{i}}{2}e^2\int dx\,dy\,\mathscr{K}_x^{-1}(0)\,\gamma_\nu K_{\nu\mu}^{-1}(y-x)\,\gamma_\mu\mathscr{K}_y^{-1}(0)\\
& -\mathrm{i}e^2\int dx\,dy\,\bar\psi_0(y)\,\gamma_\nu\psi_0(y)A_\nu^0(y)\,\mathscr{K}_x^{-1}(0)\,\gamma_\mu A_\mu^0(x)\\
& +e^2\int dx\,dy\,\bar\psi_0(y)\,\gamma_\nu\psi_0(y)\,K_{\nu\mu}^{-1}(x-y)\,\gamma_\mu\mathscr{K}_x^{-1}(0)\,. \qquad (7.1.8')
\end{aligned}
$$

In the same way as (5.2.8′) was obtained, the expressions for the scattering amplitudes of the processes in the second order of perturbation theory can be found from (7.1.8′).

In the case of electrodynamics, the expression for a matrix element will contain various combinations of functions for free particles (fermions, photons), propagators (of fermions and photons) and electromagnetic interaction vertices (of two fermions and a photon).

The propagator of a free spinor field is defined as

$$
\mathscr{G}(x-y) = \frac{\delta^2}{\delta\eta(x)\,\delta\bar\eta(y)}\,Z_0(J_\mu,\eta,\bar\eta)\bigg|_{J=\eta=\bar\eta=0}\,.
$$

Substituting into here (5.1.12) and taking into account (7.1.2, 5) yields, in the momentum representation,

$$
\mathscr{G}(p) = -\frac{1}{\not p - M} \qquad (7.1.9)
$$

or, in the coordinate representation,

$$
\mathscr{G}(x-y) = -\frac{1}{(2\pi)^4}\int dp\,\frac{1}{\not p - M}\,\mathrm{e}^{-\mathrm{i}p(x-y)}\,. \qquad (7.1.10)
$$

The propagator of the free electromagnetic field is defined as

$$
\mathscr{D}_{\mu\nu}(x-y) = \frac{\delta}{\delta J_\mu(x)}\,\frac{\delta}{\delta J_\nu(y)}\,Z_0(J_\mu,\eta,\bar\eta)\bigg|_{J=\eta=\bar\eta=0}\,.
$$

Substituting (7.1.2) and taking into account (7.1.7) yields, in the momentum representation,

$$\mathscr{D}_{\mu\nu}(k)=\frac{1}{k^2}\left[g_{\mu\nu}-(1-\alpha)\frac{k_\mu k_\nu}{k^2}\right] \tag{7.1.11}$$

and, in the coordinate representation,

$$\mathscr{D}_{\mu\nu}(x-y)=\frac{1}{(2\pi)^4}\int\frac{dk}{k^2}\left[g_{\mu\nu}-(1-\alpha)\frac{k_\mu k_\nu}{k^2}\right]\mathrm{e}^{-\mathrm{i}k(x-y)}\,. \tag{7.1.12}$$

By setting $\alpha=0$ and $\alpha=1$ in (7.1.12) we arrive at the expressions for the photon propagator in the *Landau gauge* and in the *Feynman gauge*, respectively.

In much the same way we obtain from the formula similar to (4.2.63)

$$K_{\mu\nu}(x)=\left(\Box g_{\mu\nu}-\partial_\mu\partial_\nu+\frac{1}{\alpha}g_{\mu k}g_{\nu l}\partial_k\partial_l\right).$$

Taking the limit of $\alpha\to 0$ in the inverse of this operator we obtain for the propagator of the photon in the *Coulomb gauge*

$$\mathscr{D}_{\mu\nu}(k)=-K^{-1}_{\mu\nu}(k)=\begin{cases}\mathscr{D}_{00}(k)=\dfrac{-1}{\boldsymbol{k}^2}\,,\\ \mathscr{D}_{0k}(k)=\mathscr{D}_{k0}(k)=0\,,\\ \mathscr{D}_{ij}(k)=-\left(\delta_{ij}-\dfrac{k_ik_j}{\boldsymbol{k}^2}\right)\dfrac{1}{k^2}\,.\end{cases} \tag{7.1.13}$$

The expressions for the propagator of the electromagnetic field in different gauges are summarized in Table 7.1; concerning the axial gauge cf. (5.3.16).

The interaction vertex of two fermions and one photon is found with the help of (5.1.24):

$$\begin{aligned}\Gamma_{\text{tree}}(\mathscr{A}_\mu,\bar{\Psi},\Psi)&=\int dx\,\mathscr{L}(\mathscr{A}_\mu(x),\bar{\Psi}(x),\Psi(x))\\&=\int dx[-\tfrac{1}{4}(\partial_\mu\mathscr{A}_\nu-\partial_\nu\mathscr{A}_\mu)^2\\&\quad+\mathrm{i}\bar{\Psi}\gamma_\mu\partial_\mu\Psi-M\bar{\Psi}\Psi-e\bar{\Psi}\gamma_\mu\Psi\mathscr{A}_\mu]\,,\end{aligned} \tag{7.1.14}$$

where

$$\mathscr{A}_\mu(x)=\frac{\delta Z_{\text{tree}}(J_\mu,\bar{\eta},\eta)}{\delta J_\mu(x)}\,,$$

$$\bar{\Psi}(x)=\frac{\delta Z_{\text{tree}}(J_\mu,\bar{\eta},\eta)}{\delta\eta(x)}\,,$$

**Table 7.1.** Expressions for the photon propagator in various gauges

| Gauge | Expression for $\mathscr{D}_{\mu\nu}(k)$ |
|---|---|
| $\alpha$-gauge | $\frac{1}{k^2}\left[g_{\mu\nu}-(1-\alpha)\frac{k_\mu k_\nu}{k^2}\right]$ |
| Landau gauge | $\frac{1}{k^2}\left(g_{\mu\nu}-\frac{k_\mu k_\nu}{k^2}\right)$ |
| Feynman gauge | $\frac{1}{k^2}g_{\mu\nu}$ |
| Coulomb gauge | $\mathscr{D}_{00}(k)=-\frac{1}{\boldsymbol{k}^2}$, $\mathscr{D}_{0k}(k)=\mathscr{D}_{k0}(k)=0$, $\mathscr{D}_{ij}(k)=-\left(\delta_{ij}-\frac{k_i k_j}{\boldsymbol{k}^2}\right)\frac{1}{k^2}$ |
| Axial gauge | $\frac{1}{k^2}\left(g_{\mu\nu}+\frac{n^2 k_\mu k_\nu-(nk)k_\mu n_\nu-(nk)n_\mu k_\nu}{(nk)^2}\right)$ |

$$\Psi(x)=\frac{\delta Z_{\text{tree}}(J_\mu,\bar{\eta},\eta)}{\delta\bar{\eta}(x)};$$

$Z_{\text{tree}}(J_\mu,\bar{\eta},\eta)$ is determined by (5.1.21) and (5.1.23).

Using the definition

$$\Gamma_{0\mu}(x,y,z)=(-\mathrm{i})^2\frac{\delta}{\delta\mathscr{A}_\mu(x)}\frac{\delta}{\delta\Psi(z)}\frac{\delta}{\delta\bar{\Psi}(y)}\Gamma_{\text{tree}}(\mathscr{A}_\mu,\bar{\Psi},\Psi)\Big|_{\mathscr{A}_\mu=\bar{\Psi}=\Psi=0}$$

and substituting (7.1.14), we obtain for the electromagnetic interaction vertex omitting the momentum conservation function $\delta(p+p'+k)$:

$$\Gamma_{0\mu}(p,p',k)=e\gamma_\mu\,. \tag{7.1.15}$$

The correspondence rules for the amplitudes in the momentum representation for quantum electrodynamics are summarized in Table 7.2.

It should be emphasized that, as distinct from the operator method, the path integral technique makes it possible to construct the formalism of quantum electrodynamics without explicitly introducing an indefinite metric.

The Feynman diagrams in the first non-vanishing order of the perturbation theory for two basic electrodynamic processes are depicted in Figs. 7.1, 2. The corresponding expressions for the matrix elements of the two processes which are obtained with the aid of Table 7.2 have the form

**Table 7.2.** Correspondence rules for the amplitudes in quantum electrodynamics

| Physical state | Mathematical expression | Graphical representation |
|---|---|---|
| Electron in the initial state | $v_r^{(-)}(p)$ | |
| Positron in the initial state | $\bar{v}_r^{(-)}(p)$ | |
| Electron in the final state | $\bar{v}_{r'}^{(+)}(p')$ | |
| Positron in the final state | $v_{r'}^{(+)}(p')$ | |
| Photon in the initial or the final state | $\varepsilon_\mu^\lambda$ | |
| Motion of virtual electron from 1 to 2 | $\dfrac{1}{m-\not{p}}$ | 1 2 |
| Motion of virtual positron from 1 to 2 | $\dfrac{1}{m+\not{p}}$ | 1 2 |
| Motion of virtual photon between vertices with summation indices $\mu$ and $\nu$ (Feynman gauge) | $g_{\mu\nu}\cdot\dfrac{1}{k^2}$ | $\mu$ $\nu$ |
| Electromagnetic interaction vertex | $e\gamma_\mu$ | $k$ $p_1$ $p_2$ |

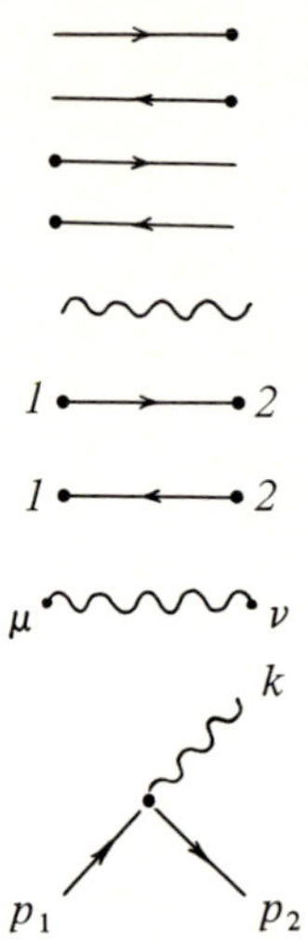

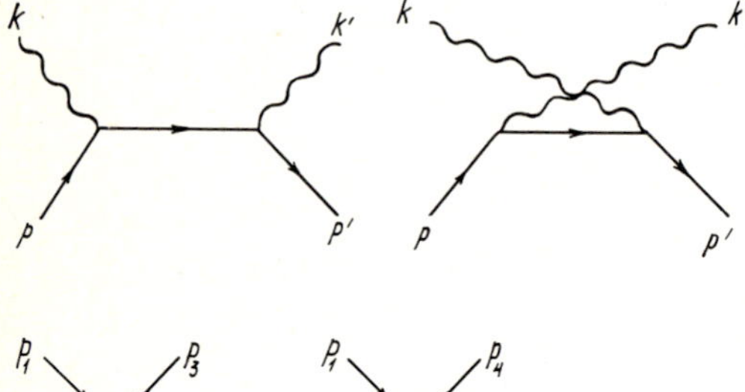


**Fig. 7.1.** Diagrams for the Compton effect on the electron

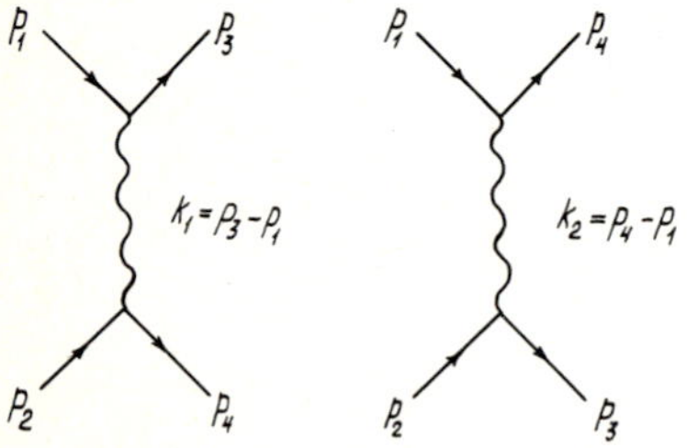

**Fig. 7.2.** Diagrams for the electron-electron scattering

i) $\gamma(k)+e^-(p)\rightarrow\gamma(k')+e^-(p')$ (Compton scattering on an electron)

$$M=(-\mathrm{i})^2e^2\left\{\bar{v}_{r'}^{(+)}(p')\not{\varepsilon}^{\lambda'}\frac{\not{p}+\not{k}+m}{(p+k)^2-m^2}\not{\varepsilon}^{\lambda}v_r^{(-)}(p)\right.$$

$$\left.+\bar{v}_{r'}^{(+)}(p')\not{\varepsilon}^{\lambda}\frac{\not{p}-\not{k}'+m}{(p-k')^2-m^2}\not{\varepsilon}^{\lambda'}v_r^{(-)}(p)\right\},\qquad \not{\varepsilon}=\varepsilon_\mu\gamma_\mu\,. \tag{7.1.16}$$

ii) $e^-(p_1)+e^-(p_2)\to e^-(p_3)+e^-(p_4)$ (electron scattering on an electron)

$$M=(-\mathrm{i})^2e^2\left\{\bar{v}_{r'}^{(+)}(p_3)\gamma_\mu v_r^{(-)}(p_1)\frac{1}{(p_3-p_1)^2}\bar{v}_{r'}^{(+)}(p_4)\gamma_\mu v_r^{(-)}(p_2)\right.$$

$$\left.-\bar{v}_{r'}^{(+)}(p_4)\gamma_\mu v_r^{(-)}(p_1)\frac{1}{(p_4-p_1)^2}\bar{v}_{r'}^{(+)}(p_3)\gamma_\mu v_r^{(-)}(p_2)\right\}. \quad (7.1.17)$$

## 7.2 Differential Cross-Sections

### 7.2.1 Formula for the Differential Cross-Section

With the help of the matrix element, the probability and thus the differential cross-section of a process can be determined. Let us first consider the process

$$1(p_1)+2(p_2)\to 3(p_3)+4(p_4), \quad (7.2.1)$$

in which two free stable particles collide and transform into two other particles. The matrix element $(S-I)_{\mathrm{fi}}\equiv \mathrm{i}\,T_{\mathrm{fi}}$ of such a process can be expressed as a product, one of the factors being the function $\delta(p_3+p_4-p_1-p_2)$ which corresponds to the energy-momentum conservation law, i.e.,

$$\mathrm{i}\,T_{\mathrm{fi}}=N_{\mathrm{e}}(2\pi)^4\delta(p_3+p_4-p_1-p_2)M_{\mathrm{fi}}, \quad (7.2.2)$$

where $N_{\mathrm{e}}$ is a normalization factor.

The probability of transition from the initial state into the final state is given by the square of the absolute value of $T_{\mathrm{fi}}$:

$$W' = |T_{\mathrm{fi}}|^2=N_{\mathrm{e}}^2[(2\pi)^4\delta(p_3+p_4-p_2-p_1)]^2|M_{\mathrm{fi}}|^2. \quad (7.2.3)$$

To obtain the probability of a transition upon which the momenta of the final particles 3 and 4 fall within the intervals $(\boldsymbol{p}_3, \boldsymbol{p}_3+d\boldsymbol{p}_3)$ and $(\boldsymbol{p}_4, \boldsymbol{p}_4+d\boldsymbol{p}_4)$, (7.2.3) is to be multiplied by the element of the phase space volume

$$\frac{d\boldsymbol{p}_3 V}{(2\pi)^3}\,\frac{d\boldsymbol{p}_4 V}{(2\pi)^3}.$$

The transition probability per unit time is then written (with $V=1$) as

$$W=(2\pi)^4N_{\mathrm{e}}^2|M_{\mathrm{fi}}|^2\delta(p_3+p_4-p_2-p_1)\frac{d\boldsymbol{p}_3}{(2\pi)^3}\,\frac{d\boldsymbol{p}_4}{(2\pi)^3}. \quad (7.2.4)$$

By definition, the differential cross-section $d\sigma_p$ of the process (7.2.1) is equal to the probability (7.2.4) divided by the flux $j_0$ of the initial particles:

$$d\sigma_p = N_e^2 \frac{(2\pi)^4}{j_0} |M_{fi}|^2 \delta(p_3+p_4-p_1-p_2) \frac{dp_3}{(2\pi)^3} \frac{dp_4}{(2\pi)^3} . \tag{7.2.5}$$

With the help of this formula the magnitude of the differential cross-section can be found for both polarized and unpolarized particles. In the latter case, a summation over the projections of spins of final particles and averaging over the projections of spins of the initial particles have to be carried out:

$$d\sigma = \sum_{\text{spin}} d\sigma_p . \tag{7.2.6}$$

(We denote both operations by a combined symbol $\sum_{\text{spin}}$.)

If more than two particles are present in the final state (7.2.5), can be written in the form

$$d\sigma_p = N_e^2 \frac{(2\pi)^4}{j_0} |M_{fi}|^2 \delta\left( \sum_f p_f - p_1 - p_2 \right) \prod_f \frac{dp_f}{(2\pi)^3} ,$$

where $M_{fi}$ is the matrix element of the process, $\prod dp_f/(2\pi)^3$ is the product of the phase space volumes of the particles in the final state, and $\sum_f p_f$ is the sum of the 4-momenta of the particles in the final state.

Let us find the expression for

$$|M_{fi}|^2 = M_{fi}^* M_{fi} .$$

Consider a process with one fermion (electron or positron) in the initial and in the final state. The expression for the matrix element can be written in the following general form

$$M_{fi} = \bar{u}(p_2) \mathrm{O} u(p_1) , \tag{7.2.7}$$

where O is an operator containing the matrices $\gamma$. Since

$$[\bar{u}(p_2) \mathrm{O} u(p_1)]^* = \bar{u}(p_1) \bar{\mathrm{O}} u(p_2)$$

(where $\bar{\mathrm{O}} = \gamma_0 \mathrm{O}^\dagger \gamma_0$), it follows

$$|M_{fi}|^2 = \mathrm{Tr}\{\Lambda(p_2) \mathrm{O} \Lambda(p_1) \bar{\mathrm{O}}\} .$$

Here

$$\Lambda(p_1) = u(p_1)\bar{u}(p_1) , \quad \Lambda(p_2) = u(p_2)\bar{u}(p_2) .$$

Hence,

$$d\sigma_p = \frac{(2\pi)^4}{j_0} N_e^2 \mathrm{Tr}\{\Lambda(p_2) \mathrm{O} \Lambda(p_1) \bar{\mathrm{O}}\} \delta(p_3+p_4-p_1-p_2) \frac{dp_3}{(2\pi)^3} \frac{dp_4}{(2\pi)^3} . \tag{7.2.8}$$

If the particles are not polarized the following relations hold:

i) for a spin-1/2 particle with mass $m$

$$\Lambda(p) = \not{p} + m ; \tag{7.2.9}$$

ii) for a spin-1/2 anti-particle

$$\Lambda(p) = \not p - m\,; \tag{7.2.10}$$

iii) for a spin-1 particle

$$\Lambda(p_\mu, p_\nu) = \sum_s u^*_\mu(s)\, u_\nu(s) = -\left(g_{\mu\nu} - \frac{p_\mu p_\nu}{m^2}\right); \tag{7.2.11}$$

iv) for a photon

$$\Lambda(k_\mu, k_\nu) = \sum_s A^*_\mu(s) A_\nu(s) = -g_{\mu\nu}\,. \tag{7.2.12}$$

For polarized particles one has the following relations:

i) for a spin-1/2 particle

$$\Lambda(p) = \tfrac{1}{2}(\not p + m)(1 - \gamma_5 \not s)\,, \tag{7.2.13}$$

where

$$s_\mu = (s_0, \boldsymbol{s})\,, \qquad s_0 = \frac{1}{m}(\boldsymbol{p}\,\boldsymbol{\zeta})\,, \qquad \boldsymbol{s} = \boldsymbol{\zeta} + (\boldsymbol{p}\,\boldsymbol{\zeta})\boldsymbol{p}\,\frac{1}{m(m+E)}\,;$$

$\boldsymbol{\zeta}$ is a unit vector along the polarization direction of the particle in the coordinate system where it is at rest;

ii) for a spin-1/2 anti-particle

$$\Lambda(p) = \tfrac{1}{2}(\not p - m)(1 - \gamma_5 \not s)\,.$$

In carrying out the summation over the photon polarizations, it is convenient to use the relations

$$\gamma_\nu \gamma_\nu = 4\,, \tag{7.2.14}$$

$$\gamma_\nu \not a \gamma_\nu = -2\not a\,, \tag{7.2.15}$$

$$\gamma_\nu \not a \not b \gamma_\nu = 4ab\,, \tag{7.2.16}$$

$$\gamma_\nu \not a \not b \not c \gamma_\nu = -2\not c \not b \not a\,. \tag{7.2.17}$$

Calculating the traces of the matrices $\gamma$ is based on the following formulae:

i) the relations

$$\gamma_\mu \gamma_\nu + \gamma_\nu \gamma_\mu = 2g_{\mu\nu}\,, \qquad \gamma_\mu \gamma_5 = -\gamma_5 \gamma_\mu\,, \qquad \gamma_5^2 = +1 \tag{7.2.18}$$

obeyed by the $\gamma$-matrices;

ii) the invariance of the trace of the product of any two matrices (in particular, of the $\gamma$-matrices) with respect to their cyclic permutation, i.e.,

$$\mathrm{Tr}\{AB\} = \sum_{i,j} A_{ij} B_{ji} = \sum_{i,j} B_{ji} A_{ij} = \mathrm{Tr}\{BA\}\,. \tag{7.2.19}$$

From (7.2.18) it follows:

1) The trace of the product of an odd number of $\gamma$-matrices is zero. In fact,

$$\mathrm{Tr}\{\gamma_\mu \gamma_\nu \ldots \gamma_\varrho\} = -\mathrm{Tr}\{\gamma_5 \gamma_\mu \gamma_\nu \ldots \gamma_\varrho \gamma_5\} = -\mathrm{Tr}\{\gamma_\mu \gamma_\nu \ldots \gamma_\varrho\} = 0 .$$

2) The trace of the product of $n$ $\gamma$-matrices does not change on reversing their sequence:

$$\mathrm{Tr}\{\gamma_\mu \gamma_\nu \gamma_\alpha \gamma_\beta \ldots\} = \mathrm{Tr}\{\ldots \gamma_\beta \gamma_\alpha \gamma_\nu \gamma_\mu\} .$$

3) $\mathrm{Tr}\{I\} = 4$, where $I$ is the unit matrix.

4) $\mathrm{Tr}\{\gamma_5\} = 0$ since
$\mathrm{Tr}\{\gamma_5\} = \mathrm{Tr}\{\gamma_0 \gamma_0 \gamma_5\} = -\mathrm{Tr}\{\gamma_0 \gamma_5 \gamma_0\} = -\mathrm{Tr}\{\gamma_0 \gamma_0 \gamma_5\} = -\mathrm{Tr}\{\gamma_5\} = 0$
(note that $\gamma_0 \gamma_0 = I$).

5) $\mathrm{Tr}\{\gamma_\mu \gamma_\nu\} = 4 g_{\mu\nu}$.

6) $\mathrm{Tr}\{\not{a}\not{b}\} = 4(ab)$, where $a$ and $b$ are arbitrary 4-vectors.

7) $$\mathrm{Tr}\{\gamma_\mu \gamma_\nu \gamma_\alpha \gamma_\beta\} = 4(g_{\mu\nu} g_{\alpha\beta} - g_{\mu\alpha} g_{\nu\beta} + g_{\mu\beta} g_{\nu\alpha}) . \tag{7.2.20}$$

8) $$\mathrm{Tr}\{\not{a}\not{b}\not{c}\not{d}\} = 4[(ab)(cd) + (ad)(bc) - (ac)(bd)] , \tag{7.2.21}$$

where $a$, $b$, $c$, and $d$ are arbitrary 4-vectors.

9) $$\begin{aligned}\mathrm{Tr}\{\not{a}\not{b}\not{c}\not{d}\not{e}\not{f}\} = 4[&(ab)(cd)(ef) + (af)(bc)(de) \\ &+ (ab)(cf)(de) + (ad)(bc)(ef) + (af)(be)(cd) + (ac)(be)(df) \\ &+ (ad)(bf)(ce) + (ae)(bd)(cf) - (af)(bd)(ce) - (ad)(be)(cf) \\ &- (ab)(ce)(df) - (ac)(bd)(ef) - (ac)(bf)(de) - (ae)(bc)(df) \\ &- (ae)(bf)(cd)] ,\end{aligned} \tag{7.2.22}$$

where $a$, $b$, $c$, $d$, $e$, and $f$ are arbitrary 4-vectors.

10) $$\tfrac{1}{4}\mathrm{i}\,\mathrm{Tr}\{\gamma_5 \gamma_\mu \gamma_\nu \gamma_\alpha \gamma_\beta\} = \varepsilon_{\mu\nu\alpha\beta} . \tag{7.2.23}$$

11) $$\begin{aligned}\tfrac{1}{4}\mathrm{Tr}\{&(\not{A}_1 + a_1)(\not{A}_2 + a_2)(\not{A}_3 + a_3)(\not{A}_4 + a_4)\} \\ &= (A_1 A_2 + a_1 a_2)(A_3 A_4 + a_3 a_4) + (A_1 A_4 + a_1 a_4)(A_2 A_3 + a_2 a_3) \\ &\quad - (A_1 A_3 - a_1 a_3)(A_2 A_4 - a_2 a_4) ,\end{aligned} \tag{7.2.24}$$

where $A_i$ and $a_i$ are quantities not containing the $\gamma$-matrices.

### 7.2.2 Cross-Section of the Compton Scattering

Let us calculate as an example the differential cross-section for the photon scattering on free electrons in the second order of the perturbation theory. The Feynman diagrams for this process are depicted in Fig. 7.1; the corresponding matrix element is given by (7.1.16). The squares of the sums of momenta entering the denominators in (7.1.16) are

$$\begin{aligned}(p+k)^2 - m^2 &= 2(pk) \equiv m^2 \varkappa_1 , \\ (p-k')^2 - m^2 &= -2(pk') \equiv m^2 \varkappa_2 .\end{aligned} \tag{7.2.25}$$

Let us replace $\not{e}^{\lambda}$ and $\not{e}^{\lambda'}$ in (7.1.16) by $\gamma_\nu$ and $\gamma_\mu$, respectively, which implies the effect of summation over polarization states of the photon. By substituting the resulting expression into (7.2.8) and by taking into account that for any 4-vector $\bar{\not{a}} = \gamma_0(a_0\gamma_0 - \boldsymbol{a}\gamma)^\dagger\gamma_0 = \not{a}$ we find, with the use of (7.2.9) and (7.2.25), the following expression for the differential cross-section for the case when all particles are unpolarized:

$$d\sigma = \frac{N'}{4}\mathrm{Tr}\{\mathrm{O}(\not{p}+m)\bar{\mathrm{O}}(\not{p}'+m)\}\delta(p+k-p'-k')\,d\boldsymbol{p}'\,d\boldsymbol{k}' \,. \tag{7.2.26}$$

Here

$$\mathrm{O} = \frac{1}{m^2\varkappa_1}\gamma_\mu(\not{f}_1+m)\gamma_\nu + \frac{1}{m^2\varkappa_2}\gamma_\nu(\not{f}_2+m)\gamma_\mu\,, \quad N' = \frac{e^4}{16\,\omega\omega'\varepsilon\varepsilon'}\,\frac{1}{j_0(2\pi)^4}\,;$$

$j_0 = \varrho(|v| - |v_s|\cos\theta)$ is the flux, $\varrho$ and $|v|$ are the density and the velocity of the scattered particles, $|v_s|$ is the velocity of the initial electron, $\theta$ is the angle between $\boldsymbol{k}$ and $\boldsymbol{v}$; $\omega$ and $\varepsilon$ are the energies of the photon and the electron, respectively, and $f_1 = p+k$, $f_2 = p-k'$.

It is convenient to rewrite (7.2.26) in the form

$$d\sigma = N'\tfrac{1}{4}\mathrm{Tr}\{F\}\delta(p+k-p'-k')\,d\boldsymbol{p}'\,d\boldsymbol{k}'\,, \tag{7.2.27}$$

where

$$\begin{aligned}\mathrm{Tr}\{F\} = {}& \frac{1}{m^4}\mathrm{Tr}\left\{\gamma_\mu\frac{\not{f}_1+m}{\varkappa_1}\gamma_\nu + \gamma_\nu\frac{\not{f}_2+m}{\varkappa_2}\gamma_\mu\right\}(\not{p}+m)\gamma_\nu\frac{\not{f}_1+m}{\varkappa_1}\gamma_\mu(\not{p}'+m) \\ &+ \frac{1}{m^4}\mathrm{Tr}\left\{\gamma_\mu\frac{\not{f}_2+m}{\varkappa_1}\gamma_\nu + \gamma_\nu\frac{\not{f}_2+m}{\varkappa_2}\gamma_\mu\right\}(\not{p}+m)\gamma_\mu\frac{\not{f}_2+m}{\varkappa_2}\gamma_\nu(\not{p}'+m)\,,\end{aligned} \tag{7.2.28}$$

Let us carry out an integration over $\boldsymbol{p}'$ and $\boldsymbol{k}'$ in (7.2.27) using the presence of the $\delta$-function. The integration over $\boldsymbol{p}'$ yields

$$d\sigma = N'\int d\boldsymbol{k}'\,\tfrac{1}{4}\mathrm{Tr}\{F\}\delta(\varepsilon+\omega-\varepsilon'-\omega')\,. \tag{7.2.29}$$

To perform the integration over $\boldsymbol{k}'$, we express the volume $d\boldsymbol{k}'$ in the spherical coordinates:

$$d\boldsymbol{k}' = |\boldsymbol{k}'|^2 d|\boldsymbol{k}'|\,d\Omega = \omega'^2\frac{\partial\omega'}{\partial E_\mathrm{f}}\,d\Omega\,dE_\mathrm{f}\,,$$

where $d\Omega$ is an element of the solid angle enclosing the vector $\boldsymbol{k}'$; $E_\mathrm{f}$ is the total energy of the final state. Then

$$d\sigma = N'\frac{1}{4}\mathrm{Tr}\{F\}\omega'^2 d\Omega\frac{\partial\omega'}{\partial E_\mathrm{f}}\,, \tag{7.2.30}$$

where

$$\varepsilon' = \sqrt{m^2+(\boldsymbol{p}+\boldsymbol{k}-\boldsymbol{k}')^2}$$
$$= \sqrt{\varepsilon^2+\omega^2+2p\omega\cos\theta_1+\omega'^2-2p\omega'\cos\theta_2-2\omega\omega'\cos\theta}$$

and

$$\frac{\partial E_{\mathrm{f}}}{\partial\omega'} = \frac{\partial(\varepsilon'+\omega')}{\partial\omega'} = \frac{p'k'}{\varepsilon'\omega'} = \frac{m^2\varkappa_1}{2\varepsilon'\omega'}.$$

With the use of the latter relation, (7.2.30) is rewritten as

$$d\sigma = N'\frac{1}{4}\mathrm{Tr}\{F\}\frac{2\varepsilon'\omega'}{m^2\varkappa_1}d\Omega. \tag{7.2.31}$$

Let us now calculate the trace (7.2.28). It is recognized that the second term in (7.2.28) can be obtained from the first one by means of the interchange $k\to -k'$, $k'\to -k$, which is equivalent to the interchange $f_1\to f_2$, $f_2\to f_1$, $\varkappa_1\to\varkappa_2$, $\varkappa_2\to\varkappa_1$. Consequently, $\mathrm{Tr}\{F\}$ can be expressed as

$$\mathrm{Tr}\{F\} = P(\varkappa_1,\varkappa_2)+P(\varkappa_2,\varkappa_1), \quad P(\varkappa_1,\varkappa_2) = h_1(\varkappa_1,\varkappa_2)+h_2(\varkappa_1,\varkappa_2),$$

where

$$h_1(\varkappa_1,\varkappa_2) = \frac{1}{m^4\varkappa_1^2}\mathrm{Tr}\{\gamma_\mu(\not f_1+m)\gamma_\nu(\not p+m)\gamma_\nu(\not f_1+m)\gamma_\mu(\not p'+m)\},$$

$$h_2(\varkappa_1,\varkappa_2) = \frac{1}{m^4\varkappa_1\varkappa_2}\mathrm{Tr}\{\gamma_\nu(\not f_2+m)\gamma_\mu(\not p+m)\gamma_\nu(\not f_1+m)\gamma_\mu(\not p'+m)\}.$$

Performing the summation over $\mu$ and $\nu$ in the expression for $h_1(\varkappa_1,\varkappa_2)$, for which purpose (7.2.14 – 17) is used, and utilizing (7.2.24) we find

$$h_1(\varkappa_1,\varkappa_2) = \frac{4}{m^4\varkappa_1^2}\mathrm{Tr}\{\not f_1\not p\not f_1\not p'+4m^2(\not f_1\not p+\not f_1\not p'-f_1^2)-m^2\not p\not p'+4m^4\}$$
$$= 8\frac{1}{\varkappa_1^2}(4+2\varkappa_1-\varkappa_1\varkappa_2).$$

Analogously, the expression

$$h_2(\varkappa_1,\varkappa_2) = 8\frac{1}{\varkappa_1\varkappa_2}(4+\varkappa_1+\varkappa_2)$$

is obtained. The trace (7.2.28) is thus given by

$$\frac{1}{8}\mathrm{Tr}\{F\} = 4\left(\frac{1}{\varkappa_1}+\frac{1}{\varkappa_2}\right)^2+4\left(\frac{1}{\varkappa_1}+\frac{1}{\varkappa_2}\right)-\left(\frac{\varkappa_1}{\varkappa_2}+\frac{\varkappa_2}{\varkappa_1}\right) \equiv F_0. \tag{7.2.32}$$

Substituting (7.2.32) into (7.2.31) yields the desired expression, corresponding to the second order of perturbation theory, for the differential cross-section of the photon scattering on free electrons:

$$d\sigma = \frac{1}{2} r_0^2 \left(\frac{\omega'}{\omega}\right)^2 F_0 d\Omega , \tag{7.2.33}$$

where $r_0^2 = e^2/4\pi m$ is the classical radius of an electron.

In terms of the invariant variables

$$s = (p+k)^2 = (p'+k')^2 ,$$
$$u = (p-k')^2 = (p'-k)^2 ,$$
$$t = (p'-p)^2 = (k'-k)^2 .$$

(7.2.25) and (7.2.32) assume the form

$$m^2 \varkappa_1 = (p+k)^2 - m^2 = s - m^2$$
$$m^2 \varkappa_2 = (p-k')^2 - m^2 = u - m^2$$
$$F_0 = 4\left[\left(\frac{m^2}{s-m^2} + \frac{m^2}{u-m^2}\right)^2 + \frac{m^2}{s-m^2} + \frac{m^2}{u-m^2} - \frac{1}{4}\left(\frac{s-m^2}{u-m^2} + \frac{u-m^2}{s-m^2}\right)\right].$$

# 8. Weak Interactions

## 8.1 Processes Caused by Neutral Weak Currents

Weak interactions of *leptons* are described in the standard model by the Lagrangians (6.1.13) and (6.1.15) and those of *quarks* by the Lagrangians (6.3.5) and (6.3.7).

Let us first consider the processes caused by the neutral weak current. This can be expressed in the form

$$J_\mu^n = J_\mu^{n,\nu} + J_\mu^{n,l} + J_\mu^{n,q} . \tag{8.1.1}$$

The particular currents involved are determined by (6.1.16, 17) and (6.3.8). According to (6.3.14), we have for the effective interaction Lagrangian for neutral weak currents

$$\mathscr{L}_{\text{eff}} = \frac{G}{\sqrt{2}} J_\mu^n J_\mu^n . \tag{8.1.2}$$

A number of predictions can be made with this Lagrangian. We shall consider some of them as examples.

The Lagrangian (8.1.2) contains two kinds of terms: diagonal terms and non-diagonal ones.

### 8.1.1 Diagonal Terms

The diagonal part of the effective Lagrangian (8.1.2) reads

$$\mathscr{L}_{\text{d}} = \frac{G}{\sqrt{2}} (J_\mu^{n,\nu} J_\mu^{n,\nu} + J_\mu^{n,l} J_\mu^{n,l} + J_\mu^{n,q} J_\mu^{n,q}) . \tag{8.1.3}$$

In this Lagrangian, the first term describes the weak interaction between neutrinos, the second one between leptons, and the third one between quarks.

Let us consider the process

$$e^+(k') + e^-(k) \to \mu^+(p') + \mu^-(p) , \tag{8.1.4}$$

which is caused by the electromagnetic interaction as well as by the neutral weak interaction. Therefore, both the virtual photon exchange (Fig. 8.1 a) and

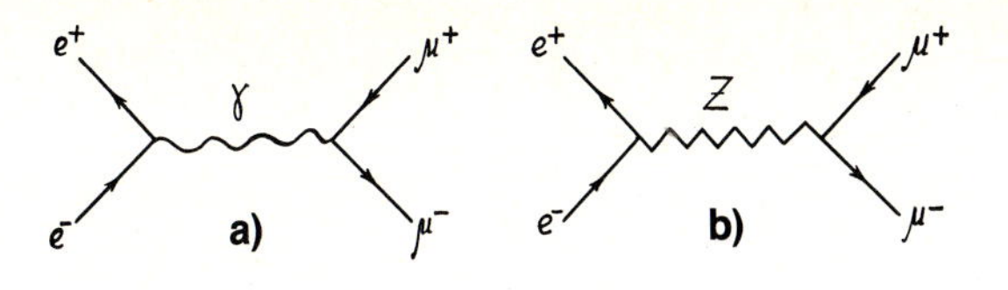


**Fig. 8.1a,b.** Diagrams for the process $e^+e^- \to \mu^+\mu^-$

the virtual $Z$-boson exchange (Fig. 8.1 b) contribute to this process, according to (6.1.12) and (6.1.17). Neglecting the masses of the electron and the muon as compared to their energies, one can express the amplitude $M$ of the process (8.1.4) in the following form[1]:

$$M = \frac{e^2}{q^2}[\bar{\mu}(p)\gamma_\mu \mu(p')][\bar{e}(k')\gamma_\mu e(k)] + \frac{\frac{1}{4}\bar{g}^2}{q^2 - m_Z^2}$$
$$\times [(-\tfrac{1}{2}+\xi)\bar{\mu}(p)\gamma_{\mu L}\mu(p') + \xi\bar{\mu}(p)\gamma_{\mu R}\mu(p')]$$
$$\times [(-\tfrac{1}{2}+\xi)\bar{e}(k')\gamma_{\mu L}e(k) + \xi\bar{e}(k')\gamma_{\mu R}e(k)] . \tag{8.1.5}$$

For the differential cross-section at $q^2 \ll m_Z^2$ it follows, in the centre-of-mass system, that

$$\frac{d\sigma}{d\cos\theta} = \frac{\pi\alpha^2}{4q^2}[(a+b)(1+\cos^2\theta) + (a-b)2\cos\theta] , \tag{8.1.6}$$

where

$$a = \tfrac{1}{2}\{[1 + x(-\tfrac{1}{2}+\xi)^2]^2 + (1 + x\xi^2)\} ,$$
$$b = [1 + x\xi(-\tfrac{1}{2}+\xi)]^2 ,$$
$$x = \frac{\bar{g}^2}{e^2}\frac{q^2}{q^2 - m_Z^2} ,$$
$$q^2 = (k+k')^2 = (p+p')^2 ;$$

$\theta$ is the angle between the momenta of the electron and the muon in the centre-of-mass system. If $x = 0$, then $a = b = 1$, and (8.1.6) reduces to the cross-section of the annihilation $e^+e^- \to \mu^+\mu^-$ caused by the electromagnetic interaction alone. If $q^2 \ll m_Z^2$, then $x < 0$ and $a - b \sim x/4 < 0$ so that, according to (8.1.6), the negative muons will move at large angles ($\theta \sim 180°$), in the direction of the incident positron. This will give rise to charge asymmetry. Such a circumstance makes it possible to single out the effect due to neutral weak currents.

---

[1] In this chapter $\mu$, $e$, $\nu$, $u$, $d$ denote the spinor functions of the initial particles and $\bar{\mu}$, $\bar{e}$, $\bar{\nu}$, $\bar{u}$, $\bar{d}$ those of final state particles. Explicit forms of these spinors, and the propagators of the particles, are given in Table 7.2. The factors $1/(q^2 - m_{W,Z}^2)$ stem from the propagators of the $W$- and $Z$-bosons.

### 8.1.2 Non-Diagonal Terms

The non-diagonal part of the Lagrangian (8.1.2) is written as

$$\mathscr{L}_{\text{nd}} = \frac{2G}{\sqrt{2}} (J_\mu^{n,\nu} J_\mu^{n,l} + J_\mu^{n,\nu} J_\mu^{n,q} + J_\mu^{n,l} J_\mu^{n,q}) . \tag{8.1.7}$$

The first term of this Lagrangian describes the weak interaction between neutrinos and leptons, the second between neutrinos and quarks, and the third between charged leptons and quarks.

#### Neutrino-Lepton Processes

An example of such a process is the $\nu_\mu$ scattering on electrons

$$\nu_\mu(k) + \mathrm{e}^-(p) \to \nu_\mu(k') + e^-(p') . \tag{8.1.8}$$

This process is caused by the neutral weak current only. Therefore, the only contribution in the lowest order of the perturbation theory is that from the virtual $Z$-boson exchange (Fig. 8.2). According to (6.1.16) and (6.1.17), this diagram is associated with the matrix element

$$M = \frac{G}{\sqrt{2}} (\bar{\nu}_\mu \gamma_{\mu\mathrm{L}} \nu_\mu) [(-\tfrac{1}{2} + \xi) \bar{e} \gamma_{\mu\mathrm{L}} e + \xi \bar{e} \gamma_{\mu\mathrm{R}} e]$$

*leading to the following total cross-section:*

$$\sigma_{\nu_\mu e} = \frac{G^2 s}{\pi} \left[ \left( -\frac{1}{2} + \xi \right)^2 + \frac{1}{3} \xi^2 \right], \tag{8.1.9}$$

where

$$s = (p+k)^2$$

#### Neutrino-Nucleon Processes

Let us consider neutrino scattering on nucleons caused by the neutral weak current:

$$\nu(k) + N(p) \to \nu(k') + X(p') , \tag{8.1.10}$$

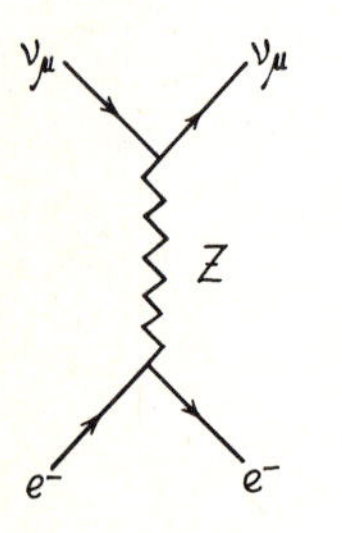


**Fig. 8.2.** Diagram for the process $\nu_\mu e^- \to \nu_\mu e^-$

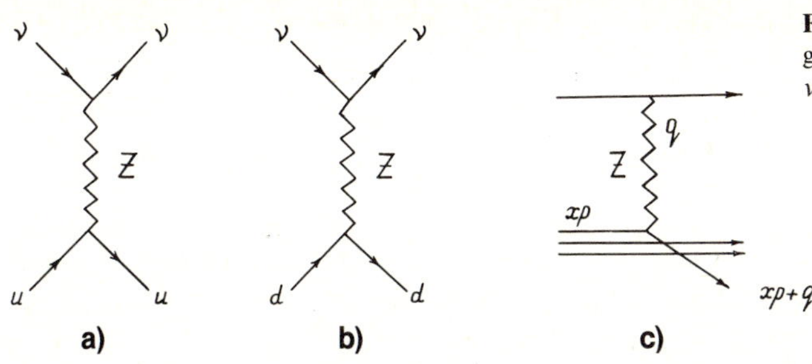


**Fig. 8.3a–c.** Quark diagrams for the process $\nu N \to \nu X$

where $N$ denotes the nucleon and $X$ stands for all other particles; $q = k - k' = p' - p$ is the transferred momentum. Let us introduce the variables $\nu = pq/M$ and $q^2$. The region where the conditions $\nu \gg M$, $-q^2 \gg M^2$ are fulfilled and $x \equiv -q^2/(2M\nu)$ is fixed is called the deep-inelastic region ($M$ is the nucleon mass). We limit the analysis of the process (8.1.10) to the deep-inelastic region.

The nucleons are made out of quarks. Hence, the neutrino-nucleon interaction is reduced to the neutrino-quark interaction. Let us neglect the contribution of the quarks $s$, $\bar{s}$, $\bar{u}$, and $\bar{d}$. Then, by virtue of (6.1.16) and (6.3.8), the matrix element for the process of neutrino scattering on $u$-quarks and $d$-quarks is written, in the limit of small transferred momenta ($|q^2| \ll m_Z^2$), as (Fig. 8.3a, b):

$$M = \frac{G}{\sqrt{2}} (\bar{\nu}\gamma_{\mu L}\nu)[(\tfrac{1}{2} - \tfrac{2}{3}\xi)\bar{u}\gamma_{\mu L}u - \tfrac{2}{3}\xi\bar{u}\gamma_{\mu R}u + (-\tfrac{1}{2} + \tfrac{1}{3}\xi)\bar{d}\gamma_{\mu L}d + \tfrac{1}{3}\xi\bar{d}\gamma_{\mu R}d] \,. \tag{8.1.11}$$

From this, the cross-section $\hat{\sigma}_\nu^n$ of neutrino scattering on quarks follows:

$$\hat{\sigma}_\nu^{\mathrm{n}} = \frac{1}{2\pi} G^2 \hat{s} B_\nu^{\mathrm{n}} \,, \tag{8.1.12}$$

where

$$B_\nu^{\mathrm{n}} = (\tfrac{1}{2} - \tfrac{2}{3}\xi)^2 + \tfrac{1}{3}(\tfrac{2}{3}\xi)^2 + (-\tfrac{1}{2} + \tfrac{1}{3}\xi)^2 + \tfrac{1}{3}(\tfrac{1}{3}\xi)^2 \,; \tag{8.1.13}$$

$\hat{s} = (px+k)^2 \simeq x2pk \simeq xs$, $s = (k+p)^2$, $k$ is the neutrino, $p$ the nucleon, and $px$ the initial quark momentum; $x$ is the fraction of the total nucleon momentum carried by the quark (Fig. 8.3c). The cross-section (8.1.12) describes the process involving a quark which has a fixed momentum. As a matter of fact, in a hadron there exist quarks with different momenta. For instance, consider a distribution function of quarks over $x$, say, $u(x)$. The quantity $xu(x)\,dx$ gives the fraction of the total proton momentum carried by the $u$-quarks whose $x$-value falls within the interval $dx$. In the same manner the distribution functions $d(x)$, $s(x)$, $\bar{u}(x)$, $\bar{d}(x)$, $\bar{s}(x)$, etc. can be introduced.

The distribution functions have been determined from experimental data. For this purpose, the cross-sections for deep-inelastic processes, expressed in terms of the distribution functions, have been utilized. A number of sets of distribution functions have been obtained in this way which are in reasonable agreement with each other. (For more detail, see Sect. 11.2.)

Let us calculate the cross-section of the neutrino scattering on an "average" nucleon, i.e., $\frac{1}{2}(\sigma_p+\sigma_n)$. This cross-section can be conveniently used in analyzing the scattering on nuclei with an equal number of protons and neutrons. For the "average" nucleon, the distribution functions of the $u$-quarks and the $d$-quarks are equal. Using this circumstance and (8.1.12) we arrive at the following expression for deep-inelastic neutrino scattering on the "average" nucleon:

$$\sigma_\nu^{\mathrm{n}} = \frac{1}{2\pi} G^2 s(U+D) B_\nu^{\mathrm{n}}, \tag{8.1.14}$$

where $U = \int_0^1 dx\, x u(x)$ and $D = \int_0^1 dx\, x d(x)$ are the total relative momenta associated with the $u$-quarks and the $d$-quarks in a proton, respectively.

The cross-section of the deep-inelastic anti-neutrino ($\bar{\nu}$) scattering on the "average nucleon" has a similar form

$$\sigma_{\bar{\nu}}^{\mathrm{n}} = \frac{1}{2\pi} G^2 s(U+D) B_{\bar{\nu}}^{\mathrm{n}}, \quad \text{where} \tag{8.1.15}$$

$$B_{\bar{\nu}}^{\mathrm{n}} = \tfrac{1}{3}(\tfrac{1}{2} - \tfrac{2}{3}\xi)^2 + (\tfrac{2}{3}\xi)^2 + \tfrac{1}{3}(-\tfrac{1}{2} + \tfrac{1}{3}\xi)^2 + (\tfrac{1}{3}\xi)^2 . \tag{8.1.16}$$

An analogous procedure is used to derive the cross-sections for deep-inelastic neutrino and anti-neutrino scattering on nucleons due to charged currents. Within the same approximation, these cross-sections read

$$\sigma_\nu^{\mathrm{c}} = \frac{1}{2\pi} G^2 s(U+D), \quad \sigma_{\bar{\nu}}^{\mathrm{c}} = \frac{1}{6\pi} G^2 s(U+D) . \tag{8.1.17}$$

Introducing the ratios of cross-sections,

$$R_\nu = \frac{\sigma_\nu^{\mathrm{n}}}{\sigma_\nu^{\mathrm{c}}} = \frac{1}{2} - \xi + \frac{20}{27}\xi^2, \tag{8.1.18}$$

$$R_{\bar{\nu}} = \frac{\sigma_{\bar{\nu}}^{\mathrm{n}}}{\sigma_{\bar{\nu}}^{\mathrm{c}}} = \frac{1}{2} - \xi + \frac{20}{9}\xi^2, \tag{8.1.19}$$

we find

$$\xi = \tfrac{1}{2}(1 + R_{\bar{\nu}} - 3R_\nu) . \tag{8.1.20}$$

**Electron-Nucleon Processes**

Let us dwell on two processes of this type.

1) We consider the deep-inelastic scattering of a longitudinally polarized electron on a nucleon:

$$e^- + N \to e^- + X . \tag{8.1.21}$$

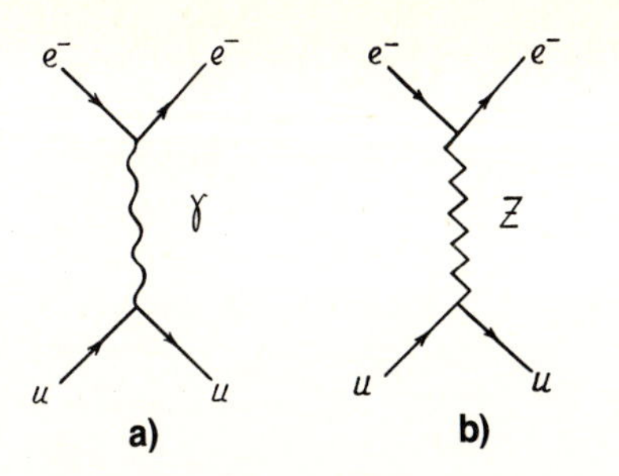


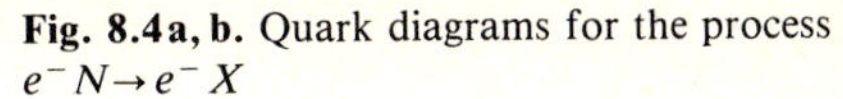
**Fig. 8.4a, b.** Quark diagrams for the process $e^- N \to e^- X$

Both the electromagnetic current (Fig. 8.4a) and the neutral weak current (Fig. 8.4b) underlie this process. According to (6.1.12), (6.1.17), (6.3.4), and (6.3.8), the matrix element of the electron scattering, e.g., on a $u$-quark of the nucleon, is written, in the limit of small transferred momenta ($|q^2| \ll m_Z^2$),

$$M = -\frac{2e^2}{3q^2}(\bar{e}\gamma_\mu e)(\bar{u}\gamma_\mu u) + \frac{2G}{\sqrt{2}}$$
$$\times [(-\tfrac{1}{2}+\xi)\bar{e}\gamma_{\mu L}e + \xi\bar{e}\gamma_{\mu R}e][(\tfrac{1}{2}-\tfrac{2}{3}\xi)\bar{u}\gamma_{\mu L}u - \tfrac{2}{3}\xi\bar{u}\gamma_{\mu R}u]\,. \quad (8.1.22)$$

The second term of (8.1.22) contains two kinds of terms non-invariant under space inversion: (i) the product of the quark axial current and the electron vector current and (ii) the product the quark vector current and the electron axial current. These terms give rise to a difference in the scattering of right-handed and left-handed electrons, that is to say, to an asymmetry in the scattering. This asymmetry is conveniently described by the quantity

$$A = \frac{d\sigma_R - d\sigma_L}{d\sigma_R + d\sigma_L},$$

where $d\sigma_R$ and $d\sigma_L$ are the scattering cross-sections of right-handed and left-handed electrons, respectively. Retaining only the contributions of $u$-quarks and $d$-quarks we have

$$A = \frac{Gq^2}{2\sqrt{2}\pi\alpha}\,\frac{9}{10}\left[\left(1-\frac{20}{9}\xi\right)+(1-4\xi)K(y)\right],$$

where

$$K(y) = \frac{1-(1-y)^2}{1+(1-y)^2}, \qquad y = 1 - E'/E\,;$$

$E$ and $E'$ denote the energy of the electron before and after the collision, respectively.

2) The presence of an additional electron-nucleon interaction originating from the neutral weak current gives rise to a term non-invariant under space inversion in the interaction potential of atom-bound electrons with the nucleons of the nucleus. The existence of this term results in a number of specific effects, such as the rotation of the plane of polarization of linearly

polarized light or more intensive absorption of right-handed photons as compared with the left-handed ones in case of the circular polarization of light. The magnitude of these effects can be calculated within the framework of the standard model.

### 8.1.3 Comparison with Experiment

Quite a number of other processes involving the neutral weak currents have been theoretically analyzed in a similar way. The expressions obtained contain the unknown parameter $\xi$. Theoretical results exhibit a reasonably good agreement with the corresponding experimental data when the value of $\xi \sim 0.23$ is used. An exception is found in experimental data on parity violation in atoms. This effect was measured by various experimental groups some of which have provided results in agreement, while others in striking disagreement with the predictions of the standard model.

### 8.1.4 Non-Standard Models

Processes originating from neutral currents have also been subject to analyses in the framework of various non-standard models (Sect. 6.4). The results

**Table 8.1.** Predictions of various models

| | Standard model | Vector model | Asymmetric model |
|---|---|---|---|
| $\dfrac{\sigma(\bar{\nu}_\mu e)}{\sigma(\nu_\mu e)}$ | $\dfrac{1-4\xi+16\xi^2}{3-12\xi+16\xi^2}$ | 1 | 1 |
| $\dfrac{\sigma(\bar{\nu}N\to\bar{\nu}X)}{\sigma(\nu N\to\nu X)}$, $B=0.8-1$ ($B=1$ for sea $=0$, cf. Sect. 11.2.2) | $\dfrac{(2-B)(1-2\xi)+\frac{40}{9}\xi^2}{(2+B)(1-2\xi)+\frac{40}{9}\xi^2}$ | 1 | as in the standard model |
| Violation of parity in heavy atoms | yes | no | no |
| Violation of parity in hydrogen | yes, $\sim 1-4\xi$ | no | yes |
| Asymmetry in polarized *ep*-scattering | yes | no | yes |
| Forward-backward asymmetry in $e^+e^-\to\mu^+\mu^-$ | yes | no | no |

of the analyses obtained with some non-standard models are summarized in Table 8.1 which can be considered as a continuation of Table 6.6. One can recognize that the predictions of the standard model and of the non-standard ones, regarding the same effects, coincide in some cases but differ in some others.

## 8.2 Processes Caused by Charged Weak Currents

### 8.2.1 Charged Weak Quark Current

This current is given by (6.3.6). It does not contain the quarks $d$, $s$, and $b$ themselves, but rather involves their linear combinations $d'$, $s'$ and $b'$, i.e., there occurs a *mixing among the quarks*. The need for mixing is dictated by experimental evidence. If a charged weak quark current had the structure

$$J_\mu^{(+),\mathrm{q}} = \bar{u}\gamma_{\mu\mathrm{L}} d + \bar{c}\gamma_{\mu\mathrm{L}} s + \bar{t}\gamma_{\mu\mathrm{L}} b\,, \tag{8.2.1}$$

the quarks $s$ and $b$ would be stable. This would imply the stability of strange particles and $b$-quark-containing hadrons as well. This, however, contradicts the experiment. The contradiction is resolved by introducing mixing among the quarks.

Quark mixing implies that weak charged quark currents can be expressed in the following general form:

$$J_\mu^{(+),\mathrm{q}} = \bar{q}_i\gamma_{\mu\mathrm{L}} V_{ij} q_j = (\bar{u},\bar{c},\bar{t})\,\gamma_{\mu\mathrm{L}} \begin{vmatrix} a_{11} & a_{12} & a_{13} \\ a_{21} & a_{22} & a_{23} \\ a_{31} & a_{32} & a_{33} \end{vmatrix} \begin{pmatrix} d \\ s \\ b \end{pmatrix}. \tag{8.2.2}$$

where $V_{ij}$ is a unitary $3\times 3$ matrix. The elements $a_{ij}$ of the matrix $V_{ij}$ can be expressed in terms of a certain number of independent parameters (i.e., the matrix $V_{ij}$ can be parametrized).

Let us find the number of these independent *mixing parameters*. For that, we consider a general case when the number of quark doublets which enter the standard model is $n$. The matrix $V$ then contains $n^2$ complex or $2n^2$ real parameters. The unitarity condition $VV^\dagger = 1$ imposes $n^2$ constraints on the parameters so that the remaining number of independent parameters will be $n^2$. The number of quarks is $2n$. The wave function of each quark is determined with the accuracy of an arbitrary (unphysical) phase; one of the phases can be chosen freely, e.g., it can be set equal to zero. That is to say, $2n-1$ phases will be unphysical and can be eliminated. Hence, the net number of physical parameters of the matrix $V$ is

$$n^2-(2n-1) = (n-1)^2\,.$$

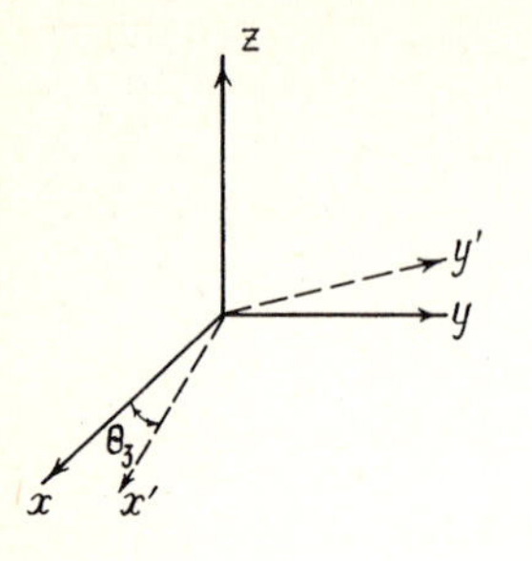


**Fig. 8.5.** To the derivation of (8.2.5)

As parameters, both phases and rotation angles can be chosen. The number of independent rotations $n_\theta$ in an $n$-dimensional space is given by

$$n_\theta = \tfrac{1}{2}(n-1)\,n\,. \tag{8.2.3}$$

For the number $n_\delta$ of physical phases it then follows

$$n_\delta = (n-1)^2 - \tfrac{1}{2}n(n-1) = \tfrac{1}{2}(n-1)(n-2)\,. \tag{8.2.4}$$

In particular,

i) for $n = 1$ (one doublet), $n_\theta = 0,\ n_\delta = 0$;
ii) for $n = 2$ (two doublets), $n_\theta = 1,\ n_\delta = 0$;
iii) for $n = 3$ (three doublets), $n_\theta = 3,\ n_\delta = 1$;
iv) for $n = 4$ (four doublets), $n_\theta = 6,\ n_\delta = 3$.

In the case of three doublets, which is of interest here, the matrix is characterized by three parameters and one phase. Usually the explicit form of these parameters is determined in the following way. Consider a triplet of coordinates $x, y, z$ (Fig. 8.5) and assign in correspondence with it a triplet of quarks $d, s, b$. Passing to mixed quarks $d'$, $s'$, $b'$ is carried out by rotation by an angle $\theta_3$ around the $z$-axis and subsequent rotations by an angle $\theta_1$ around the $x$-axis and by an angle $\theta_2$ around the new $z$-axis. The resulting rotation is described by the product of the matrices of the above rotations,

$$V = \begin{pmatrix} 1 & 0 & 0 \\ 0 & c_2 & s_2 \\ 0 & -s_2 & c_2 \end{pmatrix} \begin{pmatrix} c_1 & s_1 & 0 \\ -s_1 & c_1 & 0 \\ 0 & 0 & 1 \end{pmatrix} \begin{pmatrix} 1 & 0 & 0 \\ 0 & c_3 & s_3 \\ 0 & -s_3 & c_3 \end{pmatrix}, \tag{8.2.5}$$

where $c_i = \cos\theta_i$ and $s_i = \sin\theta_i$. Taking into account the phase $\delta$, the matrix (8.2.5) is rewritten as

$$V = \begin{pmatrix} 1 & 0 & 0 \\ 0 & c_2 & s_2 \\ 0 & -s_2 & c_2 \end{pmatrix} \begin{pmatrix} c_1 & s_1 & 0 \\ -s_1 & c_1 & 0 \\ 0 & 0 & e^{i\delta} \end{pmatrix} \begin{pmatrix} 1 & 0 & 0 \\ 0 & c_3 & s_3 \\ 0 & -s_3 & c_3 \end{pmatrix}. \tag{8.2.6}$$

After multiplication one obtains the *Kobayashi-Maskawa* matrix:

$$V = \begin{pmatrix} c_1 & s_1 c_3 & s_1 s_3 \\ -s_1 c_2 & c_1 c_2 c_3 - e^{i\delta} s_2 s_3 & c_1 c_2 s_3 + e^{i\delta} s_2 c_3 \\ s_1 s_2 & -c_1 s_2 c_3 - e^{i\delta} c_2 s_3 & -c_1 s_2 s_3 + e^{i\delta} c_2 c_3 \end{pmatrix} . \tag{8.2.7}$$

Substituting this matrix into (8.2.2) we obtain for the quark fields $d'$, $s'$, $b'$:

$$\begin{aligned} d' &= d\cos\theta_1 + s\sin\theta_1\cos\theta_3 + b\sin\theta_1\sin\theta_3 , \\ s' &= -d\sin\theta_1\cos\theta_2 + s(\cos\theta_1\cos\theta_2\cos\theta_3 - e^{i\delta}\sin\theta_2\sin\theta_3) \\ &\quad + b(\cos\theta_1\cos\theta_2\sin\theta_3 + e^{i\delta}\sin\theta_2\cos\theta_3) , \\ b' &= d\sin\theta_1\sin\theta_2 + s(-\cos\theta_1\sin\theta_2\cos\theta_3 - e^{i\delta}\cos\theta_2\sin\theta_3) \\ &\quad + b(-\cos\theta_1\sin\theta_2\sin\theta_3 + e^{i\delta}\cos\theta_2\cos\theta_3) . \end{aligned} \tag{8.2.8}$$

Assuming that $b' = b$, i.e., neglecting the transitions $ub$ and $cb$, we are left with a single angle $\theta_1 = \theta_C$ mixing the quarks $d$ and $s$:

$$d_C = d\cos\theta_C + s\sin\theta_C , \qquad s_C = -d\sin\theta_C + s\cos\theta_C . \tag{8.2.9}$$

The angle $\theta_C$ is referred to as the *Cabibbo angle*; it has been determined from the experimental data on the decay of strange particles yielding $\theta_C \sim 13°$. In this section we shall take into account only the Cabibbo mixing (8.2.9).

The charged weak current in the standard model is the sum of the lepton current and the quark current:

$$J_\mu^{(+)} = J_\mu^{(+),l} + J_\mu^{(+),q} . \tag{8.2.10}$$

The currents involved are determined by (6.1.14), (6.3.6) and (8.2.8). According to (6.3.14), we have for the effective interaction Lagrangian $\mathscr{L}_{\text{eff}}$ of charged weak currents

$$\begin{aligned} \mathscr{L}_{\text{eff}} = \frac{G}{\sqrt{2}} J_\mu^{(+)} J_\mu^{(-)} = \frac{G}{\sqrt{2}} [J_\mu^{(+),l} J_\mu^{(-),l} + J_\mu^{(+),l} J_\mu^{(-),q} \\ + J_\mu^{(+),q} J_\mu^{(-),l} + J_\mu^{(+),q} J_\mu^{(-),q}] . \end{aligned} \tag{8.2.11}$$

The first term of this Lagrangian describes the weak interaction between leptons, the second and third ones between leptons and quarks, and the fourth one between quarks.

Let us consider, as examples, some concrete processes.

### 8.2.2 Leptonic Processes

Examples of such processes are

$$\nu_\mu(k) + e^-(p) \to \nu_e(k') + \mu^-(p') , \tag{8.2.12}$$

$$\nu_e + e^- \to \nu_e + e^- , \tag{8.2.13}$$

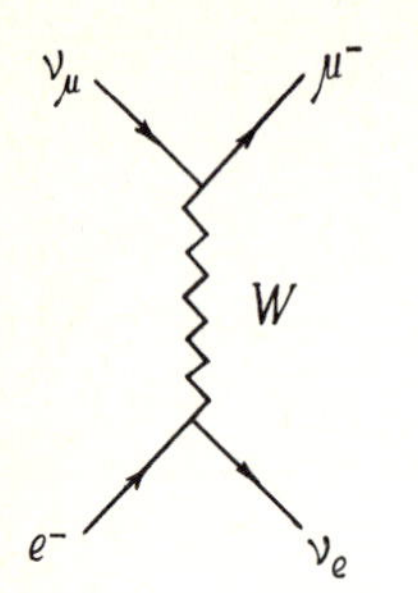


Fig. 8.6. Diagram for the process $\nu_\mu e^- \to \nu_e \mu^-$

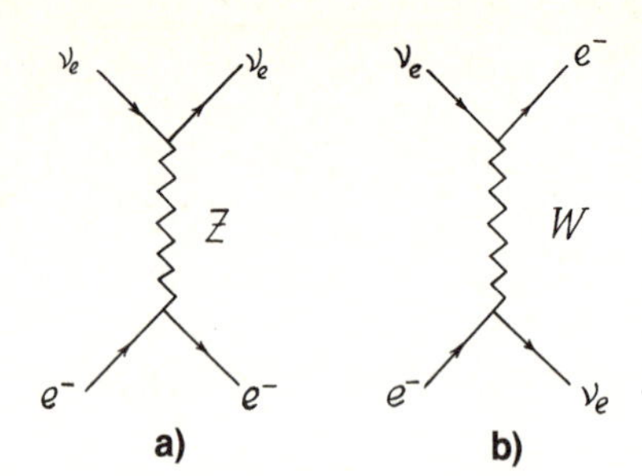


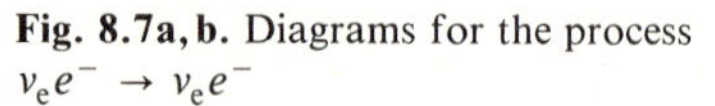

Fig. 8.7a, b. Diagrams for the process $\nu_e e^- \to \nu_e e^-$

$$\mu^- \to e^- + \bar{\nu}_e + \nu_\mu \,. \tag{8.2.14}$$

The first process originates from the charged weak lepton current only (Fig. 8.6). According to (6.1.14), the matrix element for this process reads

$$M = \frac{G}{\sqrt{2}} (\bar{\nu}_e \gamma_{\mu L} e)(\bar{\mu} \gamma_{\mu L} \nu_\mu) \,.$$

From this, we find for the total cross-section

$$\sigma = \frac{1}{\pi} G^2 \frac{(s-\mu^2)^2}{s} \,,$$

where $s = (k+p)^2$ and $\mu$ is the muon mass.

In the process (8.2.13), both neutral (Fig. 8.7a) and charged (Fig. 8.7b) currents are involved. Using (6.1.14), (6.1.16), and (6.1.17) we obtain for the matrix element of the process (8.2.13)

$$M = \frac{G}{\sqrt{2}} \{(\bar{\nu}_e \gamma_{\mu L} \nu_e)[(-\tfrac{1}{2} + \xi)\bar{e} \gamma_{\mu L} e + \xi \bar{e} \gamma_{\mu R} e] + (\bar{e} \gamma_{\mu L} \nu_e)(\bar{\nu}_e \gamma_{\mu L} e)\} \,. \tag{8.2.15}$$

With the aid of the Fierz transformation

$$(\bar{e} \gamma_{\mu L} \nu_e)(\bar{\nu}_e \gamma_{\mu L} e) = (\bar{e} \gamma_{\mu L} e)(\bar{\nu}_e \gamma_{\mu L} \nu_e) \,,$$

Eq. (8.2.15) can be rewritten as

$$M = \frac{G}{\sqrt{2}} \{\bar{e}[(\tfrac{1}{2} + \xi) \gamma_{\mu L} + \xi \gamma_{\mu R}] e\}(\bar{\nu}_e \gamma_{\mu L} \nu_e) \,.$$

This matrix element leads to the following cross-section

$$\frac{d\sigma_{\nu_e}}{dT} = \frac{2G^2 m}{\pi} \left[ \left(\frac{1}{2} + \xi\right)^2 + \xi^2 \left(1 - \frac{T}{\omega_1}\right)^2 - \left(\frac{1}{2} + \xi\right) \xi \frac{mT}{\omega_1^2} \right] , \tag{8.2.15'}$$

where $T$ is the kinetic energy of the scattered electron, $m$ is the electron mass, and $\omega_1$ is the energy of the primary neutrino.

For the process (8.2.14), (6.1.14) can be used to obtain the matrix element

$$M = \frac{G}{\sqrt{2}} (\bar{\nu}_\mu \gamma_{\mu L} \mu)(\bar{e} \gamma_{\mu L} \nu_e) ,$$

which leads to the following expression for the probability of $\mu$-decay:

$$\Gamma = \frac{G^2 \mu^5}{192 \pi^3} .$$

### 8.2.3 Semi-Leptonic Decays

These include decays involving leptons and hadrons. As examples, we consider the following semi-leptonic decays:

$$\pi^+(q) \to e^+(k_1) + \nu_e(k_2) , \tag{8.2.16}$$

$$\pi^+(q_1) \to \pi^0(q_2) + e^+(k_1) + \nu_e(k_2) , \tag{8.2.17}$$

$$n(p_1) \to p(p_2) + e^-(k_1) + \bar{\nu}_e(k_2) . \tag{8.2.18}$$

These decays are caused by the charged leptonic current $(\bar{e}\nu_e)$ and the quark current $(\bar{u}d)$. The quark diagrams for these decays are presented in Fig. 8.8. The general structure of the matrix elements for each of these decays is given by

$$M = \frac{G}{\sqrt{2}} \cos\theta_C L_\mu H_\mu , \tag{8.2.19}$$

where $L_\mu$ and $H_\mu$ are the weak leptonic and hadronic currents, respectively. For example, for the process $\nu_e + i \to e + f$ they read

$$L_\mu = \bar{e} \gamma_{\mu L} \nu_e , \tag{8.2.20}$$

$$H_\mu = \langle f | J_\mu^{(+),q} | i \rangle = \langle f | \bar{u} \gamma_\mu d | i \rangle + \langle f | \bar{u} \gamma_\mu \gamma_5 d | i \rangle \equiv V_\mu + A_\mu . \tag{8.2.21}$$

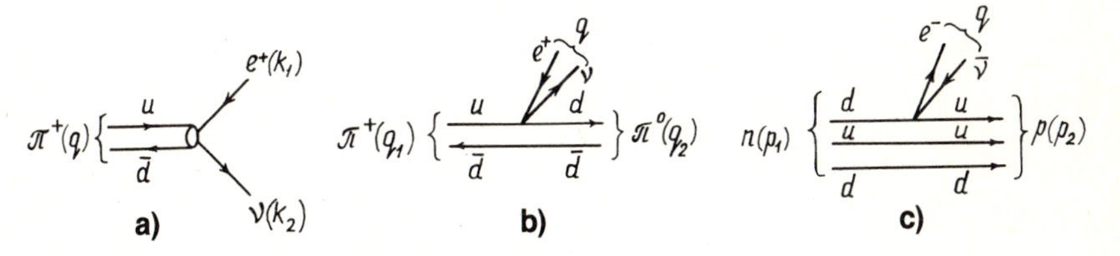


**Fig. 8.8 a – c.** Quark diagrams for the semi-leptonic decays

Here $i$ and $f$ denote the initial and the final state, respectively, $V_\mu$ and $A_\mu$ are the vector weak hadronic current and the axial weak hadronic current, respectively.

The matrix element for the leptonic current is given by (8.2.20). Because weak interactions of hadrons ought to be considered in the background of strong interactions, (8.2.21) represents the matrix element for the weak hadronic current *dressed* by strong interaction. The meaning of this is as follows: the contribution of strong interactions is taken into account completely while the weak interaction is considered within the perturbation theory. In this situation one is unable to find an explicit expression for the matrix elements of weak hadronic currents $V_\mu$ and $A_\mu$. One can only determine their kinematic structure provided that general symmetry properties are utilized.

Let us find the structure of hadronic currents associated with the processes whose diagrams are depicted in Fig. 8.8. For this purpose, a four-vector and an axial vector have to be constructed from the wave functions and the momenta of particles as well as from $\gamma$-matrices (if fermions are involved in the process). It should be kept in mind that the matrix element can depend only on the sum of the four-momenta of the created leptons but not on each momentum separately. This requirement is due to the fact that the pair of leptons is emitted at one point.

1) For the case of pseudoscalar $\pi$-meson decay into a pair of leptons (Fig. 8.8a) there are only the (pseudoscalar) wave function $\tilde{\phi}(q)$ of the $\pi$-meson and the vector of its momentum, $q=k_1+k_2$, at our disposal. Only an axial vector $A_\mu=f\tilde{\phi}(q)q_\mu$ can be constructed out of these quantities, so that $V_\mu=0$, and the weak hadronic current is

$$H_\mu=fq_\mu\tilde{\phi}(q)\,. \tag{8.2.22}$$

Here $f$ is an unknown invariant coefficient. In the case under consideration $f$ is a constant because the only invariant it may depend on, namely $q^2=m_\pi^2$, is a constant ($m_\pi$ is the $\pi$-meson mass).

2) For the pseudoscalar $\pi$-meson decay into a meson of the same kind and a pair of leptons (Fig. 8.8b) we have the wave functions $\tilde{\phi}(q_1)$ and $\tilde{\phi}(q_2)$ of the initial and the final mesons, their momenta $q_1$ and $q_2$, respectively, and the sum $q$ of the momenta of the lepton pair to construct the current vectors. Due to the momentum conservation ($q_1=q_2+q$), only two of the momenta are independent which results in the following general form of the weak hadronic current:

$$H_\mu=\tilde{\phi}(q_1)\,\tilde{\phi}(q_2)(f_1P_\mu+f_2q_\mu)\,. \tag{8.2.23}$$

Here $P=q_1+q_2$; $f_1$ and $f_2$ are unknown form factors which depend on the momenta $q_1$ and $q_2$. Only one independent invariant, which is not a constant, can be constructed from the latter quantities, for example, $q^2$. Therefore the form factors $f_1(q^2)$ and $f_2(q^2)$ will be functions of the only variable $q^2$. If the two mesons have the same parity, then $A_\mu=0$.

3) For the case of baryon decay into a baryon and a lepton pair (Fig. 8.8c) we have, in addition to the momenta of the initial and the final baryons, spinor functions and the $\gamma$-matrices. To construct the current vector, we can combine these quantities into three independent vectors: $\bar{u}\gamma_\mu u$, $(\bar{u}\sigma_{\mu\nu}u)q_\nu$, and $\bar{u}uq_\mu$ [where $q=p_2-p_1$, $\sigma_{\mu\nu}=(\gamma_\mu\gamma_\nu-\gamma_\nu\gamma_\mu)/(2\mathrm{i})$ and three independent pseudovectors: $\bar{u}\gamma_\mu\gamma_5 u$, $(\bar{u}\gamma_5 u)P_\mu$, and $\bar{u}\gamma_5 uq_\mu$ (where $P_\mu=p_1+p_2$). Consequently, we have for the weak hadronic current

$$H_\mu=A_\mu+V_\mu\,, \tag{8.2.24}$$

where

$$V_\mu=\bar{u}_p(p_2)[f_1(q^2)\gamma_\mu+f_2(q^2)\sigma_{\mu\nu}q_\nu+f_3(q^2)q_\mu]u_n(p_1)\,, \tag{8.2.25}$$

$$A_\mu=\bar{u}_p(p_2)[g_1(q^2)\gamma_\mu+g_2(q^2)P_\mu+g_3(q^2)q_\mu]\gamma_5 u_n(p_1)\,. \tag{8.2.26}$$

Here $g_i(q^2)$ and $f_i(q^2)$ are unknown form factors depending on $q^2$; $q$ is the momentum of the leptons.

It follows from the hermiticity and the *CP* (or *T*) invariance of the expressions (8.2.22–24) that the form factors involved are real functions. Indeed, these expressions will coincide with those obtained from them by Hermitian conjugation and *CP* transformation only if these form factors are real.

With the aid of (8.2.22–24) the matrix elements of the processes (8.2.16–18) can be rewritten in the form

$$M=\frac{G}{\sqrt{2}}\cos\theta_{\mathrm{C}}A_\mu L_\mu=\frac{G}{\sqrt{2}}\cos\theta_{\mathrm{C}}fq_\mu\tilde{\phi}(q)\,\bar{\nu}_{\mathrm{e}}\gamma_{\mu\mathrm{L}}e$$

$$M=\frac{G}{\sqrt{2}}\cos\theta_{\mathrm{C}}V_\mu L_\mu=\frac{G}{\sqrt{2}}\cos\theta_{\mathrm{C}}(f_1P_\mu+f_2q_\mu)\,\tilde{\phi}(q_1)\,\tilde{\phi}(q_2)\,\bar{\nu}_{\mathrm{e}}\gamma_{\mu\mathrm{L}}e$$

$$M=\frac{G}{\sqrt{2}}\cos\theta_{\mathrm{C}}(V_\mu+A_\mu)\bar{e}\gamma_{\mu\mathrm{L}}\nu_{\mathrm{e}}\,.$$

Here $\theta_{\mathrm{C}}$ is the Cabibbo angle; the quantities $V_\mu$ and $A_\mu$ in the last formula are given by (8.2.25, 26), respectively. Using these matrix elements one can find the expressions for the corresponding cross-sections.

### 8.2.4 Non-Leptonic Hadronic Decays

These consist of hadronic decays into hadrons. Consider, for example, the decay

$$\Lambda^0\rightarrow p\pi^-\,. \tag{8.2.27}$$

It is caused by the interaction of the $(\bar{u}s)$ and $(\bar{d}u)$ currents. Two kinds of quark diagrams are possible for this process: the external diagrams (Fig. 8.9a)

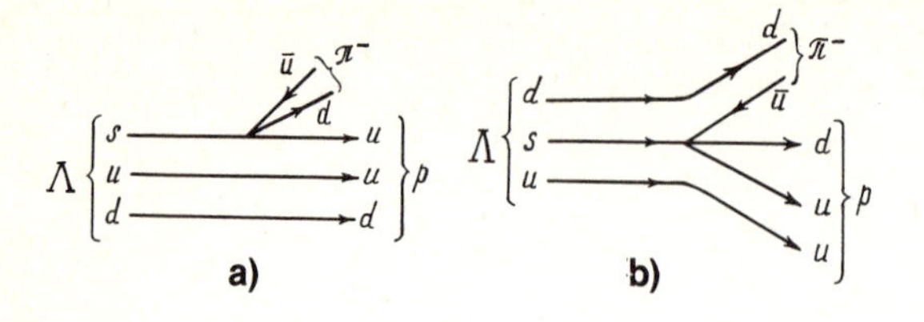


**Fig. 8.9 a, b.** Quark diagrams for the decay $\Lambda^0 \to p\pi^-$

and the internal diagrams (Fig. 8.9b). According to (6.3.6), the decay amplitude has the form

$$M = \frac{G}{\sqrt{2}} \cos\theta_C \sin\theta_C (\bar{d}\gamma_{\mu L} u)(\bar{u}\gamma_{\mu L} s) + \dots \,. \tag{8.2.28}$$

We shall neglect the internal diagrams. We further assume that the quarks which are constituents of the $\pi$-meson do not interact with the quarks contained in the initial baryon and in the final baryon. Then the matrix element (8.2.28) of the decay can be factorized to be represented as the product of two matrix elements. For example, for the term $(\bar{d}\gamma_{\mu L} u)(\bar{u}\gamma_{\mu L} s)$ we have

$$\langle \pi^- p | (\bar{d}\gamma_{\mu L} u)(\bar{u}\gamma_{\mu L} s) | \Lambda \rangle = \langle \pi^- | \bar{d}\gamma_{\mu L} u | 0 \rangle \langle p | \bar{u}\gamma_{\mu L} s | \Lambda \rangle .$$

Both matrix elements on the right-hand side have already been calculated. The first one reads

$$\langle \pi^- | \bar{d}\gamma_{\mu L} u | 0 \rangle = \langle \pi^- | \bar{d}\gamma_\mu (1+\gamma_5) u | 0 \rangle = f_\pi \phi q_\mu .$$

The second matrix element describes the $\beta$-decay of the $\Lambda$-hyperon:

$$\langle p | \bar{u}\gamma_{\mu L} s | \Lambda \rangle = \langle p | \bar{u}\gamma_\mu (1+\gamma_5) s | \Lambda \rangle = \bar{u}_p (f_1 \gamma_\mu + g_1 \gamma_\mu \gamma_5) u_\Lambda .$$

The remaining terms in (8.2.28) can be factorized in a similar way in order to calculate the cross-section of the process (8.2.27).

The decay probabilities for many other processes involving the charged currents have been calculated along the same lines. The theoretical results are in good agreement with the corresponding experimental data.

## 8.3 *CP* Violation

In discussing the processes associated with the weak interaction, we assumed the *CP* invariance to hold. However, there is experimental evidence for *CP* violation in neutral *K*-meson decays.

### 8.3.1 Neutral *K*-Meson Decay

With regard to the weak interaction, the neutral *K*-mesons are distinguished from all other elementary particles. Both the electric charge and the baryonic charge of $K^0$ and $\bar{K}^0$ are zero. They only differ in strangeness $S$ ($S = +1$ for $K^0$ and $S = -1$ for $\bar{K}^0$). Since the strangeness is not conserved in weak interactions, $K^0$ and $\bar{K}^0$ are identical with respect to them. In particular, they can convert into each other.

All particles can be classified into two groups. One group comprises particles (e.g. proton, electron, hyperon, etc.) which differ from their anti-particles by some strictly conserved quantum numbers (e.g., electric charge, baryon charge, etc.). The second group comprises particles identical with their anti-particles or truly neutral particles (e.g., photon or $\pi^0$-meson). Neutral $K^0$-mesons are on the border between these two particle groups: $K^0$ and $\bar{K}^0$ differ in strangeness but this difference is not absolute, existing with respect to strong interaction and disappearing with respect to weak interaction. While $K^0$ and $\bar{K}^0$ have a definite strangeness (or a definite hypercharge), they do not have a definite *CP* parity: under *CP* transformations, $K^0$ transforms into $\bar{K}^0$ rather than into $K^0$,

$$CP|K^0\rangle = |\bar{K}^0\rangle \,, \quad CP|\bar{K}^0\rangle = |K^0\rangle \,. \tag{8.3.1}$$

A definite *CP* parity can be attributed to the linear combination of $K^0$ and $\bar{K}^0$:

$$K_1^0 = \frac{1}{\sqrt{2}}(K^0 + \bar{K}^0) \,, \quad K_2^0 = \frac{1}{\sqrt{2}}(K^0 - \bar{K}^0) \,. \tag{8.3.2}$$

Under *CP* transformation $K^0$ converts into $\bar{K}^0$, and $\bar{K}^0$ converts into $K^0$ so that the state $K_1^0$ will transform into itself while $K_2^0$ will reverse its sign:

$$CP|K_1^0\rangle = |K_1^0\rangle \,, \quad CP|K_2^0\rangle = -\,|K_2^0\rangle \,. \tag{8.3.3}$$

Hence, $K_1^0$ has positive and $K_2^0$ has negative *CP* parity, but neither $K_1^0$ nor $K_2^0$ possess a definite value of strangeness.

If the *CP* invariance holds, the $K_1^0$-meson may only decay into two $\pi$-mesons (whose total *CP* parity is positive):

$$K_1^0 \begin{array}{l} \nearrow \pi^+ + \pi^- \\ \searrow \pi^0 + \pi^0 \,. \end{array}$$

The $K_2^0$ may only decay into 3 $\pi$-mesons (their total *CP* parity being negative):

$$K_2^0 \begin{array}{l} \nearrow \pi^+ + \pi^- + \pi^0 \\ \searrow \pi^0 + \pi^0 + \pi^0 \,. \end{array}$$

Both kinds of decay have been observed. It turns out that the life time of the $K_1^0$-meson ($0.86 \times 10^{-10}$ s) is smaller than that of the $K_2^0$-meson ($5 \times 10^{-8}$ s).

For this reason, $K_1^0$ and $K_2^0$ are also referred to as the *short-lived* $K_S^0$-meson and the *long-lived* $K_L^0$-meson, respectively.

In the absence of *CP* invariance the $K_L^0$-meson can also decay into two $\pi$-mesons, and the $K_S^0$-meson into three $\pi$-mesons:

$$K_L^0 \begin{array}{l}\nearrow \pi^+ + \pi^- \\ \searrow \pi^0 + \pi^0\end{array}, \quad K_S^0 \begin{array}{l}\nearrow \pi^+ + \pi^- + \pi^0 \\ \searrow \pi^0 + \pi^0 + \pi^0\end{array}. \tag{8.3.4}$$

Indeed, such decays have been experimentally observed which implies *CP* violation. The effect of *CP* violation is very small: for example, only 0.20% of all $K_L^0$-mesons do decay into $\pi^+ + \pi^-$; the ratio of the decay probabilities of $K_L^0 \to \pi^+ + \pi^-$ and $K_S^0 \to \pi^+ + \pi^-$ amounts to only $(3.69 \pm 0.15) \times 10^{-6}$.

In case of *P* invariance, a particle can be figuratively represented as a "nail", in case of *CP* invariance as a "screw" with a certain direction of screw thread, whose length is the same for the particle and the anti-particle, due to the fact that only a relative difference exists between them under *CP* invariance. If *CP* is violated, different "screws" have to be attributed to a particle and its anti-particle (both the sense of the screw thread and its length being different) to represent an absolute difference between the particles and anti-particles. As a consequence, the probabilities of, e.g., the following lepton decays of $K_L^0$-mesons with the production of particles or anti-particles will be different:

$$K_L^0 \begin{array}{l}\nearrow \pi^+ + e^- + \bar{\nu}_e \\ \searrow \pi^- + e^+ + \nu_e\end{array}, \quad K_L^0 \begin{array}{l}\nearrow \pi^+ + \mu^- + \bar{\nu}_\mu \\ \searrow \pi^- + \mu^+ + \nu_\mu\end{array}. \tag{8.3.5}$$

That is to say, in the case of *CP* non-invariance the probabilities of creating particles with opposite charge (e.g., $e^-$ or $e^+$, $\mu^-$ or $\mu^+$) will be different (charge asymmetry). Charge asymmetry in the mentioned leptonic decays of $K_L^0$-mesons has been experimentally observed.

### 8.3.2 *CP* Violation and Gauge Models

The origin of *CP* violation is not clarified as yet, although a great number of models have been proposed. Interesting possibilities of accounting for *CP* violation are provided by unified gauge models. In case of *CP* violation, the Lagrangian must contain *complex coupling constants.* This follows from the fact that the initial Lagrangian will differ from that obtained from it by Hermitian conjugation and *CP* transformation only if they contain complex coupling constants. In any unified gauge model, there are two interaction Lagrangians, one of them describing the interaction of fermions with gauge fields and the other the interaction of fermions with the Higgs fields (the Yukawa term). One can construct such unified gauge models in which the coupling constants entering the Lagrangians are complex.

1) We first consider the *fermion-gauge field interaction* Lagrangian. This includes charged currents containing mixed quarks (Sect. 8.2). If the mixing coefficients are complex, then the fermion-vector boson coupling constants

can become complex as well, although real coupling constants enter the initial Lagrangian. As already demonstrated in the preceding section, in a model with two quark doublets the mixing coefficients cannot be complex, implying *CP* conservation in such a model.

On the contrary, in a model with three quark doublets the mixing coefficients will be complex owing to the phase $\delta$ in (8.2.8). This will lead to complex coupling constants. In other words, in this model it is the phase $\delta$ which is responsible for *CP* violation. The Lagrangian for the model is obtained by substituting (8.2.8) into (6.3.6) which, in turn, is substituted into (6.3.14). This Lagrangian describes all the effects of *CP* violation. It contains the parameters $\theta_1$, $\theta_2$, $\theta_3$, and $\delta$. The most straightforward way of determining these parameters would be to study weak decays of hadrons containing the $b$ and $t$ quarks as constituents. Such a study, though, is in its beginnings, and therefore one attempts to deduce information on the parameters $\theta_i$ and $\delta$ from the data on the decays of nucleons, muon and strange particles. Unfortunately, the accuracy of determination of the parameters is not high enough so that effects due to *CP* violation cannot be calculated with a good accuracy. The magnitudes of these effects are listed in Table 8.2.

Let us give some explanations concerning Table 8.2. The complex quantities $\eta_{+-}$, $\eta_{00}$, and $\eta_{+-0}$ contained in it characterize the decays of $K^0_L$-mesons and $K^0_S$-mesons determined by

$$\eta_{+-}=\frac{\langle\pi^+\pi^-|K^0_L\rangle}{\langle\pi^+\pi^-|K^0_S\rangle}, \quad \eta_{00}=\frac{\langle\pi^0\pi^0|K^0_L\rangle}{\langle\pi^0\pi^0|K^0_S\rangle}, \quad \eta_{+-0}=\frac{\langle\pi^+\pi^-\pi^0|K^0_S\rangle}{\langle\pi^+\pi^-\pi^0|K^0_L\rangle}. \tag{8.3.6}$$

As follows from the requirement of relativistic invariance, a spinor particle (in particular, the neutron) can have an intrinsic electric dipole moment $d_n\sigma$,

**Table 8.2.** *CP* violation effects in various models

| Quantity \ Model | three quark doublets | $SU_{2L}\times SU_{2R}\times U_1$ | two Higgs doublets | three Higgs doublets |
|---|---|---|---|---|
| $\left\lvert\frac{\eta_{+-}-\eta_{00}}{\eta_{+-}}\right\rvert$ | $0.002-0.02$ | 0 | $\sim 0$ | 0.05 |
| $\left\lvert\frac{\eta_{+-0}}{\eta_{+-}}-1\right\rvert$ | smaller than 0.01 | O(1) | $\sim 0$ | $\sim 0$ |
| Dipole moment of neutron | $\sim 10^{-33}$ | $10^{-29}$ | $10^{-24}-10^{-30}$ | $2\times 10^{-25}$ |
| Dipole moment of $\Lambda^0$-hyperon | | | | $5\times 10^{-23}$ |
| *CP* violation in the $D^0$-decay | minor | | may be severe | minor |
| *CP* violation in the $B^0$-decay | minor to medium | | may be severe | minor |

where $\sigma$ denotes the spin vector, $d_n = el$, $e$ being the elementary electric charge and $l$ the dipole length. From present-day experimental data it is known that $l \leqslant 10^{-24}$ cm. The interaction potential $U_{\mathrm{I}} = d_n(\sigma E)$ changes its sign under $CP$ transformations, i.e., $U_{\mathrm{I}}$ is not $CP$ invariant. It can be concluded that the intrinsic electric dipole moment of a spinor particle is non-zero when $CP$ invariance is broken and is zero otherwise. That is to say, studying the electric dipole moment can provide information on $CP$ invariance. As seen from Table 8.2, the model with three quark doublets predicts a value of the electric dipole moment of neutron which does not contradict the experimental data. Other spinor particles, e.g., $\Lambda^0$-hyperons, likewise can have an intrinsic electric dipole moment.

$CP$ invariance-breaking decays will be characteristic not only of $K^0$-mesons but also of other neutral mesons, such as (charmed) $D^0$ and (heavy $b$-quarks containing) $B^0$. The investigation of these decays is just beginning.

Complex coupling constants in the interaction Lagrangian of quarks with gauge fields, i.e., $CP$ violation, can also be accounted for by a model based on the group $SU_{2\mathrm{L}} \times SU_{2\mathrm{R}} \times U_1$. This model involves both left-handed and right-handed currents. $CP$ violation originates from the phase difference $\phi$ between the invariant part $\mathscr{L}_{\mathrm{c}}$ and the non-invariant part $\mathscr{L}_{\mathrm{n}}$ of the effective weak interaction Lagrangian $\mathscr{L}_{\mathrm{I}}^{\mathrm{W}}$ (these parts of the Lagrangian are conserved and not conserved, respectively, under space inversion):

$$\mathscr{L}_{\mathrm{I}}^{\mathrm{W}} = \mathrm{e}^{\mathrm{i}\phi} \mathscr{L}_{\mathrm{c}} + \mathscr{L}_{\mathrm{n}} .$$

The $CP$ violation effects associated with this model are indicated in Table 8.2.

2) Next we pass to the *fermion-Higgs field interaction* Lagrangians (the Yukawa terms). Complex coupling constants can appear in these Lagrangians as a result of increasing the number of Higgs field doublets. We shall demonstrate the underlying mechanism by considering as an example a model based on the group $SU_2$.

Let us take a triplet

$$\psi = \begin{pmatrix} p \\ n \\ b \end{pmatrix} \tag{8.3.7}$$

which contains both left-handed and right-handed particles. The $CP$ invariant interaction Lagrangian of the fermion triplet (8.3.7) with a triplet of Higgs fields $\phi$ is written as

$$L = m \bar{\psi} \psi + \mathrm{i} h (\bar{\psi} \gamma_5 \omega \psi) \phi . \tag{8.3.8}$$

Let us consider only that component of the Higgs fields which corresponds to the neutral field $\phi^0$. Denoting the vacuum-expectation-value of the field $\phi^0$ as $\phi_{\mathrm{v}}^0$ we introduce a new field

$$\phi^{0'} = \phi^0 - \phi_{\mathrm{v}}^0 . \tag{8.3.9}$$

This brings the Lagrangian (8.3.8) to the form

$$L=\bar{p}(m+\mathrm{i}m'\gamma_5)p+m\bar{n}n+\bar{b}(m-\mathrm{i}m'\gamma_5)b+\mathrm{i}h(\bar{p}\gamma_5 p)\phi^{0'} -\mathrm{i}h(\bar{b}\gamma_5 b)\phi^{0'}\,, \quad m'=h\phi^0_{\mathrm{v}}\,. \tag{8.3.10}$$

To obtain the mass terms in the canonical form, we carry out a unitary transformation

$$p\to \mathrm{e}^{-\mathrm{i}\varepsilon\gamma_5/2}p\,,\qquad b\to \mathrm{e}^{\mathrm{i}\varepsilon\gamma_5/2}b\,,\qquad \tan\varepsilon=\frac{m'}{m}\,.$$

As a result, we get for the interaction Lagrangian of the fermions with the field $\phi^{0'}$

$$L=h\{[\bar{p}(\sin\varepsilon+\mathrm{i}\gamma_5\cos\varepsilon)p]\,\phi^{0'}+[\bar{b}(\sin\varepsilon-\mathrm{i}\gamma_5\cos\varepsilon)b]\,\phi^{0'}\}\,.$$

Although the initial Lagrangian (8.3.8) is invariant with respect to *CP* transformations, the Lagrangian obtained contains complex coupling constants, i.e., is *CP* non-invariant. *CP* violation occurs due to spontaneous symmetry-breaking.

Models with two quark doublets to which two or three Higgs' field doublets are added have also been analyzed. The resulting magnitudes of the *CP* violation effects in these models are given in Table 8.2.

Thus, as seen from Table 8.2, the unified gauge models considered do lead to *CP* violation. However, to date they only provide an estimate of the *CP* violation effect.

# 9. Higher Orders in Perturbation Theory

In the previous chapters, our consideration of the processes caused by the electromagnetic and the weak interactions was confined to the first non-vanishing order of perturbation theory. The present chapter deals with higher orders of perturbation theory.

## 9.1 Divergences of Matrix Elements

Let us consider, for example, the process of electron scattering on the Coulomb field of a nucleus with an electric charge $Z$:

$$e^{-}+Z\rightarrow e^{-}+Z\,. \tag{9.1.1}$$

The Feynman diagrams for this reaction in the third order of perturbation theory are presented in Fig. 9.1. Let us single out some elementary diagrams (Fig. 9.2) among them and analyze in some detail the corresponding matrix elements:

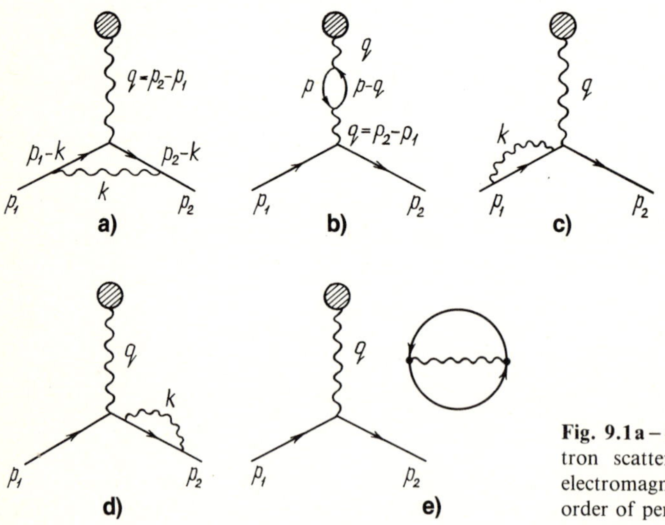


**Fig. 9.1a – e.** Diagrams for electron scattering by an external electromagnetic field in the third order of perturbation theory

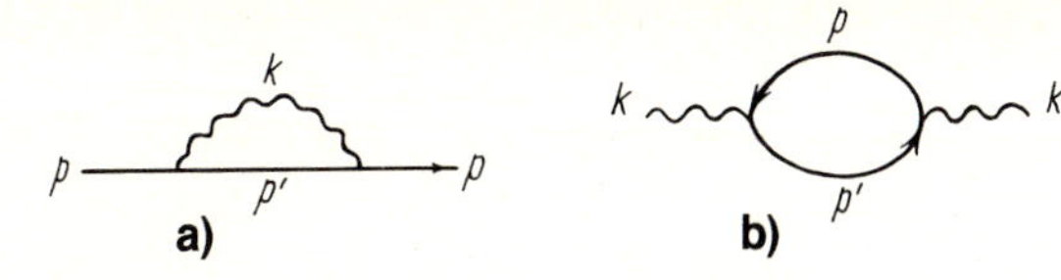


Fig. 9.2 a – c. Diagrams for (a) the electron self-energy; (b) the photon self-energy; (c) the interaction vertex

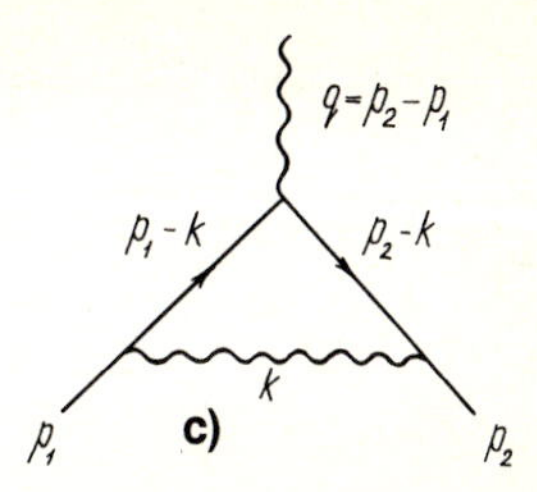


$$M = \mathscr{G}(p)\Sigma^{(2)}(p)\,\mathscr{G}(p)\,, \quad \Sigma^{(2)}(p) = \frac{e^2}{(2\pi)^3}\int dk\,\gamma_\nu \frac{(\not{p}+\not{k})+m}{(p+k)^2-m^2}\gamma_\nu \frac{1}{k^2}\,, \tag{9.1.2}$$

$$M_{\mu\nu} = \mathscr{D}_{\mu\varrho}(k)P^{(2)}_{\varrho\sigma}(k)\,\mathscr{D}_{\sigma\nu}(k)\,,$$
$$P^{(2)}_{\nu\mu}(k) = \frac{-e^2}{(2\pi)^3}\int dp\,\mathrm{Tr}\left\{\gamma_\nu \frac{(\not{p}+m)}{p^2-m^2}\gamma_\mu \frac{(\not{p}+\not{k})+m}{(p+k)^2-m^2}\right\}, \tag{9.1.3}$$

$$M = \bar{u}(p_2)\Lambda^{(3)}(p_1,p_2,q)\,u(p_1), \qquad \Lambda^{(3)}(p_1,p_2,q)$$
$$= \frac{e^3}{(2\pi)^4}\int dk\,\gamma_\nu \frac{(\not{p}_2-\not{k})+m}{(p_2-k)^2-m^2}A^e(q)\frac{(\not{p}_1-\not{k})+m}{(p_1-k)^2-m^2}\gamma_\nu \frac{1}{k^2}\,. \tag{9.1.4}$$

In (9.1.2), the fifth power of the momentum $k$, in which the integration measure is included, appears in the numerator, while the denominator contains its fourth power. The integral (9.1.2) will thus be proportional to $k$ and, since the integration is performed from 0 to $\infty$, the matrix element (9.1.2) becomes *infinite*, or *divergent*, at the upper limit $k \to \infty$. The divergence is linear in this case. Likewise, one can prove that the matrix elements (9.1.3) and (9.1.4) are quadratically and logarithmically divergent, respectively.

The graphs shown in Fig. 9.2 a – c will also enter the Feynman diagrams for higher orders of perturbation theory for the process (9.1.1). Hence, the matrix elements of the process (9.1.1) turn out to be *divergent* in higher orders of perturbation theory.

The same graphs of Fig. 9.2 a – c will be contained in higher-order Feynman diagrams for other processes as well. Consequently, the matrix elements of higher orders in perturbation theory will be divergent for all processes.

Two problems arise in connection with the occurrence of divergences in the matrix elements of higher orders in perturbation theory: (i) to find all possible types of diverging matrix elements and (ii) to elucidate whether the number of these types of divergences depends on the order of perturbation theory.

The solution to these problems depends exclusively on the type of the interaction. Let us show that, e.g. in the case of quantum electrodynamics, the number of divergent matrix element types is finite and does not depend on the order of perturbation theory.

Let us consider an arbitrary one-particle-irreducible diagram (Sect. 5.1) of quantum electrodynamics in the $n$th order of perturbation theory to find out what kinds of divergences in the matrix element are possible. In the case of one-particle-irreducible diagrams, the integrand of the matrix element does not decompose into single factors containing variables not related to each other. Therefore, the superficial divergence of the integral in the large momentum region is determined by the number $\omega$ defined as the difference between the powers of the momentum in the numerator and in the denominator; for $\omega < 0$, the integral is convergent, for $\omega \geqslant 0$ it is superficially divergent. The number $\omega$ is called the *superficial divergence index* of the diagram.

Let $F$ and $B$ be the total numbers of electron and photon lines, $F_e$ and $B_e$ those of the external electron and photon lines, and $F_i$ and $B_i$ those of the internal electron and photon lines, respectively. Let us find the quantity $\omega$. First count the number of independent integration variables. The number of momenta over which the integration is carried out is equal to the number of internal lines, i.e. to $F_i + B_i$. They are not, however, independent, since three momenta associated with each of the $n$ vertices are interrelated by the momentum conservation law. One of these conservation laws, namely that related to the process as a whole, refers to the external lines so that the internal momenta are subject to a total of $n-1$ constraints. Accordingly, the number of independent internal momenta is $F_i + B_i - (n-1)$, the number of independent integration variables being $4(F_i + B_i - n + 1)$. The propagator associated with an internal electron line contains the first power of the momentum in the denominator; the propagator associated with an internal photon line the second power. This gives rise to the appearance of the power $F_i + 2B_i$ of the momentum in the denominator of the integrand. Hence,

$$\omega = 4(F_i + B_i - n + 1) - F_i - 2B_i = 3F_i + 2B_i - 4(n-1) . \tag{9.1.5}$$

Let us now express the quantity $\omega$ in terms of the number of the external lines only. For this, we take into account that in the case of electrodynamics two electron lines and one photon line meet at each vertex. The number of the ends of electron lines is thus twice the number of vertices. Because every internal electron line has two ends and every external one has one end it follows that

$$2F_i + F_e = 2n . \tag{9.1.6}$$

Similarly, one photon line ends at each vertex so that the number of photon-line ends equals that of the vertices:

$$2B_i + B_e = n . \tag{9.1.7}$$

Substituting (9.1.6, 7) into (9.1.5) we arrive at an expression for $\omega$ in terms of the number of external lines

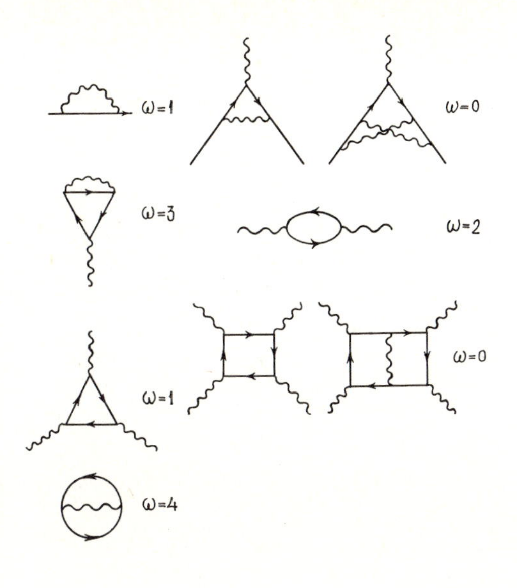


**Fig. 9.3.** Types of divergent diagrams of quantum electrodynamics

$$\omega = 4 - \tfrac{3}{2}F_e - B_e\,. \tag{9.1.8}$$

This expression enables us to count *all possible* divergent matrix elements corresponding to the one-particle-irreducible diagrams of quantum electrodynamics. It should be pointed out that an electron line cannot terminate, implying that $F_e$ is always an even number. Possible types of divergences of the matrix elements of quantum electrodynamics are depicted in Fig. 9.3.

One can see that in the case of quantum electrodynamics the number of the types of divergences is finite. According to (9.1.8), $\omega$ does not depend on $n$, i.e. the number of divergence types is independent of the order of perturbation theory.

For a gauge field, in the absence of a matter fields, we have in the $\alpha$-gauge

$$\omega = 4 - N_A - N_c\,, \tag{9.1.9}$$

where $N_A$ and $N_c$ are the numbers of external lines corresponding to the gauge field and to the ghost field $c$. Since the interaction vertex of the gauge field with the ghost field $c$ contains a derivative, the integration by parts will bring this derivative into an external line of the diagram. This gives rise to an external momentum and to lowering the divergence index of the diagram. The types of divergent diagrams, taking into account the latter circumstance, are shown in Fig. 9.4.

For a gauge field interacting with a spinor field

$$\omega = 4 - N_A - N_c - \tfrac{3}{2}N_\psi\,, \tag{9.1.10}$$

where $N_\psi$ is the number of external spinor lines. In this case the divergent diagrams shown in Fig. 9.5 appear in addition to those depicted in Fig. 9.4.

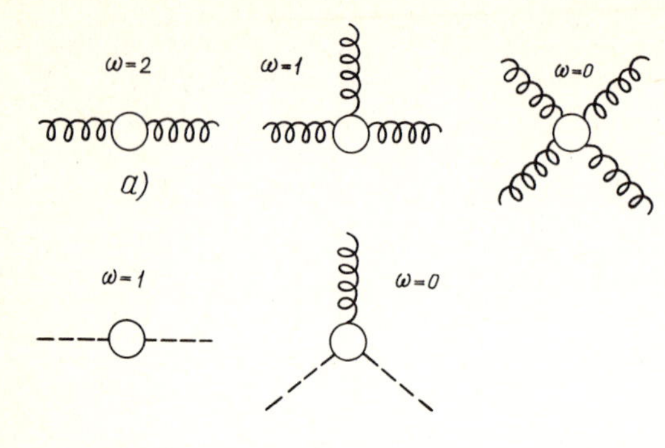


Fig. 9.4. Types of divergent diagrams for a gauge field

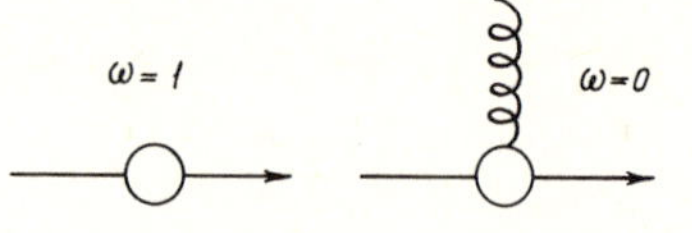


Fig. 9.5. Additional types of divergent diagrams for a gauge field interacting with a spinor field

Each of the examples considered is characterized by a finite number of divergent matrix elements, this number being independent of the order of perturbation theory. Theories possessing this property are called *renormalizable.*

Not all the interactions are renormalizable. Let us consider, for instance, the four-fermion interaction

$$L_{\mathrm{I}}^{\mathrm{F}} = \frac{G}{\sqrt{2}}\,\bar{\psi}_{\mathrm{e}}\Gamma\psi_{\mathrm{e}}\cdot\bar{\psi}_{\nu}\Gamma\psi_{\nu}\,,$$

where $\psi_{\mathrm{e}}$ is the electron wave function, $\psi_{\nu}$ is the neutrino wave function, $\Gamma$ is a matrix, and $G$ is the coupling constant. In this case, four spinor lines meet at one point, and the expression for $\omega$ is written as follows:

$$\omega = 2n + 4 - \tfrac{3}{2}F_{\mathrm{e}} - \tfrac{3}{2}L_{\mathrm{e}}\,,$$

where $F_{\mathrm{e}}$ and $L_{\mathrm{e}}$ denote the numbers of external electron and neutrino lines, respectively.

Contrary to quantum electrodynamics, for the four-fermion interaction the quantity $\omega$ depends on $n$ and in each order of perturbation theory there appear new divergences, i.e. the number of types of divergent matrix elements is infinite. Theories leading to an infinite number of divergences are called *non-renormalizable.* The four-fermion-interaction theory thus exemplifies a non-renormalizable theory.

## 9.2 Renormalization

### 9.2.1 The Renormalization Procedure

The divergences of renormalizable theories can be eliminated. A special procedure, referred to as *renormalization*, has been devised for this purpose.

The first step of this procedure is the *regularization* of divergent diagrams. It is awkward to perform calculations with divergent integrals and therefore it is instructive to temporarily modify the theory so as to make all the integrals finite. At the final stage, regularization is removed.

After regularization, one can proceed with eliminating the divergences. A special technique developed for that is called the *R-operation*. This operation enables one to obtain for the amplitudes expressions which remain finite after removing the regularization.

Additional problems arise for gauge theories. As a matter of fact, renormalization is equivalent to redefining the initial Lagrangian. Therefore, in the case of gauge fields, it is *firstly* desirable to choose such an intermediate regularization which does not violate the invariance under gauge transformations; *secondly*, performing the renormalization procedure should not violate the gauge-invariance; and, *thirdly*, the independence of the renormalized amplitude on the specific choice of the gauge should be proven. In proving the renormalizability of an amplitude, gauges are used which lead to Lagrangians containing non-physical fields (ghost fields $c$, goldstones, etc.) and to amplitudes which are not explicitly unitary. There are also gauges leading to Lagrangians not containing non-physical fields and to explicitly unitary amplitudes. The invariance of a renormalized amplitude with respect to the choice of gauge means that the amplitude is unitary as well.

### 9.2.2 Dimensional Regularization

There are several regularization methods: the Pauli-Villars procedure, the method of higher covariant derivatives, dimensional regularization etc. We shall consider in some detail the method of dimensional regularization.

The idea of *dimensional regularization* is based on the fact that the superficial divergence index of a diagram significantly depends on the dimension $n$ of the space, e.g. for QED

$$\omega = \sum_{l=1}^{L} (r_l + n - 2) - n(m-1) , \tag{9.2.1}$$

where the summation over all internal lines of the diagram is carried out; $L$ is the number of the internal lines, $r_l$ is the order of the polynomial corresponding to an internal line, and $m$ is the number of vertices. Therefore, integrals divergent in a four-dimensional space may prove to be convergent in a space of smaller dimension. The number $n$ can be thought of as being not only positive integer but also of being non-integer and even complex; the last two cases will correspond to spaces of "non-integer" and "complex" dimensions, respectively.

Before starting concrete calculations, it is necessary to formulate the rules for treating tensor quantities and the $\gamma$-matrices in an $n$-dimensional space. By definition, tensor quantities in an $n$-dimensional space obey the following relations

$$g_{\mu\nu} p_\nu = p_\mu \, , \tag{9.2.2}$$

$$p_\mu p_\mu = p^2 \, , \tag{9.2.3}$$

$$g_{\mu\nu} g_{\nu\alpha} = g_{\mu\alpha} \, , \qquad g_{\mu\nu} g_{\mu\nu} = n \, . \tag{9.2.4}$$

For theories including fermions, the matrices $\gamma_\mu$ are introduced which possess the properties

$$\gamma_\mu \gamma_\nu + \gamma_\nu \gamma_\mu = 2 g_{\mu\nu} I \, , \tag{9.2.5}$$

$$\mathrm{Tr}\{\gamma_\mu \gamma_\nu\} = 4 g_{\mu\nu} \, , \tag{9.2.6}$$

$$\mathrm{Tr}\{\gamma_\mu \gamma_\alpha \gamma_\nu \gamma_\beta\} = 4(g_{\mu\alpha} g_{\nu\beta} - g_{\mu\nu} g_{\alpha\beta} + g_{\mu\beta} g_{\alpha\nu}) \, , \tag{9.2.7}$$

$$\gamma_\mu \not{p} \gamma_\mu = 2\left(1 - \frac{n}{2}\right) \not{p} \, , \tag{9.2.8}$$

$$\gamma_\mu \not{p} \not{q} \gamma_\mu = 4pq + (n-4) \not{p} \not{q} \, , \tag{9.2.9}$$

where $I$ is the unit matrix and $\not{a} = a_\mu \gamma_\mu$.

It should be noted, that the usual definition of the matrix $\gamma_5 = \frac{\mathrm{i}}{4!} \varepsilon_{\mu\nu\varrho\sigma} \times \gamma_\mu \gamma_\nu \gamma_\varrho \gamma_\sigma$ is not applicable in a space of arbitrary dimension, because the completely antisymmetric tensor $\varepsilon_{\mu\nu\varrho\sigma}$ is defined only in the four-dimensional space. For this reason, dimensional regularization in this simple form is not applicable to the theories in which the matrix $\gamma_5$ is involved and a more refined treatment is needed.

Let us elucidate the main steps of dimensional regularization by considering the integral

$$I = \int d^n k \frac{1}{(k^2 - m^2)[(k-p)^2 - m^2]} \, , \tag{9.2.10}$$

which comes from the diagram presented in Fig. 5.5a. This integral is divergent for $n = 4$ and convergent for $n < 4$. Let us calculate the integral (9.2.10), assuming that $n < 4$.

For integrating over $k$ we use a parametrization which permits an explicit calculation, e.g. the Feynman parametrization

$$\frac{1}{ab} = \int_0^1 \frac{dx}{[ax + b(1-x)]^2} \, . \tag{9.2.11}$$

This yields, upon the change of variables $k - p(1-x) \to k$:

$$I = \int_0^1 dx \int d^n k \frac{1}{[k^2 + p^2 x(1-x) - m^2]^2} \, . \tag{9.2.12}$$

By rotating the integration contour by 90° and changing the variable $k_0 \to \mathrm{i} k_0$ we get the integral over the $n$-dimensional Euclidean space

$$I = \mathrm{i}\int_0^1 dx \int d^n k \frac{1}{[k^2+m^2-p^2x(1-x)]^2} . \tag{9.2.13}$$

We further make use of the integral

$$\int \frac{d^n k}{(k^2+M^2)^\alpha} = \pi^{n/2} M^{2(n/2-\alpha)} \frac{\Gamma(\alpha-n/2)}{\Gamma(\alpha)} , \tag{9.2.14}$$

where $\Gamma(x)$ is the gamma function. As a result, the expression for the regularized integral follows

$$I = \mathrm{i}\pi^{n/2}\Gamma\left(2-\frac{n}{2}\right)\int_0^1 dx [m^2-p^2x(1-x)]^{n/2-2} . \tag{9.2.15}$$

As expected, the removal of regularization (i.e., setting $n=4$) leads to a divergence of the integral (9.2.15), since the function $\Gamma(2-n/2)$ has a pole at this point.

More complicated divergent integrals can be regularized and calculated along the same lines.

Two important properties of dimensional regularization should be emphasized. Firstly, it does not violate the gauge-invariance of the theory, because passing from a four-dimensional space to a space of some other dimension does not affect its gauge properties. Due to this, dimensional regularization is well suited for treating gauge theories. Secondly, dimensional regularization allows a change (shift) of integration variables, $k \to k-p$, which is, in general, not allowed in divergent integrals.

### 9.2.3 The *R*-Operation

Let us explain the idea underlying renormalization by considering, as an example, the integral (9.2.10). Upon its regularization we arrived at (9.2.15). Removing the regularization, i.e., setting $n=4$ in (9.2.15), gives rise to a pole. This corresponds to the divergence of the initial integral over the four-dimensional space. Expanding the regularized integral (9.2.15) in Laurent series with respect to $n$ in the vicinity of the point $n/2=2$, we obtain

$$I = \frac{-\mathrm{i}\pi^2}{\frac{n}{2}-2} + C - \mathrm{i}\pi^2 \int_0^1 dx \ln[m^2-p^2x(1-x)] + O\left(\frac{n}{2}-2\right) , \tag{9.2.16}$$

where $C$ is a finite constant.

We then, in general, expand (9.2.16) in Taylor's series with respect to the square of the external momentum, $p^2$. Let us choose as the expansion point (also referred to as the *renormalization point*) the point $p^2=\lambda^2$. Subtracting $I(p^2=\lambda^2)$ from (9.2.16), we arrive at an expression which does not contain divergences:

$$I_{\mathrm{R}} = -\mathrm{i}\pi^2 \int_0^1 dx \ln \frac{m^2 - p^2 x(1-x)}{m^2 - \lambda^2 x(1-x)} . \tag{9.2.17}$$

This is the renormalized expression for the integral (9.2.10). The choice of the renormalization point is arbitrary. By taking a renormalization point other than $p^2 = \lambda^2$, an expression will be obtained which differs from (9.2.17) by a finite polynomial in $p^2$.

In a similar way, the renormalization of other integrals can be performed.

As seen, the renormalization procedure consists in subtracting from divergent integrals several first terms of the Taylor series expansion with respect to external momenta. The number of the subtracted terms (i.e., the minimum number required to ensure convergence of the integral) is determined by the actual divergence of the diagram. Due to the arbitrariness of the choice of the renormalization point, the subdivision of the integral into a finite part and an infinite part is not unique.

The divergent integral (9.2.10) above contains only one integration over the momentum, and renormalization is performed easily enough. For more complicated integrals, including integrations over many momenta, the simple renormalization method discussed is insufficient. In this case, the *R-operation* developed by Bogoliubov, Parasiuk, Hepp and Zimmermann is applied.

The essence of the $R$-operation is as follows. Let us consider an arbitrary diagram. It corresponds to an integral over many internal momenta. This integral may diverge not only when all momenta simultaneously go to infinity but also when a part of the momenta do, while the rest of them are fixed. In this case, the diagram is said to contain divergent subdiagrams. Recursive subtraction is applied for such diagrams: first the divergences of the subdiagrams and then those of the diagram as a whole are eliminated.

The renormalization procedure can also be formulated in a different language. The point is that replacing divergent integrals by the renormalized ones is equivalent to including some additional terms in the initial Lagrangian. These are called the *counter-terms*. Therefore, renormalization can be carried out by introducing counter-terms in the Lagrangian. Thus, to renormalize a theory, it is sufficient to follow the prescriptions of the $R$-operation.

As already mentioned, the renormalization of gauge theories brings about additional problems associated with the requirement that the gauge invariance of the theory be not violated. In this case, renormalization is also carried out by means of the $R$-operation. To prove that the gauge-invariance is not violated by that, it is convenient to utilize the *generalized Ward identities*. Therefore, we shall first dwell on these identities which have been found by Slavnov and Taylor and are also called the *Slavnov-Taylor identities*.

### 9.2.4 Generalized Ward Identities

Let us obtain the generalized Ward identities for a gauge field. We proceed from the generating functional $W(J)$ for the Green's function of the gauge field in the generalized Lorentz gauge (4.2.60),

$$W(J) = \int \prod_x \mathscr{D}A_\mu^k \mathscr{D}a^k \delta(\partial_\mu A_\mu^k - a^k) \Delta_a(A_\mu^k) \times \exp\left\{ \mathrm{i}I - \mathrm{i}\int dx \left[ \frac{1}{2\alpha} (a^k)^2 + J_\mu^k A_\mu^k \right] \right\}, \quad (9.2.18)$$

where $I$ is the gauge-invariant action.

The idea of deriving the generalized Ward identities is as follows. We change the gauge in the generating functional (9.2.18) [cf. (9.2.19) below]; as a result, it is rewritten in the form of (9.2.25) containing the function $\varrho^k(x)$. Since (9.2.25) is obtained from (9.2.18) by means of identical transformations, the expressions for the generating functional they give are equal. Next we find the functional derivatives with respect to $\varrho^k(x)$ of both sides of this equation. The functional (9.2.18) does not depend on $\varrho^k(x)$, i.e., its derivative with respect to $\varrho^k(x)$ is zero. Differentiating the functional (9.2.25) with respect to $\varrho^k(x)$ then leads to the set of generalized Ward identities (9.2.27). The gauge-invariance of the action $I$ is essentially used in this derivation. Now we pass to the calculations along the lines sketched.

Changing to the gauge

$$\partial_\mu A_\mu^{k\omega} - a^k(x) - \varrho^k(x) = 0 \quad (9.2.19)$$

in (9.2.18) is performed in the same way as before (Sect. 4.2). For that we introduce the gauge-invariant functional $\tilde{\Delta}_a(A_\mu^k)$ which is defined by

$$\tilde{\Delta}_a(A_\mu^k) \int \prod_x \mathscr{D}\omega(x) \delta[\partial_\mu A_\mu^{k\omega} - a^k(x) - \varrho^k(x)] = 1 . \quad (9.2.20)$$

Substituting (9.2.20) into (9.2.18) yields

$$W = \int \prod_{x,k} \mathscr{D}A_\mu^k \mathscr{D}a^k \mathscr{D}\omega \delta(\partial_\mu A_\mu^k - a^k) \delta(\partial_\mu A_\mu^{k\omega} - a^k - \varrho^k) \Delta_a(A_\mu^k) \times \tilde{\Delta}_a(A_\mu^k) \exp\left\{ \mathrm{i}I - \mathrm{i}\int dx \left[ \frac{1}{2\alpha} (a^k)^2 + J_\mu^k A_\mu^k \right] \right\} . \quad (9.2.21)$$

Changing the variables $A_\mu^k \to A_\mu^{k\omega}$, $\omega \to \omega^{-1}$ brings (9.2.21) to the form

$$W(J) = \int \prod_{x,k} \mathscr{D}A_\mu^k(x) \mathscr{D}a^k(x) \mathscr{D}\omega(x) \Delta_a(A_\mu^k) \tilde{\Delta}_a(A_\mu^k) \delta(\partial_\mu A_\mu^{k\omega} - a^k) \times \delta(\partial_\mu A_\mu^k - a^k - \varrho^k) \exp\left[ \mathrm{i}I - \mathrm{i}\int dx \left( \frac{1}{2\alpha} (a^k)^2 + J_\mu^k A_\mu^{k\omega} \right) \right] . \quad (9.2.22)$$

To perform the integration over $\omega(x)$ and $a^k(x)$, we make use of the fact that the term $J_\mu^k A_\mu^{k\omega}$ in (9.2.22) can be represented, due to the presence of the two

$\delta$-functionals, as $J_\mu^k A_\mu^{k\omega_0}$, where $\omega_0(x; A_\mu^k, \varrho^k)$ is the solution of the following set of equations

$$\partial_\mu A_\mu^{k\omega_0} - a^k = 0\,, \qquad \partial_\mu A_\mu^k - a^k - \varrho^k = 0\,. \tag{9.2.23}$$

Now we take into account (4.2.61) and use the relation

$$\tilde{\Delta}_a(A_\mu^k) = \det M_a = \det(\Box\,\delta_{kl} + g f_{klm} A_\mu^m \partial_\mu + g f_{klm} \partial_\mu A_\mu^m)$$

which is fulfilled on the surface $\partial_\mu A_\mu^k = a^k + \varrho^k$. The integration in (9.2.22) then yields

$$W(J) = \int \prod_{x,k} \mathscr{D} A_\mu^k(x) \det M_a \times \exp\left\{ \mathrm{i} I - \mathrm{i} \int dx \left[ \frac{1}{2\alpha} (\partial_\mu A_\mu^k - \varrho^k)^2 + J_\mu^k A_\mu^{k\omega_0} \right] \right\}. \tag{9.2.24}$$

Let us find the explicit form of $A_\mu^{k\omega_0}(x)$ by considering $\varrho^k(x)$ as infinitesimal functions. The set of equations (9.2.23) is then rewritten

$$\partial_\mu A_\mu^{k\omega_0} - a^k = \partial_\mu A_\mu^k - a^k + M_a^{kl} \varepsilon_l^0 = 0\,, \qquad \partial_\mu A_\mu^k - a^k - \varrho^k = 0\,,$$

where $\varepsilon_k^0(x)$ denotes the parameters of the infinitesimal gauge transformation $\omega_0$. It follows

$$\begin{aligned} J_\mu^k(x) A_\mu^{k\omega_0}(x) &= J_\mu^k(x) A_\mu^k(x) + J_\mu^k(x) \tilde{\nabla}_\mu^{kl} \varepsilon_l^0(x) \\ &= J_\mu^k(x) A_\mu^k(x) - J_\mu^k(x) \tilde{\nabla}_\mu^{kl} \int dy (\partial_\mu \tilde{\nabla}_\mu^{ml})^{-1}(x,y) \varrho^m(y)\,, \end{aligned}$$

where

$$\tilde{\nabla}_\mu^{kl} \equiv \delta^{kl} \partial_\mu + g f_{klm} A_\mu^m\,, \qquad M_a^{kl} = \partial_\mu \tilde{\nabla}_\mu^{kl}\,.$$

Upon substituting $A_\mu^{k\omega_0}(x)$ into (9.2.24), the expression for the generating functional assumes the form

$$\begin{aligned} W(J) = \int \prod_{x,k} \mathscr{D} A_\mu^k(x) \det M_a & \times \exp\left\{ \mathrm{i} I - \mathrm{i} \int dx \left[ \frac{1}{2\alpha} (\partial_\mu A_\mu^k - \varrho^k)^2 + J_\mu^k A_\mu^k \right] \right. \\ & \left. + \mathrm{i} \int dx\, dy\, J_\mu^k(x) \tilde{\nabla}_{\mu x}^{kl} (M_a^{ml})^{-1}(x,y) \varrho^m(y) \right\}. \end{aligned} \tag{9.2.25}$$

Let us now differentiate both sides of (9.2.25) with respect to $\varrho^k(x)$. Because the initial functional (9.2.18), coinciding with (9.2.25), does not depend on $\varrho^k(x)$, we have

$$\left. \frac{\delta W(J)}{\delta \varrho^k} \right|_{\varrho^k = 0} = 0\,. \tag{9.2.26}$$

Substituting (9.2.25) into (9.2.26) we obtain, after the differentiation, the desired set of *generalized Ward identities* for gauge fields

$$\int \prod_{x,k} \mathscr{D} A_\mu^k \det M_a \left[ \frac{1}{\alpha} \partial_\mu A_\mu^k(y) + \int dz\, J_\mu^l(z)\, \tilde{\nabla}_{\mu z}^{lm} (M_a^{km})^{-1}(z,y) \right]$$

$$\times \exp\left\{ \mathrm{i}I - \mathrm{i}\int dx \left[ \frac{1}{2\alpha} (\partial_\mu A_\mu^k)^2 + J_\mu^k A_\mu^k \right] \right\} = 0\,, \qquad (9.2.27)$$

where

$$\int dz [M_a^{kl}(x,z)]^{-1} M_a^{km}(z,y) = \delta^{lm} \delta(x-y)\,.$$

This system of identities can be rewritten in terms of functional derivatives of the generating functional $W(J)$ with respect to the currents

$$\left\{ \frac{1}{\alpha} \partial_\mu \left[ \frac{1}{\mathrm{i}} \frac{\delta}{\delta J_\mu^k(x)} \right] + \int dy\, J_\mu^l(y) \left[ \tilde{\nabla}_\mu^{lm} \left( y, \frac{1}{\mathrm{i}} \frac{\delta}{\delta J_\mu^k(y)} \right) \right. \right.$$

$$\left. \left. \times (\partial_\mu \tilde{\nabla}_\mu^{km})^{-1} \left( y, x; \frac{1}{\mathrm{i}} \frac{\delta}{\delta J_\mu^k} \right) \right] \right\} W(J) \equiv 0\,. \qquad (9.2.28)$$

In a similar way, generalized Ward identities can be found for other cases, e.g., for gauge fields interacting with (spinor or scalar) fields of matter, for theories with spontaneously broken symmetry, etc.

The generalized Ward identities stem from the physical equivalence of various gauges. As seen from (9.2.28), they lead to certain relations between the Green's functions.

The generalized Ward identities can also be obtained in a different way. Again, we consider a gauge field. Its generating functional can be written as

$$W = \int \prod_{x,k} \mathscr{D} A_\mu^k \mathscr{D} \bar{c}^k \mathscr{D} c^k \exp\{\mathrm{i}\int dx [\mathscr{L}_{\mathrm{eff}}(x) + J_\mu^k A_\mu^k + \bar{c}^k \chi^k + \bar{\chi}^k c^k]\,, \qquad (9.2.29)$$

with

$$\mathscr{L}_{\mathrm{eff}}(x) = -\frac{1}{4} (F_{\mu\nu}^k)^2 - \frac{1}{2\alpha} (\partial_\mu A_\mu^k)^2 - \partial_\mu \bar{c}^k \partial_\mu c^k - g f_{klm} \partial_\mu \bar{c}^k A_\mu^m c^l\,. \qquad (9.2.30)$$

It can be proven directly that the effective Lagrangian $\mathscr{L}_{\mathrm{eff}}$ is invariant under the following transformations of the gauge field $A_\mu^k(x)$ and the ghost fields $c^k(x)$ (the *Becchi-Rouet-Stora transformation*):

$$\begin{aligned} A_\mu^k(x) &\to A_\mu^k(x) + (\nabla_\mu c^k(x))\,\varepsilon\,, \\ c^k(x) &\to c^k(x) - \tfrac{1}{2} f_{klm} c^l(x) c^m(x)\,\varepsilon\,, \qquad (9.2.31) \\ \bar{c}^k(x) &\to \bar{c}^k(x) + \frac{1}{\alpha} (\partial_\mu A_\mu^k(x))\,\varepsilon\,, \end{aligned}$$

where $\varepsilon$ is a parameter independent of $x$ and obeying $\varepsilon^2 = 0$,

$$\varepsilon c^k + c^k \varepsilon = 0\,, \quad \varepsilon \bar{c}^k + \bar{c}^k \varepsilon = 0\,, \quad [\varepsilon, A_\mu^k] = 0\,.$$

Substituting (9.2.31) into (9.2.29) and taking into account that the Jacobian of the transformations is equal to unity, we arrive at an expression for $W$ that contains $\varepsilon$. Since the functional (9.2.29) does not depend on $\varepsilon$, we get, after differentiating with respect to $\varepsilon$,

$$\begin{aligned} 0 = \int \prod_{x,k} \mathscr{D}A_\mu^k \, \mathscr{D}\bar{c}^k \, \mathscr{D}c^k \int dy \Big\{ J_\mu^k(y) [\tilde{\nabla}_\mu c(y)]^k \\ - \frac{1}{\alpha} \partial_\mu A_\mu^k(y) \chi^k(y) - \frac{1}{2} \bar{\chi}^k(y) f_{klm} c^l(y) c^m(y) \Big\} \\ \times \exp\{ i \int dx [\mathscr{L}_{\text{eff}}(x) + J_\mu^k A_\mu^k + \bar{c}^k \chi^k + \bar{\chi}^k c^k] \}\,. \end{aligned} \qquad (9.2.32)$$

Let us differentiate this equation with respect to $\bar{\chi}$ and $\chi$, set $\bar{\chi} = \chi = 0$, and subsequently integrate over $c$ and $\bar{c}$. As a result, the generalized Ward identities (9.2.27) will follow.

Likewise, the Ward identities for models more general than gauge fields can be obtained.

### 9.2.5 Renormalization of Gauge Fields

Now we go over to renormalization of gauge fields. Its specifics will be illustrated by the instance of the Yang-Mills field. According to (5.3.2, 3), the effective Lagrangian for this case, in the $\alpha$-gauge, reads

$$\mathscr{L}_{\text{eff}} = -\frac{1}{4} F_{\mu\nu}^k F_{\mu\nu}^k - \frac{1}{2\alpha} (\partial_\mu A_\mu^k)^2 - \partial_\mu \bar{c}^k \partial_\mu c^k - g f_{lmk} \partial_\mu \bar{c}^k \cdot A_\mu^m c^l\,. \qquad (9.2.33)$$

The types of divergent diagrams are depicted in Fig. 9.4.

Let us consider, as an example, the diagram a of Fig. 9.4. It diverges quadratically ($\omega = 2$). Following the general renormalization procedure, first it has to be regularized. For what follows, it is only important that regularization does not violate the invariance of the regularized action, and we shall not write down its explicit form assuming that regularization has been carried out.

Next, according to the $R$-operation, three terms ought to be subtracted from the vertex Green's function corresponding to the diagram under consideration. The most general expression for the terms to be subtracted, which fulfills the requirements of the relativistic invariance and the $SU_2$-invariance, is written

$$\Pi'^{ab}_{\mu\nu}(p) = \delta_{ab} [b_1 g_{\mu\nu} + b_2 p_\mu p_\nu + b_3 (p^2 g_{\mu\nu} - p_\mu p_\nu)]\,, \qquad (9.2.34)$$

$b_i$ denoting arbitrary constants.

Subtracting the polynomial (9.2.34) is equivalent to introducing into the Lagrangian the counter-terms

$$\mathscr{L}' = b_1(A^k_\mu)^2 + b_2(\partial_\mu A^k_\mu)^2 + \tfrac{1}{2}b_3(\partial_\nu A^k_\mu - \partial_\mu A^k_\nu)^2 . \tag{9.2.35}$$

The expressions for other divergent diagrams in Fig. 9.4 can be written in a similar way. To renormalize all divergent diagrams, eight subtraction terms $b_i$ have to be introduced.

The constants $b_i$ introduced according to the $R$-operation provide renormalization of the Green's function. It must be proven that this does not violate the gauge-invariance of the theory. To that end, we proceed as follows. We consider, besides (9.2.33), another effective Lagrangian, whose gauge-invariance is explicitly seen. In the case under consideration, it contains only three, rather than eight, constants $Z_i$ and, therefore, the possibility of renormalizing it is not obvious. In what follows we make use of the generalized Ward identities to prove that the three constants $Z_i$ entering (9.2.37) are sufficient for all Green's functions to be renormalized. This will demonstrate the feasibility of renormalization not violating the gauge-invariance of the theory. In other words, renormalization is performed by means of the $R$-operation, while the generalized Ward identities enable one to prove that the renormalized theory remains gauge-invariant.

1) First, we shall write down the gauge-invariant Lagrangian of the gauge field. The gauge-invariance of the Lagrangian is not violated by multiplying it by a constant and by re-defining the charge $g$. Therefore the general gauge-invariant form of the Lagrangian is

$$\mathscr{L}_{\text{eff}} = -\frac{Z_2}{4}(\partial_\mu A^k_\nu - \partial_\nu A^k_\mu - gZ_1Z_2^{-1}\varepsilon_{klm}A^l_\mu A^m_\nu)^2 . \tag{9.2.36}$$

Here $\tilde{g} = Z_1Z_2^{-1}g$ plays the role of the "charge" in the gauge transformation. For the theory to be self-consistent, this same parameter ought to appear in the definition of the covariant derivative. Hence, the coupling constant $g$ in the operator

$$M^{kl} = \delta^{kl}\Box + g\varepsilon_{klm}A^m_\mu\partial_\mu + g\varepsilon_{klm}\partial_\mu A^m_\mu$$

has to be replaced by $\tilde{g}$. Let us represent $\det M(\tilde{g})$ as an integral over the ghost fields

$$\det M(\tilde{g}) = \int\prod_x \mathscr{D}\bar{c}^k\,\mathscr{D}c^k \exp[\mathrm{i}\int dx\,\tilde{Z}_2\bar{c}^k\partial_\mu(\partial_\mu c^k + \tilde{Z}_1\tilde{Z}_2^{-1}g\varepsilon_{klm}A^m_\mu c^l)] , \tag{9.2.36'}$$

where $\tilde{Z}_2^{-1}\tilde{Z}_1 = Z_2^{-1}Z_1$. Then the general expression for the gauge-invariant effective Lagrangian takes the form

$$\begin{aligned}\mathscr{L}_{\text{eff}} = &-\tfrac{1}{4}Z_2(\partial_\mu A^k_\nu - \partial_\nu A^k_\mu - Z_1Z_2^{-1}g\varepsilon_{klm}A^l_\mu A^m_\nu)^2 \\ &-\frac{1}{2\alpha}(\partial_\mu A^k_\mu)^2 + \tilde{Z}_2\bar{c}^k\partial_\mu(\partial_\mu c^k + \tilde{Z}_1\tilde{Z}_2^{-1}g\varepsilon_{klm}A^m_\mu c^l) .\end{aligned} \tag{9.2.37}$$

Since the constants $Z_1$, $Z_2$, $\tilde{Z}_1$, and $\tilde{Z}_2$ are related by $\tilde{Z}_2^{-1}\tilde{Z}_1 = Z_2^{-1}Z_1$, the gauge-invariant Lagrangian (9.2.37) contains three independent constants $Z_i$.

2) Let us now use the generalized Ward identities to prove that the three constants $Z_i$ suffice for all divergences to be eliminated from the Green's function. This is possible, since the Ward identities impose certain constraints on the Green's functions leading to a reduction in the number of the constants $b_i$.

The proof of the finiteness of the Green's functions is carried out by induction. First one considers the Green's function of order $g^2$, then of order $g^3$, and finally one proves that the Green's functions of order $g^{n+1}$ will be finite, provided that those of order $g^n$ are finite. We shall not give a detailed proof here. Instead, we confine ourselves to exemplifying the mechanism of reduction of the number of constants $b_i$ by considering the Green's function corresponding to the diagram a of Fig. 9.4. In the second order in $g^2$, this Green's function, $D^{ab}_{\mu\nu}(p)$, is related to the Green's function $\prod^{mn}_{\mu\nu}(p)$ by

$$D^{ab}_{\alpha\beta}(p) = \mathscr{D}^{am}_{\alpha\mu}(p)\prod\nolimits^{mn}_{\mu\nu}(p)\,\mathscr{D}^{nb}_{\nu\beta}(p) + \mathscr{D}^{ab}_{\alpha\beta}(p)\,.$$

One can show that the substitution of this relation into the generalized Ward identities yields

$$p_\mu p_\nu \prod\nolimits^{ab}_{\mu\nu}(p) = 0\,. \tag{9.2.38}$$

3) In a gauge-invariant theory, this requirement has to be satisfied by the function (9.2.34), too. Multiplying the latter by $p_\mu p_\nu$ we recognize that the condition (9.2.38) is fulfilled, provided that $b_1 = b_2 = 0$. From a comparison of (9.2.37) and (9.2.35) it follows that $1 + b_3 \sim Z_2$. Consequently, the constants $b_1$ and $b_2$ in (9.2.34) are zero, and the only constant, $Z_2^0$, is sufficient for the divergence of the Green's function $D^{ab}_{\mu\nu}$ to be eliminated. (Here $Z_2^0$ denotes the value of the constant $Z_2$ in the second order of perturbation theory.)

Analogously, it can be shown that the three constants $Z_i$ provide the finiteness of all Green's functions in all orders of perturbation theory.

Thus, the renormalization of a gauge field is performed without violating the gauge-invariance.

Likewise, it is possible to renormalize more complicated gauge models, viz. gauge fields interacting with matter fields, theories with spontaneously broken symmetry, etc.

### 9.2.6 Unitarity of the Amplitude

It is evident on physical grounds that the renormalized amplitude should not depend on the choice of the gauge. That the amplitudes we used do meet this requirement can be proven in the following way. We proceed from the expression for the amplitude in terms of the generating functional. Let the renormalized generating functional be given in the Lorentz gauge. Due to the gauge-invariance of the renormalized Lagrangian we can pass to another gauge, e.g., to the one with a manifestly unitary amplitude, in the usual way

(Sect. 4.2). The resulting generating functional will differ from the initial one only in the form of the source terms. However, as can be shown, this difference disappears upon substitution of these functionals into the amplitude, i.e., the renormalized amplitude does not change upon a change of the gauge. Consequently, the renormalized amplitude is unitary.

### 9.2.7 Anomalies

In deriving the unitary renormalized amplitude, we essentially made use of the fact that there exists an intermediate regularization invariant under gauge transformations. However, theories are conceivable for which there is no invariant regularization. This pertains, e.g., to theories where the gauge transformations of the fermion fields contain the matrix $\gamma_5$. No consistent definition of the matrix $\gamma_5$ can be given for a space with arbitrary dimension which renders impossible the dimensional regularization. Other known regularizations are incompatible with the $\gamma_5$-invariance as well. Theories lacking invariant regularization have certain specifics.

An illustration of this is provided by a model invariant under the simplest group $U_1$. The model is described by the Lagrangian

$$\mathscr{L} = -\tfrac{1}{4}(\partial_\mu A_\nu - \partial_\nu A_\mu)^2 + \mathrm{i}\,\bar{\psi}\gamma_\mu(\partial_\mu - \mathrm{i}gA_\mu\gamma_5)\,\psi \tag{9.2.39}$$

invariant under the following Abelian transformations involving the matrix $\gamma_5$:

$$\begin{aligned} &A_\mu(x) \to A_\mu(x) + \partial_\mu \varepsilon(x)\,, \\ &\psi(x) \to \mathrm{e}^{\mathrm{i}g\gamma_5\varepsilon(x)}\,\psi(x)\,, \qquad \bar{\psi}(x) \to \bar{\psi}(x)\,\mathrm{e}^{\mathrm{i}g\gamma_5\varepsilon(x)}\,. \end{aligned} \tag{9.2.40}$$

The invariance of the Lagrangian (9.2.39) under the transformations (9.2.40) leads to the Ward identity (in the $\alpha$-gauge)

$$\left[\frac{1}{\alpha}\partial_\mu\left(\frac{\delta W}{\delta J_\mu(x)}\right) - W\partial_\mu J_\mu(x) + \mathrm{i}g\bar{\eta}(x)\,\gamma_5\frac{\delta W}{\delta\bar{\eta}(x)} + \mathrm{i}g\frac{\delta W}{\delta\eta(x)}\gamma_5\eta(x)\right] = 0\,, \tag{9.2.41}$$

where

$$W = \int \mathscr{D}A_\mu\,\mathscr{D}\bar{\psi}\,\mathscr{D}\psi \exp\left\{\mathrm{i}\int dx\left[\mathscr{L} + \frac{1}{2\alpha}(\partial_\mu A_\mu)^2 + J_\mu A_\mu + \bar{\eta}\psi + \bar{\psi}\eta\right]\right\}.$$

Because of the lack of invariant regularization, this identity contains divergent integrals.

In the theories considered so far, the generalized Ward identities were obtained from the invariant regularized action so that the renormalized Green's functions obey these identities. On the contrary, the renormalized Green's function of the model (9.2.39) does not obey the identity (9.2.41). To verify this, let us consider, for example, the three-point vertex Green's function (shown in Fig. 9.6a) which is determined by

$$\Gamma_{\mu\nu\alpha}(p,q)G_{\mu\mu'}(p)G_{\nu\nu'}(q)G_{\alpha\alpha'}(p+q)$$
$$= \int e^{ipx}e^{iqy}\left(\frac{-i\delta^3 W}{\delta J_{\mu'}(x)\delta J_{\nu'}(y)\delta J_{\alpha'}(0)}\right)dx\,dy\,. \tag{9.2.42}$$

Taking a particular case of the relations (9.2.41),

$$\partial^{x_1}_{\mu_1}\frac{\delta^n \ln W}{\delta J_{\mu_1}(x_1)\ldots\delta J_{\mu_n}(x_n)}\bigg|_{J=\eta=0} = 0\,, \qquad n>2\,, \tag{9.2.43}$$

we obtain

$$p_\mu \Gamma_{\mu\nu\alpha}(p,q) = q_\nu \Gamma_{\mu\nu\alpha}(p,q) = (p+q)_\alpha \Gamma_{\mu\nu\alpha}(p,q) = 0\,. \tag{9.2.44}$$

A direct calculation of the function $\Gamma_{\mu\nu\alpha}(p,q)$, taking into account that it is symmetric with respect to the arguments $(p,\mu)$, $(q,\nu)$, and $(-(p+q),\alpha)$, yields

$$i(p+q)_\alpha \Gamma_{\mu\nu\alpha}(p,q) = -\frac{g^3}{6\pi^2}\varepsilon^{\mu\nu\alpha\beta}p_\alpha q_\beta\,. \tag{9.2.45}$$

The divergence index of the diagram shown in Fig. 9.6a equals unity, i.e., the function is determined with the accuracy of a polynomial of first order in $p$ and $q$. The most general expression for the renormalized vertex function $\Gamma_{R\mu\nu\alpha}(p,q)$ thus has the form

$$\Gamma_{R\mu\nu\alpha}(p,q) = \Gamma_{\mu\nu\alpha}(p,q) + c_1\varepsilon_{\mu\nu\alpha\beta}p_\beta + c_2\varepsilon_{\mu\nu\alpha\beta}q_\beta\,, \tag{9.2.46}$$

where $\Gamma_{\mu\nu\alpha}(p,q)$ is the symmetric vertex function obeying (9.2.45). The function $\Gamma_{R\mu\nu\alpha}(p,q)$ also ought to be symmetric with respect to the arguments $(\mu,p)$, $(\nu,q)$, $(\alpha,-(p+q))$; this leads to

$$c_1 = c_2 = 0\,.$$

Consequently, there is no possible choice of local counter-terms to make the renormalized functions $\Gamma_{R\mu\nu\alpha}(p,q)$ satisfy the identity (9.2.44). The cases when the right-hand side of the identities (9.2.45) is non-zero are said to have *anomalies*. The occurrence of the anomalies implies that different gauges are non-equivalent and the model described by the Lagrangian (9.2.39) is inconsistent. For the model (9.2.39), the unitary renormalized $S$-matrix does not exist, at least in the framework of perturbation theory. This difficulty is inherent in all theories which are invariant under gauge transformations involving the matrix $\gamma_5$.

The difficulty just considered can be circumvented. For this purpose, we introduce in (9.2.39), besides the field $\psi$, another field $\psi'$ which interacts with the vector field in the same manner as $\psi$, differing only in the sign of the charge $g$:

$$\mathscr{L} = -\tfrac{1}{4}(\partial_\nu A_\mu - \partial_\mu A_\nu)^2 + i\bar\psi\gamma_\mu(\partial_\mu - igA_\mu\gamma_5)\psi + i\bar\psi'\gamma_\mu(\partial_\mu + igA_\mu\gamma_5)\psi'\,. \tag{9.2.47}$$

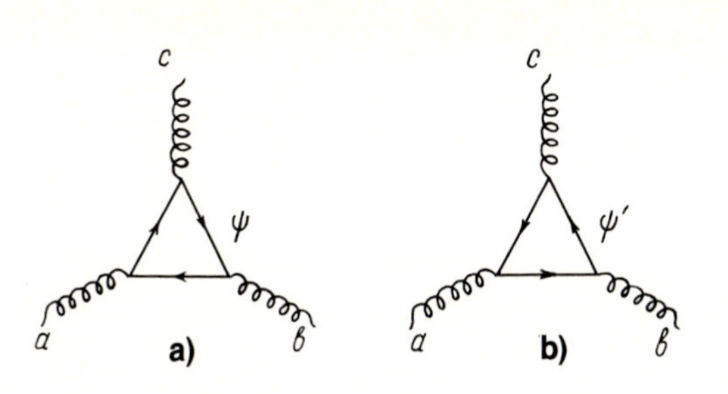


**Fig. 9.6a, b.** Anomalous triangle diagram (**a**) for the field $\psi$; (**b**) for the field $\psi'$

Then, beside the diagram in Fig. 9.6a, there appears an analogous diagram (Fig. 9.6b) whose internal lines correspond to the field $\psi'$. As follows from (9.2.45), the divergence of a vertex function is proportional to $g^3$. The contributions from the diagrams associated with the fields $\psi$ and $\psi'$ are therefore equal in magnitude, but different in sign. As a result, the total vertex function $\Gamma_{\mu\nu\alpha}$ will satisfy the identities (9.2.44). Furthermore, it can be proven by direct calculation that the rest of the one-loop diagrams satisfy the identities (9.2.44) as well. As can be demonstrated in a general form, diagrams containing more than one loop exhibit no anomalies.

The above mechanism for compensating anomalies can also be used in models with spontaneously broken symmetry. In such models the interaction Lagrangian of spinor and vector fields giving rise to anomalies remains unchanged, and all the above considerations concerning the anomaly cancellation remain applicable.

Anomalies may occur also in non-Abelian gauge theories. In this case, the divergence of a three-point vertex Green's function can be calculated in the same way as in an Abelian theory. The specifics of the non-Abelian theory will manifest itself only in an additional factor in the expression for the divergence, analogous to (9.2.45):

$$\mathrm{i}(p+q)_\alpha \Gamma^{abc}_{\mu\nu\alpha}(p,q) = \mathrm{Tr}\{\gamma_5 [T_a, T_b]_+ T_c\} \varepsilon_{\mu\nu\alpha\beta} p_\alpha q_\beta \,. \tag{9.2.48}$$

Here $T$ denotes the generators of the gauge group which may include the matrix $\gamma_5$. It follows from (9.2.48) that anomalies are absent and the theory is renormalizable if

$$A_{abc} = \mathrm{Tr}\{\gamma_5 [T_a, T_b]_+ T_c\} = 0 \,. \tag{9.2.49}$$

### 9.2.8 Renormalization of the Standard Model

The Lagrangian of the standard model (Chap. 6) is invariant under gauge transformations; the above renormalization procedure is therefore applicable to them.

As has been shown (Sects. 8.1, 2), the weak currents of the standard model involve the matrix $\gamma_5$, i.e., the standard model contains anomalies. It can be shown that taking into account only the leptonic sector of the standard model gives rise to a non-zero factor $A_{abc}$, determined by (9.2.49), implying that the model is not renormalizable. However, when both the lepton sector and the quark sector of the standard model are considered the anomalies associated with these sectors are mutually cancelled and the factor $A_{abc}$ vanishes. Thus, the *standard model* as a whole is *renormalizable*.

# Part IV

## Gauge Theory of Strong Interactions

In this part we shall discuss the possibility of applying gauge fields to describe the *strong interactions* of particles.

The theory of strong interactions is one of the most complicated problems of elementary particle physics. Following the discovery of the $\pi$-mesons, there were attempts to construct a theory of strong interactions based on the perturbation theory within the Lagrangian formalism (patterned upon quantum electrodynamics). These attempts were based on the assumption that strong interactions are mediated by $\pi$-mesons. It soon became clear, however, that such a treatment cannot even lead to qualitative agreement between theoretical and experimental results. Therefore, other theoretical approaches were put forward which do not make use of perturbation theory but rather are based on general principles of analyticity and unitarity (e.g., dispersion-relation methods, etc.).

However, recently our understanding of the structure of hadrons and of their interactions has changed substantially. It has been established in deep-inelastic lepton-hadron scattering experiments that at high momentum transfers, or at short distances, hadrons behave like systems consisting of weakly interacting particles – quarks and gluons – which, in the first approximation, may be considered free. The same result follows from the theory in which all interactions between the quarks are mediated by gauge fields: at small distances the coupling constant is quite small. Such theories are said to be *asymptotically free*. These theories predict, on the other hand, that the interaction between particles increases with increasing distance.

Thus, we arrive at the following qualitative picture: hadrons have a complicated structure, the interaction of the constituent particles tending to zero at small distances (much smaller than the hadron size). At large distances (of the order of the hadron size) the interaction becomes strong and the particles cannot escape the hadron; one speaks of quark and gluon *confinement*. This picture underlies the model of strong interactions referred to as *quantum chromodynamics*. In quantum chromodynamics, hadrons are considered as bound-states of quarks, while the strong interaction between the hadrons is reduced to the interaction of their constituent quarks mediated by the gauge fields of the *colour group* $SU_3$. These latter fields are called the *gluon fields*.

Since strong, electromagnetic, and weak interactions are all mediated by gauge fields, there appears a possibility of unifying all these types of interactions into a single one. This is called the *grand unification*.

In this part we shall discuss the asymptotically free theories, the dynamical structure of hadrons following from the deep-inelastic lepton-hadron scattering experiments, chromodynamics, and the grand unification.

# 10. Asymptotically Free Theories

In constructing a theory, information about the properties of the coupling constants between the fields is of primary importance. In a general case, the coupling constants depend on the momenta. To emphasize this circumstance, the notion of *effective* (or *running*) *coupling constant*, or of *effective charge,* has been introduced. Of special interest is investigating the behaviour of the effective charge at large momentum transfers ($|p|^2 \to \infty$), in the so called *asymptotic* region, because it is this region where the physically important hard processes occur.

Asymptotic properties of the effective charges of Abelian and non-Abelian fields are essentially different. In the case of Abelian fields, the effective coupling constant grows with increasing $p^2$, while for non-Abelian theories it decreases. It is said that non-Abelian theories are *asymptotically free.* Asymptotically free fields provide new means of developing a consistent quantum field theory and play an important role in constructing a gauge theory of strong interactions.

A special mathematical method has been devised to investigate the asymptotic properties of the effective charges. It is based on the *renormalization group equations.*

We shall start by deriving the renormalization group equation and obtain from it an equation for the effective charge. The latter equation will be used to analyze the asymptotic properties of the effective charges of some fields and models.

## 10.1 Renormalization Group Equations and Their Solutions

### 10.1.1 Multiplicative Renormalizations and Their Groups

Let us consider renormalizable theories. For them, equations can be obtained which are a convenient aid in studying their asymptotic properties.

A well-known example of renormalizable theory is provided by quantum electrodynamics. To remove the divergences in quantum electrodynamics, it is sufficient to renormalize only three Green's functions, namely, the electron propagator $G(p, e_0, m_0)$, the photon propagator $D_{\mu\nu}(k, e_0, m_0)$, and the vertex function $\tilde{\Gamma}_\mu(p, k, q; e_0, m_0)$, where $p, k, q$ are 4-momenta and $m_0$ and $e_0$ are the bare (or unrenormalized) electron mass and charge, respectively. For this reason we shall consider the renormalization of these Green's functions only.

The Green's functions containing all orders of perturbation theory are divergent. To regularize them, we introduce a cut-off parameter $\Lambda$. On removing the regularization ($\Lambda \to \infty$), the Green's functions become divergent again.

Due to the renormalizability of the theory, finite renormalized parts $G_R(p, e_R, m_R)$, $D_{\mu\nu R}(k, e_R, m_R)$, and $\tilde{\Gamma}_{\mu R}(p, k, q; e_R, m_R)$ can be singled out in the divergent Green's functions. This is done by multiplying each of them by an appropriate renormalization factor

$$G(p, e_0, m_0, \Lambda) = Z_2(e_0, m_0, \Lambda, \mu)\, G_R(p, e_R, m_R, \mu)\,, \tag{10.1.1}$$

$$D_{\mu\nu}(k, e_0, m_0, \Lambda) = Z_3(e_0, m_0, \Lambda, \mu)\, D_{\mu\nu R}(k, e_R, m_R, \mu)\,, \tag{10.1.2}$$

$$\tilde{\Gamma}_\mu(p, k, q; e_0, m_0, \Lambda) = Z_1^{-1}(e_0, m_0, \Lambda, \mu)\, \tilde{\Gamma}_{\mu R}(p, k, q; e_R, m_R, \mu)\,, \tag{10.1.3}$$

where $e_R$ and $m_R$ are the renormalized electron charge and mass, respectively, and $\mu$ is an arbitrary parameter with the dimension of mass. We shall assume in this chapter that the vertex functions $\tilde{\Gamma}$ do not involve the charge, i.e., $\Gamma = e\tilde{\Gamma}$, so that the renormalization of the charge $e$ is carried out separately. The renormalized charge is then related to the bare charge by

$$e_R = Z_1^{-1} Z_2 Z_3^{1/2} e_0 \tag{10.1.4}$$

or, taking into account the Ward identity ($Z_1 = Z_2$), by

$$e_R = Z_3^{1/2} e_0\,. \tag{10.1.5}$$

The subdivision of a Green's function into a finite and an infinite part cannot be made in a unique way. Therefore, the renormalized Green's functions depend on the parameter $\mu$, which is the renormalization point (cf. Sect. 9.2). Commonly, the value of this parameter is additionally fixed in a theory. (It is matched in such a way that the renormalized mass $m_R$ and charge $e_R$ be equal to the physically observed mass and charge values $m$ and $e$.)

The renormalization factors $Z_i$ enter (10.1.1 – 3) as multipliers, and quantum electrodynamics is said to be *multiplicatively renormalizable.* Any renormalizable theory can be represented in such a form.

The relations of the type of (10.1.1 – 3) can be written, in a general case, as the relations for an arbitrary Green's function

$$\tilde{\Gamma}(p, g_0, m_0, \Lambda) = Z_\Gamma^{-1}(g_0, m_0, \Lambda, \mu)\, \tilde{\Gamma}_R(p, g_R, m_R, \mu)\,, \tag{10.1.6}$$

where $p$ denotes the set of all momenta the Green's function depends on; $Z_\Gamma$ is the renormalization constant for the Green's function $\tilde{\Gamma}$; the indices specifying the fields, the type of vertex, etc. are omitted.

The transformations of multiplicative renormalization of the kind (10.1.6) possess the group property. The corresponding group is called the *renormalization group.*

### 10.1.2 Dependence of the Factors $Z_i$ on the Dimensionless Parameters

It can be demonstrated that all the renormalization factors in renormalizable theories may be chosen to be dependent on dimensionless parameters.

Using this, (10.1.6) is rewritten as

$$\tilde{\Gamma}(p, g_0, m_0, \Lambda) = Z_\Gamma^{-1}\left(g_0, \frac{\Lambda}{\mu}\right)\tilde{\Gamma}_{\mathrm{R}}(p, g_{\mathrm{R}}, m_{\mathrm{R}}, \mu) \tag{10.1.7}$$

or

$$\tilde{\Gamma}_{\mathrm{R}}(p, g_{\mathrm{R}}, m_{\mathrm{R}}, \mu) = Z_\Gamma\left(g_0, \frac{\Lambda}{\mu}\right)\tilde{\Gamma}(p, g_0, m_0, \Lambda)\,, \tag{10.1.8}$$

where, by definition,

$$g_{\mathrm{R}} = Z_g\left(g_0, \frac{\Lambda}{\mu}\right)g_0\,, \qquad m_{\mathrm{R}} = Z_m\left(g_0, \frac{\Lambda}{\mu}\right)m_0\,. \tag{10.1.9}$$

Here $Z_m$ and $Z_g$ are the mass and charge renormalization constants, respectively.

### 10.1.3 The Renormalization Group Equations

The parameter $\mu$ entering (10.1.6) is arbitrary. By changing it, the renormalized charges, masses, and the values of the function $\tilde{\Gamma}_{\mathrm{R}}$ also change ($g_0$ and $m_0$ being fixed). Considering these changes leads to certain functional equations, known as the *renormalization group equations.* In deriving them, we shall proceed from (10.1.8).

1) Let us differentiate (10.1.8) with respect to $\mu$ at fixed $g_0$ and $m_0$ and multiply both sides of the resulting equation by $\mu$:

$$\mu\frac{\partial\tilde{\Gamma}_{\mathrm{R}}}{\partial\mu} + \mu\frac{\partial g_{\mathrm{R}}}{\partial\mu}\,\frac{\partial\tilde{\Gamma}_{\mathrm{R}}}{\partial g_{\mathrm{R}}} + \mu\frac{\partial m_{\mathrm{R}}}{\partial\mu}\,\frac{\partial\tilde{\Gamma}_{\mathrm{R}}}{\partial m_{\mathrm{R}}} = \mu\frac{\partial Z_\Gamma}{\partial\mu}\tilde{\Gamma}\,. \tag{10.1.10}$$

By virtue of the relation

$$\frac{\partial Z_\Gamma}{\partial\mu}\tilde{\Gamma} = \frac{1}{Z_\Gamma}\,\frac{\partial Z_\Gamma}{\partial\mu}Z_\Gamma\tilde{\Gamma} = \frac{\partial\ln Z_\Gamma}{\partial\mu}\tilde{\Gamma}_{\mathrm{R}}\,,$$

Eq. (10.1.10) takes the form

$$\left[\mu\frac{\partial}{\partial\mu} + \beta(g_{\mathrm{R}})\frac{\partial}{\partial g_{\mathrm{R}}} + m_{\mathrm{R}}\gamma'(g_{\mathrm{R}})\frac{\partial}{\partial m_{\mathrm{R}}} - \gamma(g_{\mathrm{R}})\right]\tilde{\Gamma}_{\mathrm{R}} = 0\,, \tag{10.1.11}$$

where

$$\beta(g_R) = \mu \frac{\partial}{\partial \mu} g_R\left(g_0, \frac{\Lambda}{\mu}\right),$$

$$m_R \gamma'(g_R) = \mu \frac{\partial m_R}{\partial \mu} = \mu m_0 \frac{\partial Z_m}{\partial \mu} = \mu Z_m m_0 \frac{\partial \ln Z_m}{\partial \mu} = \mu m_R \frac{\partial \ln Z_m}{\partial \mu},$$

$$\gamma(g_R) = \mu \frac{\partial \ln Z_\Gamma}{\partial \mu}.$$

2) Assume that $\tilde{\Gamma}_R$ is a homogeneous function of the dimensional variables $p$, $m$, $\mu$ of order $d$, i.e.,

$$\tilde{\Gamma}_R(lp, lm, l\mu) = l^d \tilde{\Gamma}_R(p, m, \mu),$$

(which is fulfilled in perturbation theory). Then, by using the Euler theorem for homogeneous functions we find

$$\left[\mu \frac{\partial}{\partial \mu} + m_R \frac{\partial}{\partial m_R} + p \frac{\partial}{\partial p}\right] \tilde{\Gamma}_R = d \tilde{\Gamma}_R. \tag{10.1.12}$$

The quantity $d$ is called the *normal dimension* of the function $\tilde{\Gamma}_R(p, g_R, m_R, \mu)$.

3) Determining from this the quantity $\mu(\partial/\partial\mu)\tilde{\Gamma}_R$ and substituting it into (10.1.12) we get

$$\left\{p \frac{\partial}{\partial p} + m_R[1 - \gamma'(g_R)] \frac{\partial}{\partial m_R} - d - \beta(g_R) \frac{\partial}{\partial g_R} + \gamma(g_R)\right\} \tilde{\Gamma}_R = 0. \tag{10.1.13}$$

4) Introduce the function $\tilde{\Gamma}(kp, g_R, m_R, \mu)$ all momenta of which are equally scaled. This function satisfies (10.1.11) while (10.1.13) transforms to

$$\left\{k \frac{\partial}{\partial k} - \beta(g_R) \frac{\partial}{\partial g_R} + [1 - \gamma'(g_R)] m_R \frac{\partial}{\partial m_R} + \gamma(g_R) - d\right\}$$
$$\times \tilde{\Gamma}_R(kp, g_R, m_R, \mu) = 0. \tag{10.1.14}$$

Let us introduce a new variable $t = \ln k$, so that

$$k \frac{\partial}{\partial k} = k \frac{\partial t}{\partial k} \frac{\partial}{\partial t} = \frac{\partial}{\partial t}.$$

The final form of the *renormalization group equation* then follows from (10.1.14):

$$\left\{\frac{\partial}{\partial t} - \beta(g_R) \frac{\partial}{\partial g_R} + m_R[1 - \gamma'(g_R)] \frac{\partial}{\partial m_R} - [d - \gamma(g_R)]\right\}$$
$$\times \tilde{\Gamma}_R(kp, g_R, m_R, \mu) = 0, \tag{10.1.15}$$

where

$$\beta(g_R) = -g_0 \frac{\partial}{\partial \ln \Lambda/\mu} Z_g\left(g_0, \frac{\Lambda}{\mu}\right). \tag{10.1.16}$$

$$\gamma'(g_R) = -\frac{\partial}{\partial \ln \Lambda/\mu} \ln Z_m\left(g_0, \frac{\Lambda}{\mu}\right), \tag{10.1.17}$$

$$\gamma(g_R) = -\frac{\partial}{\partial \ln \Lambda/\mu} \ln Z_\Gamma\left(g_0, \frac{\Lambda}{\mu}\right). \tag{10.1.18}$$

The functions $\beta(g_R)$, $\gamma'(g_R)$, and $\gamma(g_R)$ are unknown. One can see that they only depend on the dimensionless coupling constants. The quantity $\gamma(g_R)$ is called the *anomalous dimension* of the Green's function.

### 10.1.4 Effective Charge and Asymptotic Freedom

The solution of (10.1.15) reads (the index $R$ being omitted hereafter)

$$\tilde{\Gamma}(kp, g, m, \mu) = \exp\left\{\int_0^t [d - \gamma(G(t'))]\, dt'\right\} \tilde{\Gamma}(p, G(t), \mathcal{M}(t), \mu), \tag{10.1.19}$$

$$\frac{\partial G(t)}{\partial t} = \beta(G(t)), \quad G(0) = g, \tag{10.1.20}$$

$$\frac{\partial \mathcal{M}(t)}{\partial t} = -\mathcal{M}(t)[1 - \gamma'(G(t))], \quad \mathcal{M}(0) = m. \tag{10.1.21}$$

Expression (10.1.19) makes it possible to reduce investigating the functions $\tilde{\Gamma}$ which depend on the momenta, to investigating the functions $\tilde{\Gamma}$ which depend on charges and masses, with the fixed momenta.

The quantities $G(t)$ and $\mathcal{M}(t)$ are called the *effective charge* and the *effective* (or *running*) *mass*, respectively. As can be seen, the parameter characterizing the interaction in the asymptotic region is the effective charge.

In this chapter we shall study the properties of the effective charge $G(t)$ in the range of large $t$, i.e., in the asymptotic region. These properties are described by (10.1.20).

The solution of (10.1.20) is determined by an unknown function $\beta(G(t))$. In practice, it can be calculated only by means of perturbation theory. The properties of the solution of (10.1.20) basically depend on the sign of the function $\beta(G(t))$. For example, let it be positive and have, in the first order of perturbation theory, the form $\beta(G(t)) = aG^3(t)/2$, where $a > 0$. Equation (10.1.20) for the effective charge then takes the form

$$\frac{\partial G(t)}{\partial t} = \frac{1}{2} aG^3(t), \quad G(0) = g, \tag{10.1.22}$$

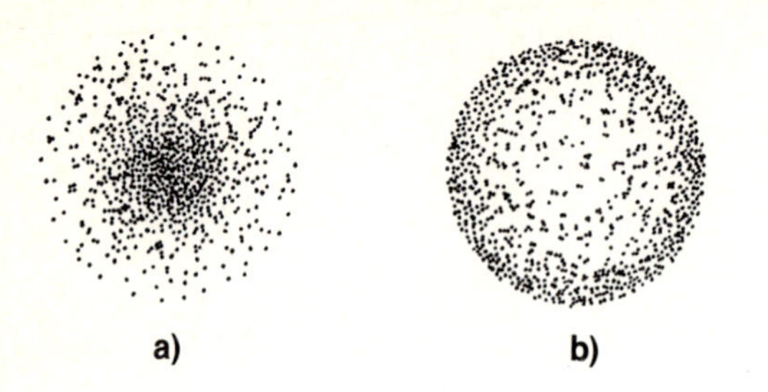


**Fig. 10.1a, b.** Vacuum polarization (effective charge): (**a**) in the asymptotically non-free theories; (**b**) in the asymptotically free theories

and its solution is

$$G^2(t) = \frac{g^2}{1 - ag^2 t} . \tag{10.1.23}$$

If the function $\beta(G(t))$ is negative and its form in the first order of perturbation theory is $\beta(G(t)) = -aG^3(t)/2$, then the solution of (10.1.20) is written as

$$G^2(t) = \frac{g^2}{1 + ag^2 t}, \quad a > 0 . \tag{10.1.24}$$

The main difference between (10.1.23) and (10.1.24) is that the signs in the denominator are different.

As follows from (10.1.23), the effective charge increases with growing $t$ (Fig. 10.1 a) and becomes infinite (i.e., has a pole) at $t \sim 1/ag^2$ indicating that in such theories we, in fact, leave the applicability domain of perturbation theory. The appearance of such a pole means that, in view of the relation between the effective charge and the Green's function, the physical Green's functions also have a pole. This fact contradicts the unitarity condition and is referred to as the "fictitious pole" difficulty. This difficulty can be expressed in other terms as follows. It is seen from (10.1.23) that there is no pole if the charge $g$ is zero. This leads to the absence of all physical processes and is, of course, unreasonable. Therefore one also speaks of the "zero-charge" difficulty.

The opposite situation is encountered in the case corresponding to (10.1.24) when the effective charge tends to zero in the asymptotic region (Fig. 10.1 b). Hence, in theories with a negative function $\beta(G(t))$ there is no "zero-charge" difficulty: the effective charge decreases with increasing momentum and vanishes in the asymptotic limit. Theories in which the effective charge $G(t)$ goes to zero as the momentum tends to infinity are referred to as *asymptotically free*.

Let us find the conditions under which there exist asymptotically free solutions of the equation. To asymptotically free solutions correspond the points at which the effective charge and thus also the function $\beta(G(t))$ vanish. Let us call the real points where $\beta(G(t))$ is zero the *fixed points* of the equation for $G(t)$.

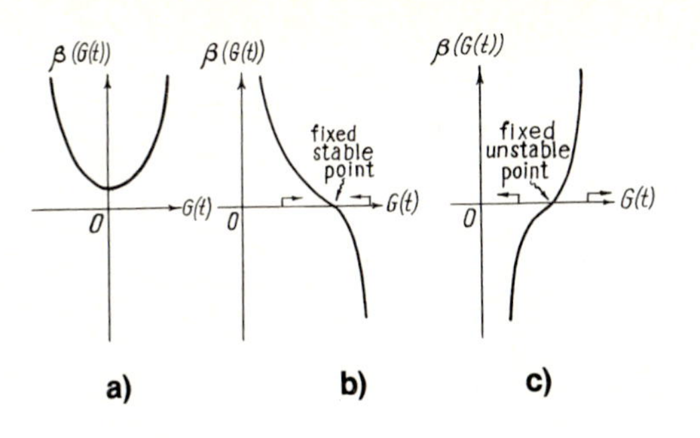


**Fig. 10.2a–c.** Fixed points: **(a)** absent fixed point; **(b)** stable fixed point; **(c)** unstable fixed point

If there are no fixed points for the function $\beta(G(t))$ (Fig. 10.2a), then (10.1.20) has no asymptotically free solutions.

One should distinguish between two types of fixed points: the stable fixed points and the unstable ones. Let $g_\infty$ be a fixed point of the function $\beta(G(t))$. Then, if the derivative $\partial\beta(G(t))/\partial G(t) < 0$ (Fig. 10.2b), the effective charge $G(t)$ is increasing, provided that its initial value lies left to $g_\infty$ and is decreasing otherwise. Therefore, at $t \to \infty$ the charge $G(t) \to g_\infty$ (Fig. 10.2b). A *fixed point* $g_\infty$ for which $\partial\beta(G)/\partial G < 0$ is called *stable.* If $\partial\beta(G(t))/\partial G(t) > 0$, then $G(t)$ does not tend to $g_\infty$, and such a *fixed point* $g_\infty$ is said to be *unstable* (Fig. 10.2c). The function $\beta(G(t))$ may have several fixed points; stable and unstable points alternate.

Are there asymptotically free theories? To answer this question, a great number of various models have been investigated. Both asymptotically free and asymptotically non-free ones have been found among them. We shall consider some of these models.

### 10.1.5 Method of Investigation

The major stages of investigating the models are as follows.

1) Specifying the model, i.e., (i) its symmetry group (e.g., $U_1$, $SU_2$, $SU_3$, etc.), (ii) its constituents (that is, field multiplets), and (iii) the expression for the Lagrangian invariant under the group chosen.
2) Specifying (i) the Green's functions and the renormalization factors and (ii) a relation between the bare and the renormalized charges in terms of the factors $Z_i$.
3) Calculating the factors $Z_i$ in the one-loop approximation. For this purpose one should (i) draw the Feynman diagrams for the propagators and the vertex functions whose renormalization factors are involved in the relation $g_R = Z_g g_0$, where $Z_g$ is the charge renormalization constant; (ii) write down the Feynman integrals associated with these diagrams; (iii) calculate the integrals retaining the terms $\sim \ln \Lambda/\mu$ and find $Z_g$.
4) Finding the function $\beta(G(t))$ or a set of functions $\beta_i$ in case of several charges.
5) Formulating equations for the effective charges.

6) Finding the fixed points of the functions $\beta_i$ and determining their type.
7) Investigating the equations for the effective charges.

It should be taken into account that the equations for the charges $G(t)$ are independent of the mass and therefore the asymptotic freedom is independent of the particle mass. In other words, the question of asymptotic freedom is to be solved in a massless theory. Thus, one may take as a basis the Lagrangian of massless particles. To make the matter fields (which are spinor, scalar, etc.) massive, one has to add to the Lagrangian appropriate mass, or Yukawa terms. The massiveness of the gauge fields is achieved by adding a necessary number of scalar Higgs multiplets.

Now we shall consider some specific models.

## 10.2 The Models

### 10.2.1 Models Without Non-Abelian Gauge Fields

One example of such models is quantum electrodynamics.

1) Quantum electrodynamics is invariant under the Abelian group $U_1$ of phase transformations. The starting field is that of charged spinor particles (the electromagnetic field being a gauge field). The Lagrangian of quantum electrodynamics is written as (cf. Sect. 2.3)

$$\mathscr{L} = -\tfrac{1}{4}F_{\mu\nu}F_{\mu\nu} + \mathrm{i}\,\bar{\psi}\gamma_\mu\nabla_\mu\psi\,,$$

where $\nabla_\mu = \partial_\mu + \mathrm{i}e_0\mathrm{A}_\mu$ is the covariant derivative.

2) Renormalization of the Green's functions is determined by (10.1.1 – 3); here $e = Z_3^{1/2}e_0$, i.e., only the factor $Z_3$ entering the expression (10.1.2) for the photon Green's function has to be calculated.

3) The Feynman diagrams for the photon Green's function $D_{\mu\nu}(k)$ to one-loop approximation are shown in Fig. 10.3. The corresponding Feynman integral has the form

$$D_{\mu\nu}(k) = \mathscr{D}_{\mu\nu}(k) + \frac{\mathrm{i}e_0^2}{(2\pi)^4}\,\mathscr{D}_{\mu\lambda}(k)P_{\lambda\varrho}(k)\,\mathscr{D}_{\varrho\nu}(k)\,, \qquad (10.2.1)$$

where

$$P_{\lambda\varrho}(k) = \int dp\,\mathrm{Tr}\left\{\gamma_\lambda\frac{\not{p}+\not{k}+m}{(p+k)^2-m^2}\gamma_\varrho\frac{\not{p}+m}{p^2-m^2}\right\}.$$

Let us calculate this integral retaining only the terms which contain $\ln(\Lambda/\mu)$. This yields

$$P_{\lambda\varrho}(k) = \frac{16\mathrm{i}\pi^2}{3}\ln\frac{\Lambda}{\mu}\,k^2\left(g_{\lambda\varrho} - \frac{k_\lambda k_\varrho}{k^2}\right). \qquad (10.2.2)$$

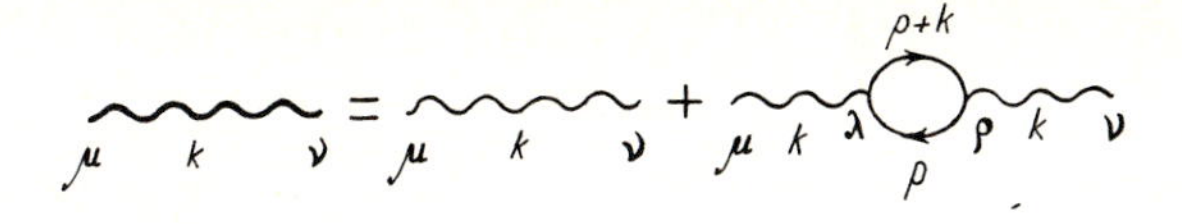


**Fig. 10.3.** Green's function of photon up to one-loop approximation

Choose the photon propagator in the form (cf. Table 7.1)

$$\mathscr{D}_{\mu\nu}(k) = \frac{1}{k^2}\left(g_{\mu\nu} - \frac{k_\mu k_\nu}{k^2}\right). \tag{10.2.2$'$}$$

Substituting (10.2.2) and (10.2.2′) into (10.2.1) we find

$$D_{\mu\nu}(k) = \left(1 - \frac{e_0^2}{6\pi^2}\ln\frac{\Lambda}{\mu}\right)\mathscr{D}_{\mu\nu}(k)\,,$$

i.e.,

$$Z_3 = 1 - \frac{e_0^2}{6\pi^2}\ln\frac{\Lambda}{\mu}.$$

4) From this, using (10.1.5, 16) we obtain for the function $\beta(e)$

$$\beta(e) = -\frac{\partial}{\partial \ln \Lambda/\mu} Z_3^{1/2} e_0 \bigg|_{e_0=\mathrm{const}} \simeq \frac{e^3}{12\pi^2}.$$

5) The equation for the effective charge $E(t)$ is written in the form of (10.1.22), where $a = 1/(6\pi^2)$.

6) The solution of this equation coincides with (10.1.23):

$$E^2(t) = \frac{e^2}{1 - ae^2 t}.$$

Consequently, quantum electrodynamics is not an asymptotically free theory $[\beta(e) > 0]$.

In a similar way other models not including non-Abelian gauge fields (a scalar field, a fermion field, their combination, etc.) have been analyzed. It has been found that they are not asymptotically free. In fact, one can show that *only non-Abelian* gauge field theories *can* be asymptotically free (the *Coleman-Gross theorem*). Let us consider some models containing non-Abelian gauge fields.

### 10.2.2 Models with Non-Abelian Gauge Fields

We shall investigate three such models in the case of $SU_2$-symmetry.

#### The Yang-Mills Field

1) According to (5.3.2, 3), the effective Lagrangian for the Yang-Mills field has the form

$$\mathscr{L}_{\mathrm{YM}}(A_\mu^k(x), c^k(x)) = -\frac{1}{4}\left[\partial_\mu A_\nu^k - \partial_\nu A_\mu^k - \frac{g_0}{2}\,\varepsilon_{klm}(A_\mu^l A_\nu^m - A_\nu^l A_\mu^m)\right]^2$$

$$-[\partial_\mu \bar{c}^k(x)][\partial_\mu c^k(x)] - g_0 \varepsilon_{lmk}(\partial_\mu \bar{c}^k) A_\mu^m c^l , \quad (10.2.3)$$

where $\varepsilon_{klm}$ are the structure constants, $g_0$ is the bare (or unrenormalized) coupling constant of the Yang-Mills field, and $c^k(x)$ are the ghost fields.

2) The renormalized ($g$) and the bare ($g_0$) coupling constants are related by

$$g = Z_1^{-1} Z_2^{3/2} g_0 . \quad (10.2.4)$$

Here $Z_2$ and $Z_1$ are the renormalization constants of the propagator $D_{\mu\nu}^{kl}(p)$ and of the vertex $\tilde{\Gamma}_{\mu\nu\varrho}^{klm}(p,k,q)$, respectively (Fig. 10.4a, b)

$$D_{\mu\nu}^{kl}(p) = Z_2^{-1} D_{B\mu\nu}^{kl}(p) , \qquad \tilde{\Gamma}_{\mu\nu\varrho}^{klm}(p,q,r) = Z_1 \tilde{\Gamma}_{B\mu\nu\varrho}^{klm}(p,q,r) , \quad (10.2.5)$$

where the subscript $B$ refers to the bare Green's function.

3) The Feynman diagrams for both Green's functions (10.2.5) are depicted, in the one-loop approximation, in Fig. 10.5. Let us calculate, for instance, the Feynman integral corresponding to the diagram a of Fig. 10.5. Let the Greek indices refer to the spatial and the Latin indices to the $SU_2$ indices of propagators and vertices. The integral under consideration reads

$$I_a = \mathscr{D}_{\mu\delta}^{kn}(p)\left\{\frac{\mathrm{i} g_0^2}{2(2\pi)^4}\int dq\, \varepsilon_{nji}[(-p-2q)_\delta g_{\varrho\varkappa} + (2p+q)_\varrho g_{\delta\varkappa} + (q-p)_\varkappa g_{\delta\varrho}]\right.$$

$$\times \frac{\delta_{ie}}{(p+q)^2}\left(g_{\varkappa\pi} - \frac{(p+q)_\varkappa (p+q)_\pi}{(p+q)^2}\right)\varepsilon_{efm}[(q-p)_\pi g_{\lambda\sigma}$$

$$\left. + (2p+q)_\lambda g_{\pi\sigma} + (-p-2q)_\sigma g_{\pi\lambda}]\,\frac{\delta_{fj}}{q^2}\left(g_{\lambda\varrho} - \frac{q_\lambda q_\varrho}{q^2}\right)\right\}\mathscr{D}_{\sigma\nu}^{ml}(p) . \quad (10.2.6)$$

Here $\varepsilon_{klm}$ are the structure constants of the Lie algebra of the group $SU_2$, $\delta_{kl}$ are the Kronecker symbols, and

$$\mathscr{D}_{\mu\nu}^{kl}(p) = \frac{\delta_{kl}}{p^2}\left(g_{\mu\nu} - \frac{p_\mu p_\nu}{p^2}\right)$$

is the propagator of the Yang-Mills field determined by (5.3.14). The calculation of the integral (10.2.6), retaining only the terms that contain $\ln(\Lambda/\mu)$, yields

$$I_a = \mathscr{D}_{\mu\delta}^{kn}(p)\,\frac{g_0^2}{16\pi^2}\,\varepsilon_{nji}\varepsilon_{ijm}\,\frac{25}{6}\left(\ln\frac{\Lambda}{\mu}\right)\left(g_{\delta\sigma} - \frac{p_\delta p_\sigma}{p^2}\right)p^2\,\mathscr{D}_{\sigma\nu}^{ml}(p) .$$

The combination of the $SU_2$ structure constants involved in this expression is determined by the formula $\sum_{ij} \varepsilon_{nij}\varepsilon_{mij} = 2\delta_{mn}$. Calculating, in a similar way,

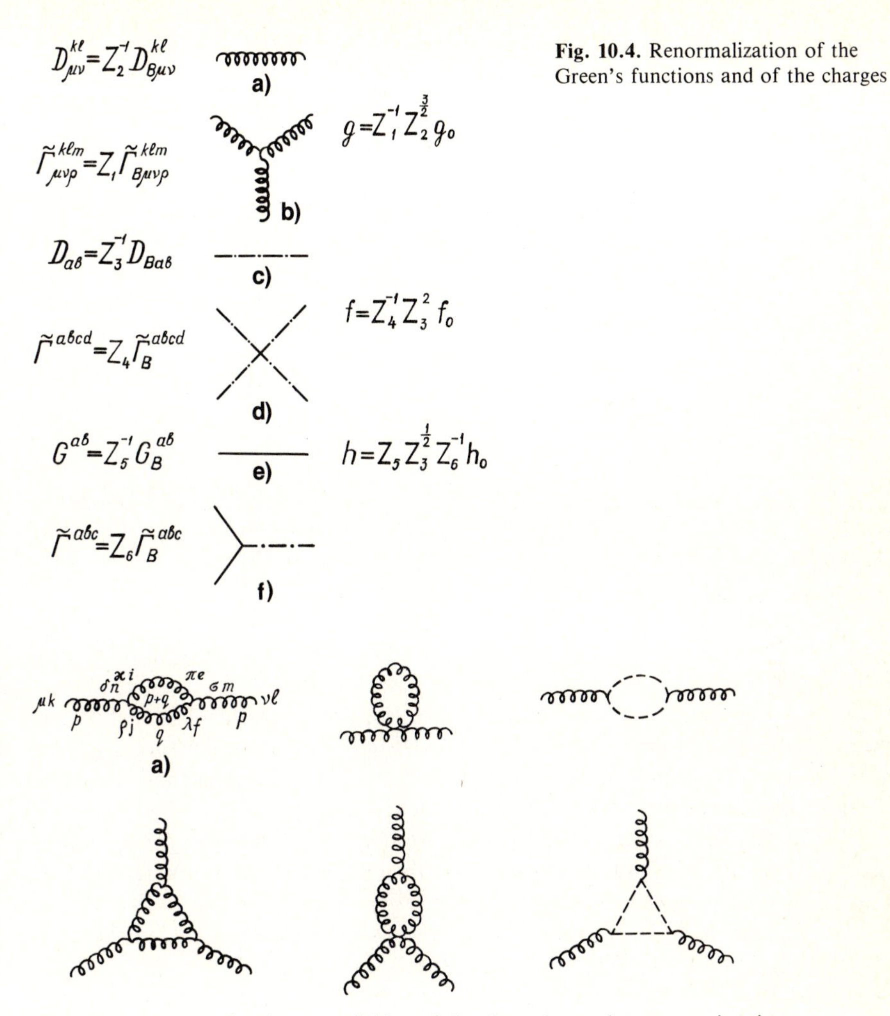


**Fig. 10.4.** Renormalization of the Green's functions and of the charges

**Fig. 10.5.** Diagrams for the gauge fields and the ghosts in one-loop approximation

the integrals corresponding to all of the diagrams of Fig. 10.5 and subsequently summing the results obtained yields, with the aid of (10.2.1) and (10.2.5),

$$Z_2 = 1 + \frac{g_0^2}{16\pi^2}\,\frac{26}{3}\ln\frac{\Lambda}{\mu},$$

$$Z_1^{-1} = 1 - \frac{g_0^2}{16\pi^2}\,\frac{17}{3}\ln\frac{\Lambda}{\mu}.$$

4) For the function $\beta(G)$ the expression $\beta(G) = -G^3/(16\pi^2)\cdot 22/3$ then follows.

5) The equation for the effective charge $G(t)$ is written in the form of (10.1.22), where $a = -1/(8\pi^2) \cdot 22/3$.

6) The solution of this equation coincides with (10.1.24), where $a = 1/(8\pi^2) \cdot 22/3$. Thus, the (non-Abelian) Yang-Mills field is asymptotically free: $\beta(G) < 0$.

**The Yang-Mills Field with Spontaneously Broken Symmetry**

1) The model is invariant under the group $SU_2$. It includes a triplet of the Yang-Mills fields $A_\mu^k(x)$ and a scalar triplet $\phi^a(x)$ $(a = 1, 2, 3)$. The model is described by the Lagrangian

$$\mathscr{L} = \mathscr{L}_{\rm YM} + \frac{1}{2}(\nabla_\mu^{ab}\phi^b)^2 - \frac{f_0}{4!}(\phi^a\phi^a)^2 + m^2\phi^a\phi^a, \tag{10.2.7}$$

where $\mathscr{L}_{\rm YM}$ is the Lagrangian of the Yang-Mills field, $f_0$ is the bare self-interaction constant of the scalar field $\phi^a(x)$, $\nabla_\mu^{ab} = \partial_\mu\delta_{ab} + \mathrm{i}g_0(\omega_k)_{ab}A_\mu^k$ is the covariant derivative, and $-\mathrm{i}(\omega_k)_{ab}$ are the generators of the triplet representation of the group $SU_2$, as given by (1.2.7).

The last term in the Lagrangian has been introduced to bring about spontaneous symmetry-breaking. The Higgs mechanism makes two components of the Yang-Mills field to become massive (i.e., to convert to massive vector bosons), while one component remains massless (and may be identified with a photon). In other words, adding a scalar triplet permits converting some of massless particles into massive ones. Unfortunately, the model then ceases to be asymptotically free, as we shall prove below.

2) The relations between the renormalized $(g, f)$ and bare $(g_0, f_0)$ constants involve the renormalization constants of four Green's functions depicted in Fig. 10.4a – d.

3) The Feynman diagrams for all Green's functions are given, in the one-loop approximation, in Figs. 10.5 and 10.6.

In the expression for the diagrams containing the scalar field lines there appear the generators $(-\mathrm{i}\omega_k)$ of the triplet representation of the group. The combinations of generators for typical diagrams of Fig. 10.6a, b have the form

$$(\omega_n)_{ji}(\omega_m)_{ef}\delta_{ie}\delta_{fj}\delta_{nk}\delta_{ml} = \mathrm{Tr}\{\omega_k\omega_l\} = 2\delta_{kl},$$
$$(\omega_k)_{dc}(\omega_l)_{ef}\delta_{kl}\delta_{df}\delta_{ac}\delta_{be} = (\omega_k\omega_k)_{ba} = 2\delta_{ab}.$$

Calculating the Feynman integrals associated with these diagrams leads to the following result:

$$Z_1^{-1} = 1 - 5\frac{g_0^2}{16\pi^2}\ln\frac{\Lambda}{\mu},$$

$$Z_2 = 1 + 8\frac{g_0^2}{16\pi^2}\ln\frac{\Lambda}{\mu},$$

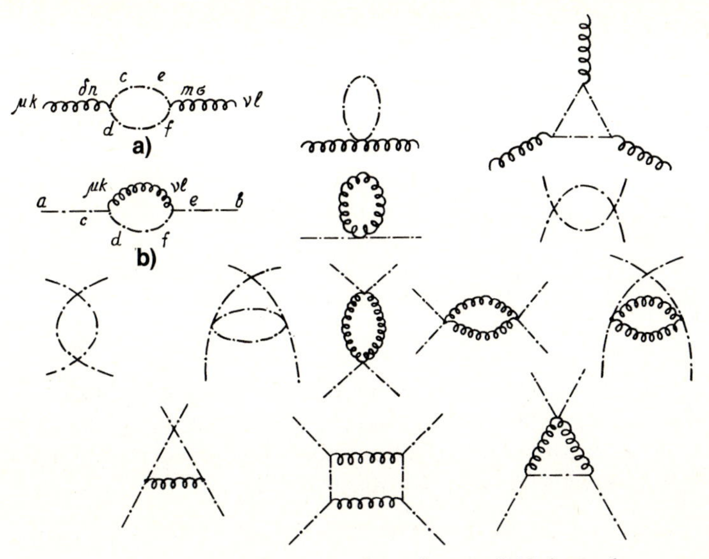


**Fig. 10.6.** Diagrams containing the scalar and gauge fields in one-loop approximation

$$Z_3 = 1 + 12\frac{g_0^2}{16\pi^2}\ln\frac{\Lambda}{\mu},$$

$$Z_4^{-1} = 1 - \frac{1}{16\pi^2}\left(\frac{11}{3}f_0 + 72\frac{g_0^4}{f_0}\right)\ln\frac{\Lambda}{\mu}.$$

4) For the functions $\beta_g(G)$ and $\beta_f(G,F)$ which enter the equations for the effective charges $G(t)$ and $F(t)$, it follows:

$$\beta_g(G(t)) = -7\frac{G^3(t)}{16\pi^2},$$

$$\beta_f(G(t),F(t)) = \frac{1}{16\pi^2}\left[\frac{11}{3}F^2(t) - 24G^2(t)F(t) + 72G^4(t)\right].$$

5) Using these formulae, the set of equations for the charges $G(t)$ and $F(t)$ is written

$$16\pi^2\frac{dG^2(t)}{dt} = -14G^4, \quad G(0) \equiv g, \tag{10.2.8}$$

$$16\pi^2\frac{dF(t)}{dt} = \frac{11}{3}F^2 - 24G^2F + 72G^4, \quad F(0) \equiv f. \tag{10.2.9}$$

6) As follows from (10.2.8), the model is asymptotically free with respect to the charge $G$, since the inequality $\beta_g(G)<0$ holds. The solution of (10.2.8) has the form

$$G^2(t)=\frac{g^2}{1+ag^2t}, \tag{10.2.10}$$

where $a=7/(8\pi^2)$. Substituting (10.2.10) into (10.2.9) and changing the variables $F=\bar{F}G^2$ and $x=[\ln(1+ag^2t)]/a$, we find

$$16\pi^2\frac{d\bar{F}}{dx}=\frac{11}{3}\bar{F}^2-10\bar{F}+72\,.$$

This equation has no fixed points, i.e., its right-hand side has no real roots. Thus, the model is not asymptotically free with respect to the charge $F$.

This result is true all the more if two scalar triplets are introduced to make all components of the Yang-Mills field become massive.

## The Yang-Mills and the Matter Fields

Let us consider a Yang-Mills field interacting with matter fields. The choice of a specific model is rather arbitrary because various multiplets of particles and symmetry-breaking terms may be chosen. Therefore, quite a number of different models have been analyzed. We shall concern ourselves with one of them.

1) The model is invariant under the group $SU_2$. It is composed of a triplet of Yang-Mills fields $A_\mu^k(x)$, $m$ spinor triplets $\psi_j^a(x)$ $(j=1,2,\ldots,m)$, and a scalar triplet $\phi^a(x)$ that interacts with one of the spinor triplets $(a=1,2,3$ throughout). The model is described by the massless Lagrangian

$$\begin{aligned}\mathscr{L}=\mathscr{L}_{\mathrm{YM}}+\sum_{j=1}^{m}\mathrm{i}\,\bar{\psi}_j^a\gamma_\mu\nabla_\mu^{ab}\psi_j^b+\frac{1}{2}(\nabla_\mu^{ab}\phi^b)^2-\frac{f_0}{4!}(\phi^a\phi^a)^2\\ -\mathrm{i}\,h_0\bar{\psi}_1^a\varepsilon_{acb}\psi_1^b\phi^c\,,\end{aligned} \tag{10.2.11}$$

where $\mathscr{L}_{\mathrm{YM}}$ is the Lagrangian for the Yang-Mills field determined by (10.2.3); $\nabla_\mu^{ab}=\partial_\mu\delta_{ab}-g_0\varepsilon_{kab}A_\mu^k$ is the covariant derivative; $g_0$ is the bare coupling constant of the Yang-Mills field with itself and with the spinor and scalar fields; $h_0$ is the bare coupling constant for the interaction of the spinor field with the scalar field (the Yukawa interaction); and $f_0$ is the bare self-interaction constant of the scalar field.

The first term in (10.2.11) is associated with the Yang-Mills field and with the ghost fields $c$; the second term with the spinor field and its interaction with the Yang-Mills field; the third term with the scalar field and its interaction with the Yang-Mills field; the fourth term with the self-interaction of scalar fields; and the fifth term with the interaction between the spinor and the scalar fields.

For the spinor and scalar matter fields to acquire mass, we add to the Lagrangian a mass term

$$\mathscr{L}_{\mathrm{M}} = -M_1 \bar{\psi}_1^a \psi_1^a - \sum_{j=2}^{m} M_j \bar{\psi}_j^a \psi_j^a + \frac{1}{2}\mu^2 \phi^a \phi^a . \tag{10.2.12}$$

In the Lagrangian (10.2.12), the last term corresponding to the scalar field has been introduced for the sake of spontaneous symmetry-breaking. Due to it, two components of the Yang-Mills field acquire the mass $g\mu(f/6)^{1/2}$; the masses of the components of the first spinor triplet split into $M_1$, $M_1 \pm \mu h(6/f)^{1/2}$; there arise a scalar meson of mass $M = \sqrt{2}\mu$ and two massless (unphysical) scalar mesons.

2) To determine the charge renormalization constants, we use the Green's functions whose diagrams are presented, together with the corresponding relations between the renormalized and the bare coupling constants, in Fig. 10.4a – f.

3) The Feynman diagrams for the Green's functions are shown, in the one-loop approximation, in Figs. 10.5 – 7. Calculating the Feynman integrals for these diagrams we obtain the expressions for $Z_i$.

4) By means of these expressions, the form of the functions $\beta_g(G)$, $\beta_h(G,H)$, and $\beta_f(G,H,F)$ is readily found. Since only the diagrams containing no other constants than $g$ contribute to the factors $Z_1$ and $Z_2$, the function $\beta_g(G)$ depends on $G(t)$ only. The factors $Z_3$, $Z_5$ and $Z_6$ are determined by the contributions from the diagrams involving $G(t)$ and $H(t)$, while the factor $Z_4$ has contributions from the diagrams involving $G(t)$, $H(t)$ and $F(t)$. Accordingly, $\beta_h(G,H)$ depends on $G(t)$ and $H(t)$ while $\beta_f(G,H,F)$ depends on $G(t)$, $H(t)$ and $F(t)$.

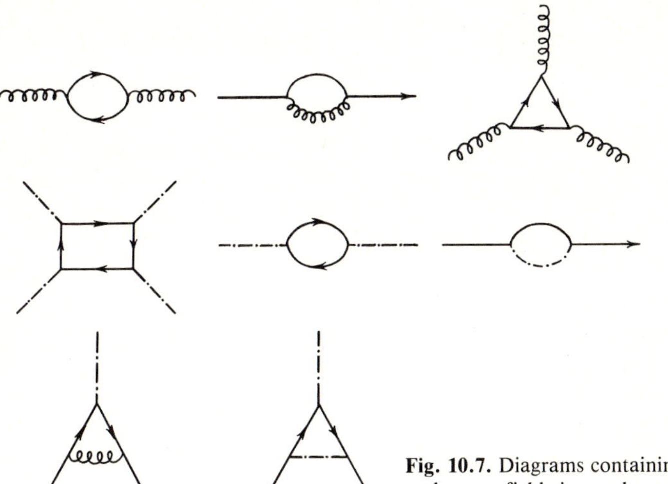

**Fig. 10.7.** Diagrams containing the spinor, scalar, and gauge fields in one-loop approximation

5) The set of equations for the effective coupling constants $G(t)$, $H(t)$ and $F(t)$ is written

$$16\pi^2 \frac{dG^2(t)}{dt} = -\left(14 - \frac{16}{3} m\right) G^4(t)\,, \qquad G(0) \equiv g\,, \tag{10.2.13}$$

$$16\pi^2 \frac{dH^2(t)}{dt} = 16H^4(t) - 24H^2(t)G^2(t)\,, \qquad H(0) \equiv h\,, \tag{10.2.14}$$

$$16\pi^2 \frac{dF(t)}{dt} = \frac{11}{3} F^2(t) - 24G^2(t)F(t) + 72G^4(t)$$
$$+ 16F(t)H^2(t) - 96H^4(t)\,, \qquad F(0) \equiv f\,. \tag{10.2.15}$$

6) Let us find out whether there exist asymptotically free solutions to this set of equations, i.e., whether there are solutions for which all three effective charges simultaneously go to zero as $t \to \infty$. It follows from (10.2.13) that the model will be asymptotically free with respect to the coupling constant $G$ if $\beta_g < 0$, or $14 - \frac{16}{3}m > 0$, i.e., only if the model includes either one or two spinor triplets. With a larger number of spinor triplets $\beta_g > 0$, and the model ceases to be asymptotically free. In other words, the requirement of asymptotic freedom imposes certain restrictions on the possible number of spinor triplets.

The solution of (10.2.13) reads

$$G^2(t) = \frac{g^2}{1 + g^2 a t}\,, \tag{10.2.16}$$

where

$$a = \frac{1}{16\pi^2}\left(14 - \frac{16}{3} m\right).$$

Let us find under which conditions the model will be asymptotically free with respect to the coupling constant $H$. For this purpose we consider (10.2.14). After the substitution $H = \bar{H}G$ and passing to the new variable $x = [\ln(1 + ag^2 t)]/a$, it assumes the form

$$16\pi^2 \frac{d\bar{H}^2}{dx} = 16\bar{H}^4 - \left(10 + \frac{16}{3} m\right) \bar{H}^2\,. \tag{10.2.17}$$

Let us write this equation in the general form

$$\frac{dy(x)}{dx} = \beta(y) = Ay^2 + By + C \tag{10.2.18}$$

which recovers (10.2.17) for

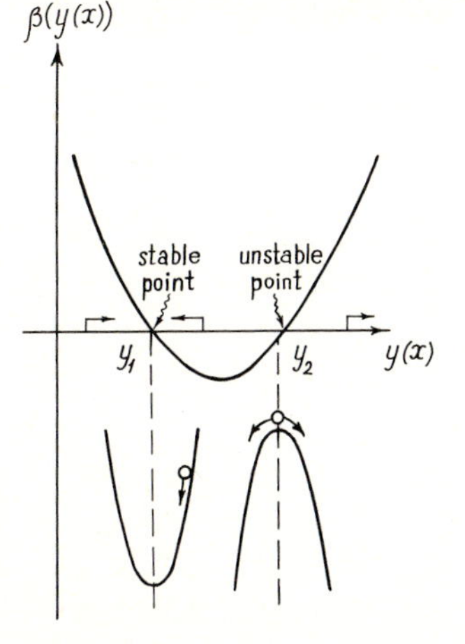


**Fig. 10.8.** Investigating the fixed points

$$y=\bar{H}^2\,,\quad A_h=\frac{1}{\pi^2}>0\,,\quad B_h=-\frac{1}{16\pi^2}\left(10+\frac{16}{3}m\right),\quad C_h=0\,. \tag{10.2.19}$$

Asymptotically free solutions of (10.2.18) are determined by the zeroes of the function $\beta(y)$ (i.e., by the fixed points). The roots of $Ay^2+By+C=0$, which correspond to the fixed points, are $y_{1,2}=(-B\mp\sqrt{B^2-4AC})/(2A)$, i.e., the function $\beta(y)$ will have two zeroes, provided that $B^2-4AC\geqslant 0$ (Fig. 10.8) and will have no zeroes otherwise (Fig. 10.2a).

The zeroes of the function $\beta(y)$ may be stable as well as unstable. In a stable point ($y_s$),

$$d\beta(y)/dy\,|_{y=y_s}<0\,,$$

in an unstable point ($y_u$),

$$d\beta(y)/dy\,|_{y=y_u}>0\,.$$

In Fig. 10.8, the point $y_1$ is stable and the point $y_2$ is unstable.

The effective coupling constants may be both positive and negative. However, taking into account (10.2.19), we should consider only the case of a positive constant, $y\geqslant 0$.

For the solution to asymptotically tend to the stable point $y_1$, the initial value $y(0)$ ought to be chosen smaller than the root $y_2$: $y(0)<y_2$. In case of an unstable point $y_2$, the initial value has to be taken equal to this root: $y(0)=y_2$.

Summarizing, we state that (10.2.17) can have both a stable and an unstable fixed point. Either will give rise to an asymptotically free solution, provided that the following conditions are fulfilled:

For stable points:

$$\left.\begin{aligned} &B^2-4AC\geqslant 0\,,\\ &y_{\mathrm{s}}\geqslant 0\,,\\ &d\beta(y)/dy\,|_{y=y_{\mathrm{s}}}<0\,,\\ &y(0)<y_{\mathrm{u}}\,.\end{aligned}\right\}\quad(10.2.20)$$

For unstable points:

$$\left.\begin{aligned} &B^2-4AC\geqslant 0\,,\\ &y_{\mathrm{u}}\geqslant 0\,,\\ &d\beta(y)/dy\,|_{y=y_{\mathrm{u}}}>0\,,\\ &y(0)=y_{\mathrm{u}}\,.\end{aligned}\right\}\quad(10.2.21)$$

**Stable Points.** Let us demonstrate that within stable solutions only, the model under consideration is not asymptotically free with respect to the coupling constant $F$.

By substituting (10.2.19) into (10.2.20) we obtain

$$\left[\frac{1}{16\pi^2}\left(10+\frac{16}{3}m\right)\right]^2>0\,,$$

$$y_1=0\,,$$

$$-\frac{1}{16\pi^2}\left(10+\frac{16}{3}m\right)<0\,,$$

$$h^2<\left(\frac{5}{8}+\frac{1}{3}m\right)g^2\,.$$

These conditions are fulfilled both for $m=1$ and for $m=2$, i.e., the model is asymptotically free with respect to the coupling constant $H$. Besides, at $t\to\infty$ we have $\bar{H}\to 0$, and $H=\bar{H}G$ goes to zero faster than $G$. However, with respect to the coupling constant $F$ the model turns out not to be asymptotically free. To prove this, we consider (10.2.15). Substituting (10.2.16) and performing a change of variables $F=\bar{F}G^2$, $x=[\ln(1+ag^2t)]/a$, we obtain

$$16\pi^2\frac{d\bar{F}}{dx}=\frac{11}{3}\bar{F}^2-\left(10+\frac{16}{3}m\right)\bar{F}+72+16\bar{F}\bar{H}^2-96\bar{H}^4\,. \qquad (10.2.22)$$

Since $\bar{H}\to 0$ as $t\to\infty$, the last two terms in (10.2.22), which characterize the interaction of the scalar field with fermions, can be neglected at asymptotic values of $t$:

$$16\pi^2\frac{d\bar{F}}{dx}=\frac{11}{3}\bar{F}^2-\left(10+\frac{16}{3}m\right)\bar{F}+72\,, \qquad (10.2.23)$$

where

$$A_f=\frac{11}{48\pi^2}\,,\qquad B_f=-\frac{1}{16\pi^2}\left(10+\frac{16}{3}m\right),\qquad \text{and}\qquad C_f=\frac{9}{2\pi^2}\,.$$

It is recognized that with these values the condition for the existence of a fixed point ($B^2 - 4AC \geqslant 0$) is fulfilled only for $m \geqslant 5$. That is to say, a solution asymptotically free with respect to the coupling constant $F$ only exists when the number of spinor triplets exceeds 4. In this case, however, the model becomes asymptotically non-free with respect to the coupling constant $G$. Due to this circumstance the model as a whole cannot be asymptotically free. This implies that it is not possible to introduce into the model a triplet of scalar fields to bring about spontaneous symmetry-breaking. Thus, with regard to stable points only, the model under consideration is not asymptotically free.

**Unstable Points.** Let us demonstrate now that, when unstable points are considered, the set of equations (10.2.13 – 15) has an asymptotically free solution, even though a spontaneously symmetry-breaking scalar triplet $\phi^a(x)$ is present in the model.

After substituting (10.2.19) into (10.2.21) we have

$$\left[\frac{1}{16\pi^2}\left(10 + \frac{16}{3}m\right)\right]^2 > 0\,,$$

$$\left(\frac{5}{8} + \frac{m}{3}\right) > 0\,,$$

$$\frac{1}{\pi^2}\left(\frac{5}{8} + \frac{1}{3}m\right) > 0$$

$$h^2 = \left(\frac{5}{8} + \frac{1}{3}m\right)g^2\,.$$

The first three conditions hold for $m = 1,2$; the last one indicates that the constants $h$ and $g$ are interrelated: $h^2 = (5/8 + m/3)\,g^2$, so that $\bar{H}$ proves to be a constant, $\bar{H} = (5/8 + m/3)^{1/2}$. Now, at $t \to \infty$, the effective charge $H = \bar{H}G$ tends to zero as $G$. Hence, for $m = 1,2$, the model is still asymptotically free with respect to the coupling constant $H$.

$\bar{H}$ being a constant, one cannot neglect the last two terms in (10.2.22). By setting $\bar{H} = (5/8 + m/3)^{1/2}$ in (10.2.22) it can be rewritten as

$$16\pi^2 \frac{d\bar{F}}{dx} = \frac{11}{3}\bar{F}^2 + 72 - 96\left(\frac{5}{8} + \frac{m}{3}\right)^2, \tag{10.2.24}$$

where now

$$A_f = \frac{11}{48\pi^2}\,, \qquad B_f = 0\,, \qquad \text{and} \qquad C_f = \frac{1}{16\pi^2}\left(\frac{69}{2} - 40m - \frac{32m^2}{3}\right).$$

Substituting (10.2.24) into (10.2.20) yields, in the stable point,

i) $$\frac{32m^2}{3} + 40m - \frac{69}{2} \geqslant 0\,,$$

ii) $$-\sqrt{\frac{3}{11}\left(\frac{32m^2}{3} + 40m - \frac{69}{2}\right)} \geqslant 0\,.$$

Condition (ii) is not fulfilled, and we do not consider the corresponding solution.

Substituting (10.2.24) into (10.2.21) yields, in the unstable point,

$$\frac{32m^2}{3} + 40m - \frac{69}{2} \geqslant 0\,,$$

$$y_2 = c \geqslant 0\,,$$

$$\frac{11}{24\pi^2}c > 0\,,$$

$$f = g^2 c\,,$$

where

$$c = \sqrt{\frac{32m^2}{11} + \frac{120m}{11} - \frac{207}{22}}\,.$$

The first three conditions hold for $m = 1, 2$; the last one shows that the constants $g$ and $f$ are interrelated, $f^2 = c^2 g^4$, implying the constancy of $\bar{F}$: $\bar{F} = f/g^2$. Correspondingly, at $t \to \infty$, the effective charge $F = \bar{F}G^2$ tends to zero as $G^2$. That is to say, for $m = 1, 2$ the model is asymptotically free with respect to the coupling constant $F$, too.

Consequently, in unstable points the model is asymptotically free with respect to all three coupling constants only in one case, namely, when the constraints $h^2 = (5/8 + m/3)\,g^2$ and $f^2 = c^2 g^4$ are imposed on them. In this case, the effective charge $G(t)$ tends to zero at $t \to \infty$, and so do the effective charges $H(t)$ [which goes to zero as $G(t)$] and $F(t)$ [which goes to zero as $G^2(t)$]. In other words, asymptotic freedom is only possible at strictly fixed values of the coupling constants, and the coupling constants cannot be "moved" in the vicinity of these values.

It should be emphasized that the existence of definite relations between the coupling constants leads to some definite relations between the particle masses. Thus, in the model under consideration, the vector meson mass $m_{\mathrm{V}}$ and the scalar meson mass $m_{\mathrm{S}}$ are related by $3\,m_{\mathrm{S}}^2 = c m_{\mathrm{V}}^2$.

Hence, (10.1.20) permits studying the asymptotic properties of the effective charges for different models in order to find whether a model is asymptotically free.

# 11. Dynamical Structure of Hadrons

## 11.1 Experimental Basis for Scaling

Let us consider the inelastic processes

$$e^- N \to e^- X\,, \qquad \mu^- N \to \mu^- X\,, \qquad \nu N \to \mu^- X\,, \qquad \bar{\nu} N \to \mu^+ X\,,$$

where $N$ denotes a nucleon and $X$ stands for all other particles. These processes exhibit in the deep-inelastic region an interesting and important property referred to as *scaling*. It is this property whose discovery gave a strong impact on the development of contemporary views on the *hadron structure*. The above processes will be given here a detailed analysis.

1) We start with the inelastic scattering of electrons by protons

$$e^- p \to e^- X\,. \tag{11.1.1}$$

The diagram for this process is given, in the one-photon approximation, in Fig. 11.1.

To this diagram corresponds the following matrix element

$$S_{\rm fi} = -\frac{4\pi\alpha}{q^2}(2\pi)^4 \mathrm{i}\,\delta(p_X + k' - p - k)\, j_\mu \langle X | J_\mu(0)\, | p \rangle\,,$$

where $j_\mu = \bar{u}(k')\,\gamma_\mu u(k)$ is the electromagnetic current in the vertex (a), $J_\mu$ is the electromagnetic current in the vertex (b), and $q = k - k'$ is the virtual photon momentum.

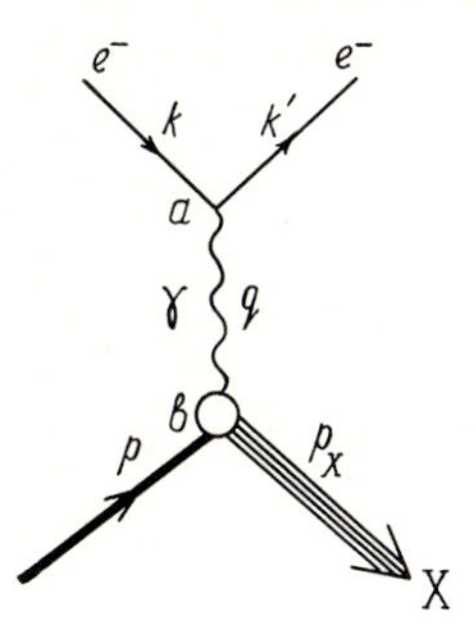


**Fig. 11.1.** Diagram for inelastic scattering of electron on proton in the one-photon approximation

The differential cross-section $d\sigma$ for the process (11.1.1), summed over all final hadron states, is

$$E' \frac{d\sigma}{dk'} = \frac{4\alpha^2}{q^4} \frac{1}{E} L_{\mu\nu} W_{\mu\nu}, \tag{11.1.2}$$

where

$$L_{\mu\nu} = \sum_{\text{spin}} j_\mu j_\nu^\dagger = \frac{1}{2}[k_\mu k'_\nu + k_\nu k'_\mu - g_{\mu\nu}(kk')],$$

$$W_{\mu\nu} = \frac{1}{4\pi} \sum_{X,\text{spin}} \int dx\, e^{i(q+p-p_X)x} \langle p|J_\mu(0)|X\rangle\langle X|J_\nu(0)|p\rangle . \tag{11.1.3}$$

Here $E$ and $E'$ denote the electron energy before and after the collision, respectively, and the lepton mass has been set equal to zero ($m_e = 0$). The function $W_{\mu\nu}$ contains all the information on the dynamics relevant to the vertex (b).

Taking into account the translation invariance and the completeness of the states of $X$, (11.1.3) is rewritten as

$$W_{\mu\nu}(p,q) = \frac{1}{4\pi} \sum_{\text{spin}} \int dx\, e^{iqx} \langle p|[J_\mu(x), J_\nu(0)]_-|p\rangle . \tag{11.1.4}$$

The requirements of the Lorentz invariance and of the current conservation lead to the following expression for the function $W_{\mu\nu}$:

$$W_{\mu\nu}(\nu, q^2) = \left(-g_{\mu\nu} + \frac{q_\mu q_\nu}{q^2}\right) W_1(\nu, q^2) + \frac{1}{M^2}\left(p_\mu - \frac{pq}{q^2} q_\mu\right)\left(p_\nu - \frac{pq}{q^2} q_\nu\right) W_2(\nu, q^2), \tag{11.1.5}$$

where $\nu = pq/M$ ($M$ denoting the nucleon mass); $W_1(\nu, q^2)$ and $W_2(\nu, q^2)$ are unknown functions referred to as the *structure functions*.

The region where the variables satisfy the conditions

$$\nu \gg M, \; -q^2 \gg M^2, \quad x = -q^2/(2\nu M) \text{ and finite}, \tag{11.1.6}$$

is called the *deep-inelastic region*. The magnitude of $x$ ranges from 0 to 1 and is taken not too close to its limiting values.

In the laboratory frame,

$$\nu = \frac{pq}{M} = (E - E'), \quad q^2 = -4EE' \sin^2\frac{\theta}{2} < 0, \tag{11.1.7}$$

where $\theta$ is the electron scattering angle.

For the differential cross-section we have, in the laboratory frame,

$$\frac{d^2\sigma}{dE'\,d\Omega} = \frac{EE'}{\pi}\,\frac{d^2\sigma}{dq^2 d\nu}$$

$$= \frac{4\alpha^2}{q^4}E'^2\left[2\,W_1(\nu,q^2)\sin^2\frac{\theta}{2} + W_2(\nu,q^2)\cos^2\frac{\theta}{2}\right]. \qquad (11.1.8)$$

The functions $W_1$ and $W_2$ are related in a simple way to the total absorption cross-sections of the longitudinal ($\sigma_L$) and the transversal ($\sigma_T$) virtual photons:

$$W_1(\nu,q^2) = \frac{K}{4\pi^2\alpha M}\,\sigma_T(\nu,q^2)\,, \qquad (11.1.9)$$

$$W_2(\nu,q^2) = \frac{K}{4\pi^2\alpha M}\,\frac{-q^2}{\nu^2-q^2}\,[\sigma_T(\nu,q^2)+\sigma_L(\nu,q^2)]\,, \qquad K = M\nu + \frac{q^2}{2}\,. \qquad (11.1.10)$$

One can also introduce a parameter

$$R = \sigma_L/\sigma_T \qquad (11.1.11)$$

giving the ratio of the two cross-sections.

The formulae obtained are also applicable to the process of inelastic scattering of $\mu$-mesons on proton.

As an example, Fig. 11.2a shows the experimentally measured dependence of the structure function $\nu W_2(x,q^2)$ on $q^2$ at fixed values of $x$. As can

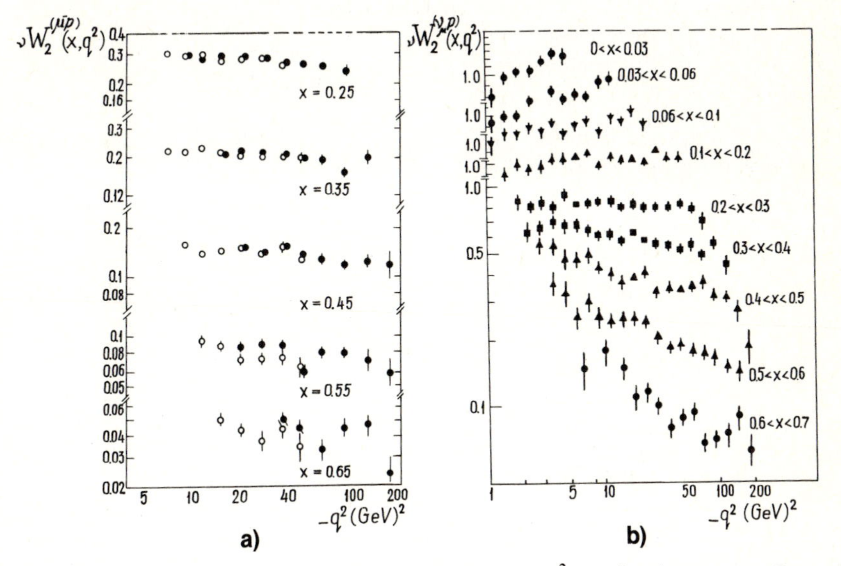


**Fig. 11.2a, b.** Dependence of the structure functions on $q^2$: **(a)** for the process $\mu^- p \to \mu^- X$; **(b)** for the process $\nu_\mu p \to \mu^- X$

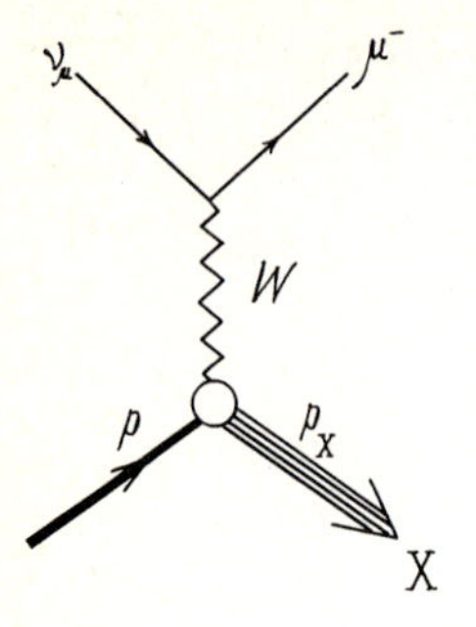


**Fig. 11.3.** Diagram for inelastic scattering of neutrino on proton

be seen, the structure function $\nu W_2(x, q^2)$ depends weakly on the momentum-transfer $q^2$, so that in the first approximation it may be considered independent of $q^2$. The property of independence of the structure functions of $q^2$ is referred to as *exact* scaling. Thus, in deep-inelastic lepton scattering experiments on proton, a slight *scaling violation* is observed.

2) Let us now consider the process of inelastic scattering of neutrino $\nu_\mu$ and anti-neutrino $\bar{\nu}_\mu$ on proton accompanied by production of charged leptons (e.g., $\mu^\pm$-mesons):

$$\nu_\mu + p \to \mu^- + X\,, \tag{11.1.12}$$

$$\bar{\nu}_\mu + p \to \mu^+ + X\,. \tag{11.1.13}$$

The diagrams for these processes are depicted in Fig. 11.3. The processes stem from the weak interaction, and therefore their matrix elements can be written as the product of two charged weak currents (Sect. 8.2):

$$M^{\nu/\bar{\nu}} = \frac{G}{\sqrt{2}} \langle X | J_\mu^{(+),\mathrm{q}}(0) | p \rangle J_\mu^{(-),l}\,.$$

Here $J_\mu^{(-),l}$ denotes the charged weak lepton currents, and $J_\mu^{(+),\mathrm{q}}$ the charged weak hadron currents. From this, we find for the differential cross-sections of the processes (11.1.12) and (11.1.13) the following expressions, each containing three structure functions:

$$\frac{d^2\sigma^{\nu/\bar{\nu}}}{dx\,dy} = \frac{G^2 ME}{\pi}\left[xy^2 M\, W_1(\nu, q^2) + \left(1 - y - \frac{Mxy}{2E}\right)\nu\, W_2(\nu, q^2)\right.$$
$$\left. \pm y\left(1 - \frac{y}{2}\right) x\nu\, W_3(\nu, q^2)\right]. \tag{11.1.14}$$

Here $y = \nu/E$, $G$ is the weak interaction constant, $E$ is the energy of the incident neutrino, $W_i$ denote the structure functions; the upper signs refer to the process (11.1.12), and the lower ones to the process (11.1.13).

Experiment shows that violation of scaling in the processes (11.1.12, 13) is minute (Fig. 11.2b).

Thus, for the deep-inelastic processes scaling is violated. However, this violation is small, and in the first approximation one can consider scaling as exact. Therefore, in this chapter we will explain how exact scaling can be understood within the quark-parton model. The problem of scaling violation in the framework of quantum chromodynamics will be treated in Chap. 12.

## 11.2 Exact Scaling and the Parton Structure of Hadrons

### 11.2.1 The Quark-Parton Model

The idea underlying the *parton model* is rather simple. Let us suppose that (i) a hadron is composed of point-like particles – partons – whose form and number are not specified; (ii) the partons within hadrons are free particles, i.e., the scattering amplitude of a particle by a hadron is the sum of the scattering amplitudes of this particle by individual partons.

The functions $W_i$ then possess scaling properties. To demonstrate this, the functions $\hat{W}_i^{(r)}$ have to be found which correspond to the scattering by a free point-like parton of the type $r$ (the hat denotes a quantity at the *parton level*, as distinct from the corresponding quantity at the *hadron level*). The $\hat{W}_i^{(r)}$ thus found have to be integrated over all possible parton momenta and have to be summed over all parton types $r$. In this way, one obtains the structure functions $W_i$ of the hadron.

For a particle scattered off a free parton $r$, the functions $\hat{W}_i^{(r)}$, $i = 1, 2, \ldots$, can be calculated exactly (in a given order of perturbation theory). For the case of the virtual photon scattering ($i = 1, 2$) by a spin$-1/2$ parton, for instance, the tensor $\hat{W}_{\mu\nu}^{(r)}$ of (11.1.5) has, in the lowest order in the electromagnetic coupling constant, and in the deep-inelastic kinematical region, the form

$$\hat{W}_{\mu\nu}^{(r)} = \frac{1}{x_r M} Q_r^2 \delta(q^2 + 2p^{(r)}q)(2p_\mu^{(r)} p_\nu^{(r)} + p_\mu^{(r)} q_\nu + p_\nu^{(r)} q_\mu - p^{(r)} q g_{\mu\nu}) ,$$

where $q$ is the photon momentum, $Q_r$ is the parton charge and $p^{(r)}$ is the parton momentum which is a fraction $x_r$ of the hadron momentum $p$, i.e.,

$$p_\mu^{(r)} = x_r p_\mu \quad \text{with} \quad 0 \leqslant x_r \leqslant 1 .$$

Using the relation $(p^{(r)}q) = x_r(pq) = x_r M\nu$ and comparing the last equation with the corresponding one of (11.1.5) term by term, we find that the functions $\hat{W}_1^{(r)}$ and $\hat{W}_2^{(r)}$ are related as

$$2Mx_r \hat{W}_1^{(r)} = \nu \hat{W}_2^{(r)} ,$$

and are determined by

$$2Mx_r\hat{W}_1^{(r)} = \nu\hat{W}_2^{(r)} = Q_r^2 2M\nu x_r\delta(q^2+2x_rM\nu)$$
$$\equiv Q_r^2 x_r\delta\left(x_r+\frac{q^2}{2M\nu}\right)$$
$$\equiv Q_r^2 x_r\delta(x_r-x)\,, \qquad (11.2.1)$$

where $x = -q^2/(2M\nu)$. In other words, the structure functions $\hat{W}_1^{(r)}$ and $\nu\hat{W}_2^{(r)}$, due to a parton $r$, depend on a single scaling variable $x$.

Let $q^r(x_r)$ be the probability of finding a parton of the type $r$ carrying the fraction $x_r$ of the hadron momentum. Then the structure functions of the hadron are obtained by multiplying the parton structure functions $\hat{W}_i^{(r)}$ of (11.2.1) by the probability $q^r(x_r)$, integrating over all possible $x_r$ and finally summing over all the types of partons:

$$W_1(\nu,q^2) = \sum_r\int_0^1 dx_r q^r(x_r)\,\hat{W}_1^{(r)} = \sum_r\int_0^1 dx_r q^r(x_r)\frac{Q_r^2}{2M}\delta(x_r-x)$$
$$=\frac{1}{2M}\sum_r Q_r^2 q^r(x)\,,$$

$$W_2(\nu,q^2) = \sum_r\int_0^1 dx_r q^r(x_r)\,\hat{W}_2^{(r)} = \sum_r\int_0^1 dx_r q^r(x_r)\frac{Q_r^2}{\nu}x_r\delta(x_r-x)$$
$$=\frac{x}{\nu}\sum_r Q_r^2 q^r(x)\,,$$

i.e.,

$$2xMW_1(\nu,q^2) = \nu W_2(\nu,q^2) = x\sum_r Q_r^2 q^r(x)\,. \qquad (11.2.2)$$

One can see from (11.2.2) that the *deep-inelastic structure functions* $MW_1(\nu,q^2)$ and $\nu W_2(\nu,q^2)$ of a hadron have the *scaling property*, i.e., they are functions of a single variable $x = -q^2/(2M\nu)$. This property is referred to as the *Bjorken scaling*.

Thus, in the deep-inelastic region we have

$$MW_1(\nu,q^2)\to F_1(x)\,,$$
$$\nu W_2(\nu,q^2)\to F_2(x)\,,$$

and (11.2.2) can be written in the form

$$2xF_1(x) = F_2(x) = x\sum_r Q_r^2 q^r(x)\,. \qquad (11.2.2')$$

In other words, the hadron structure functions $F_1(x)$ and $F_2(x)$ are expressed, in the parton model, in terms of parton probabilities or *parton distribution functions* $q^r(x)$ inside the hadron[1].

---

[1] In the same way as $MW_1$ and $\nu W_2$ scale, in the case of neutrino scattering on a hadron, one can show that the parton model gives scaling also for the third structure function $\nu W_3$, i.e.,

$$\nu W_3(\nu,q^2)\to F_3(x)\,.$$

### 11.2.2 Sum Rules

Integrating (11.2.2′), one obtains the relation

$$\int_0^1 \frac{dx}{x} F_2(x) = \sum_r Q_r^2 N_r \,, \qquad \text{where} \tag{11.2.3}$$

$N_r = \int_0^1 dx\, q^r(x)$ is the total number of partons of the kind $r$.

More concrete and economical is the quark-parton model according to which the fermions consist of three quarks and the bosons of a quark and an anti-quark. These are referred to as the *valence quarks*; they provide a hadron with its quantum numbers (the charge, the iso-spin, the baryon number, strangeness, charm, beauty). Hadrons may also contain an admixture of quark-anti-quark pairs; these form a *quark sea* and do not affect the quantum numbers of hadrons. Therefore, for example, for the proton, taking into account its quantum numbers, the following relations have to hold:

$$\int_0^1 dx\,[\tfrac{2}{3}(u(x)-\bar{u}(x)) - \tfrac{1}{3}(d(x)-\bar{d}(x)) - \tfrac{1}{3}(s(x)-\bar{s}(x)) + \tfrac{2}{3}(c(x)-\bar{c}(x))] = 1\,, \qquad \text{(the charge)}$$

$$\int_0^1 dx\,\tfrac{1}{3}[u(x)-\bar{u}(x)+d(x)-\bar{d}(x)+s(x)-\bar{s}(x)+c(x)-\bar{c}(x)] = 1\,,$$
(the baryon number)

$$\int_0^1 dx\,[s(x)-\bar{s}(x)] = 0\,, \qquad \text{(strangeness)}$$

$$\int_0^1 dx\,[c(x)-\bar{c}(x)] = 0\,. \qquad \text{(charm)}$$

These relations lead to the following *sum rules*

$$\int_0^1 dx(u(x)-\bar{u}(x)) = 2\,, \qquad \int_0^1 dx(d(x)-\bar{d}(x)) = 1\,,$$
$$\int_0^1 dx(s(x)-\bar{s}(x)) = 0\,, \qquad \int_0^1 dx(c(x)-\bar{c}(x)) = 0\,. \tag{11.2.4}$$

Let us assume that

1) the contributions due to the sea, $q_s^r(x)$, and the valence quarks, $q_v^r(x)$, are independent, i.e.,

$$q^r(x) = q_v^r(x) + q_s^r(x)\,; \tag{11.2.5}$$

2) the iso-spin of the sea is zero and its charge parity is positive:

$$u_s = d_s = \bar{u}_s = \bar{d}_s = s_1 \,, \qquad s_s = \bar{s}_s = s_2 \,, \quad c_s = \bar{c}_s = c \,. \tag{11.2.5'}$$

The sum rules (11.2.4) are then rewritten

$$\int_0^1 dx\, u_v(x) = 2 \,, \quad \int_0^1 dx\, d_v(x) = 1 \,. \tag{11.2.4'}$$

### 11.2.3 Relations Between the Structure Functions

With the help of these sum rules, certain relations between the structure functions can be established.

1) From (11.2.2′) we have

$$\begin{aligned} 2F_1^{\mathrm{ep}} &= \frac{1}{x}F_2^{\mathrm{ep}} = \frac{4}{9}(u+\bar{u}) + \frac{1}{9}(d+\bar{d}) + \frac{1}{9}(s+\bar{s}) + \frac{4}{9}(c+\bar{c}) \,, \\ 2F_1^{\mathrm{en}} &= \frac{1}{x}F_2^{\mathrm{en}} = \frac{1}{9}(u+\bar{u}) + \frac{4}{9}(d+\bar{d}) + \frac{1}{9}(s+\bar{s}) + \frac{4}{9}(c+\bar{c}) \,. \end{aligned} \tag{11.2.6}$$

From this it follows

$$F_2^{\mathrm{ep}} - F_2^{\mathrm{en}} = \tfrac{1}{3}x(u_v - d_v)$$

or, using (11.2.4′),

$$\int_0^1 \frac{dx}{x}(F_2^{\mathrm{ep}}(x) - F_2^{\mathrm{en}}(x)) = \frac{1}{3} \,. \tag{11.2.7}$$

2) From the relations (11.2.5), (11.2.5′), and (11.2.6), neglecting $c$-quarks, we get

$$\frac{F_2^{\mathrm{en}}}{F_2^{\mathrm{ep}}} = \frac{\frac{1}{9}u_v + \frac{4}{9}d_v + \frac{10}{9}s_1 + \frac{2}{9}s_2}{\frac{4}{9}u_v + \frac{1}{9}d_v + \frac{10}{9}s_1 + \frac{2}{9}s_2} \,.$$

If we neglect the contribution due to the *sea* and set $d_v = 0$ (which may be the case as $x \to 1$), then

$$\frac{F_2^{\mathrm{en}}}{F_2^{\mathrm{ep}}} = \frac{1}{4} \,. \tag{11.2.8}$$

3) Taking into account the expression (6.3.6) for the charged weak quark current and the formula (11.2.2′), we find (above the charm threshold):

$$F_1^{\nu p} = (\bar{u}+d+s+\bar{c}) , \qquad F_1^{\bar{\nu} p} = (u+\bar{d}+\bar{s}+c) ,$$
$$F_2^{\nu p} = 2x(\bar{u}+d+s+\bar{c}) , \qquad F_2^{\bar{\nu} p} = 2x(u+\bar{d}+\bar{s}+c) , \tag{11.2.9}$$
$$F_3^{\nu p} = 2(-\bar{u}+d+s-\bar{c}) , \qquad F_3^{\bar{\nu} p} = 2(u-\bar{d}-\bar{s}+c) .$$

The corresponding relations for the neutron structure functions are obtained by the interchange $u \rightleftarrows d$ and $\bar{u} \rightleftarrows \bar{d}$. Equation (11.2.9) implies that neutrino processes obey the relations

$$2xF_1^{\nu/\bar{\nu}}(x) = F_2^{\nu/\bar{\nu}}(x) .$$

Let us introduce, for convenience, the notations

$$q = u+d+s+c ,$$
$$\bar{q} = \bar{u}+\bar{d}+\bar{s}+\bar{c} . \tag{11.2.9'}$$

For the scattering on an iso-scalar target $[N = (p+n)/2]$, we then find

$$F_1^{\nu N} = F_1^{\bar{\nu} N} = \tfrac{1}{2}(q+\bar{q}) ,$$
$$F_2^{\nu N} = F_2^{\bar{\nu} N} = x(q+\bar{q}) , \tag{11.2.10}$$
$$F_3^{\nu N|\bar{\nu} N} = q-\bar{q} \mp (c+\bar{c}-s-\bar{s}) .$$

Neglecting in (11.2.9, 10) small contributions from the quarks $s$, $\bar{s}$, $c$, and $\bar{c}$, we obtain

$$\frac{d^2\sigma^{\nu N}}{dx\,dy} = \frac{G^2ME}{\pi} x[q+(1-y)^2\bar{q}] ,$$
$$\frac{d^2\sigma^{\bar{\nu} N}}{dx\,dy} = \frac{G^2ME}{\pi} x[\bar{q}+(1-y)^2 q] , \qquad y = \frac{\nu}{E} . \tag{11.2.11}$$

The relations (11.2.9 – 11) lead to the following results.

1) Integrating (11.2.11) over $x$ and $y$ yields

$$R_\nu = \frac{\sigma^{\bar{\nu} N}}{\sigma^{\nu N}} \geqslant \frac{1}{3} .$$

This ratio is exactly equal to 1/3, provided that the hadron contains no anti-quarks. The experimental value of $R_\nu$ is somewhat larger than 1/3 implying that anti-quarks are contained in the hadron in a small quantity.

2) Let us integrate (11.2.11) over $x$ and take into account that the contribution due to anti-quarks is relatively small. The resulting expression indicates that the cross-section of neutrino scattering depends weakly on $y$, while the corresponding dependence for anti-neutrino is given by $(1-y)^2$. These relations are confirmed by experiment.

3) Since

$$F_2^{eN} = \frac{5}{18}x(q+\bar{q}) + \frac{1}{6}x(c-s+\bar{c}-\bar{s}) , \quad F_2^{\nu N} = x(q+\bar{q})$$

we have

$$F_2^{eN} = \frac{5}{18}F_2^{\nu N} . \tag{11.2.12}$$

4) Making use of (11.2.4), we arrive at the sum rule

$$\frac{1}{2}\int_0^1 dx\,[F_3^{\nu p}(x) + F_3^{\bar{\nu} p}(x)] = 3 , \tag{11.2.13}$$

i.e., the integral on the left-hand side is equal to the number of valence quarks. If the contributions from the quarks $s$ and $c$ are neglected the formula

$$\frac{1}{2}\int_0^1 dx\,[F_3^{\nu p}(x) + F_3^{\bar{\nu} p}(x)] = \int_0^1 dx\, F_3^{\nu N}(x)$$

$$= \int_0^1 dx \frac{3\pi}{2G^2ME}\left(\frac{d\sigma^{\nu N}(x)}{dx} - \frac{d\sigma^{\bar{\nu} N}(x)}{dx}\right)$$

is obtained. (11.2.14)

5) From the formula (11.1.14) follows a relation which is independent of $F_3$:

$$\frac{d^2\sigma^{\nu N}}{dx\,dy} + \frac{d^2\sigma^{\bar{\nu} N}}{dx\,dy} = \frac{G^2ME}{\pi}F_2^{\nu N}(x)[1+(1-y)^2 - y^2R'] ,$$

where

$$R' = \frac{F_2^{\nu N}(x) - 2xF_1^{\nu N}(x)}{F_2^{\nu N}(x)} \tag{11.2.15}$$

and

$$F_i^{\bar{\nu} N}(x) = F_i^{\nu N}(x) , \quad i = 1, 2 .$$

### 11.2.4 Comparison with Experiment

The relations (11.2.7, 8), (11.2.12 – 15) and, as well, other relations which follow from the quark-parton model were subjected to experimental checks. Theoretical and experimental results are in good agreement which verifies the hypothesis of the composite structure of hadrons.

For free spin-1/2 particles, the ratio $R = \sigma_L/\sigma_T \sim 0$ is predicted, while for spin-zero particles, $R \sim \infty$. Experiment gives $R \sim 0$, i.e., the parton spin is 1/2. According to the relation (11.2.13), the protons include three valence quarks; from (11.2.12) it follows that the quarks have fractional charges.

It should be recalled that the deep-inelastic processes involving neutral currents already studied (Sect. 8.1) also confirm the quark structure of hadrons.

### 11.2.5 Quark and Gluon Distribution Functions

In the model with non-interacting quarks, the cross-section of scattering on a hadron is equal to the sum of scattering cross-sections on its constituent quarks. The cross-section of neutrino and anti-neutrino scattering by a quark, caused by charged weak currents, is given by the expressions

$$d\sigma^{\nu q} = \frac{G^2 \hat{s}}{\pi} dy \quad \text{and} \quad d\sigma^{\bar{\nu} q} = \frac{G^2 \hat{s}}{\pi}(1-y)^2 dy\,, \tag{11.2.16}$$

respectively, where $\hat{s} = (k+px)^2 \simeq 2xkp \cong xs(=2xME_\nu$ in the laboratory system).

The cross-section of the electron scattering on a quark is expressed by

$$d\sigma^{eq} = N_q \frac{2\pi\alpha^2}{\hat{s}} \frac{1}{y^2}[1+(1-y)^2]\,dy\,, \tag{11.2.17}$$

where

$$N_q = \begin{cases} 4/9 \text{ for the quarks } u, \bar{u} \\ 1/9 \text{ for the quarks } d, s, \bar{d}, \bar{s}\,. \end{cases}$$

With the aid of (11.2.16, 17) we obtain the cross-sections of electron, neutrino, and anti-neutrino scattering on a proton, expressed in terms of the distribution functions $q(x)$:

$$d\sigma^{ep} = \frac{2\pi\alpha^2}{s} \frac{1+(1-y)^2}{x^2 y^2} x\left[\frac{4}{9}(u+\bar{u}) + \frac{1}{9}(d+\bar{d}) + \frac{1}{9}(s+\bar{s})\right] dx\,dy\,,$$

$$d\sigma^{\nu p} = \frac{G^2 s}{\pi} x[d+s+\bar{u}(1-y^2)]\,dx\,dy\,, \tag{11.2.18}$$

$$d\sigma^{\bar{\nu} p} = \frac{G^2 s}{\pi} x[u(1-y^2)+\bar{d}+\bar{s}]\,dx\,dy\,,$$

where $s = (k+p)^2$.

The corresponding scattering cross-sections on a neutron are obtained by interchanging $u \rightleftarrows d$, $\bar{u} \rightleftarrows \bar{d}$ in (11.2.18).

Equation (11.2.18) can be utilized to find the distribution functions $q(x)$. For that, experimental values of the corresponding cross-sections (taken at fixed $q^2$) have to be substituted into the left-hand sides of (11.2.18). As already mentioned (Sect. 8.1), several sets of the functions $q_i(x)$, which differ slightly, have been obtained by such an analysis. One of such sets is presented in Fig. 11.4 for $xq_i(x)$. By integrating each of these distributions over $x$, we find the total relative momentum carried by a given quark:

$$U \equiv \int_0^1 dx\, xu(x) \approx 0.28\,, \quad D \equiv \int_0^1 dx\, xd(x) \approx 0.15\,,$$

$$\bar{U} \approx \bar{D} \approx 0.03\,, \quad S \approx \bar{S} \approx 0.01\,.$$

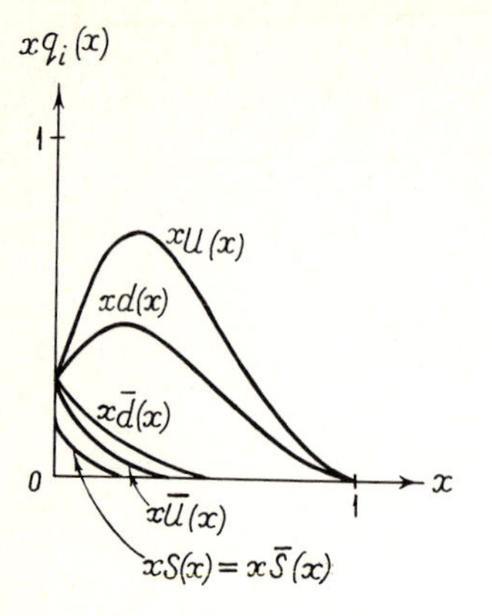


**Fig. 11.4.** Quark distribution functions

Thus, the total relative momenta of all quarks and all anti-quarks are equal to

$$Q = U + D + S \approx 0.44 \quad \text{and} \quad \bar{Q} = \bar{U} + \bar{D} + \bar{S} = 0.07 ,$$

respectively. Hence, the share of the quarks and anti-quarks in the proton momentum is about a half. The relation between the momenta of quarks and anti-quarks is characterized by one of the following quantities:

$$\alpha = \frac{\bar{Q}}{Q + \bar{Q}}, \qquad B = \frac{Q - \bar{Q}}{Q + \bar{Q}} = 1 - 2\alpha . \tag{11.2.19}$$

High-energy neutrino scattering experiments give the values of

$$\alpha \approx 0.15 \quad \text{and} \quad B \approx 0.7 .$$

It further follows from these experiments that the distribution functions $q(x)$ may be represented as polynomials with respect to $x$, of the type

$$x^a (1 - x)^b ,$$

where $a$ and $b$ are parameters to be determined from experimental data.

With only one half of the hadron momentum being carried by the quarks, it is to be concluded that the other half is associated with gluons. Let us introduce the *distribution function* $g(x)$ *for gluons* in a hadron. This function can be determined indirectly, e.g., from the distribution of strange and charmed particles produced by the gluons. Let $G = \int_0^1 dx\, x g(x)$ be the total relative momentum carried by the gluons. Then

$$Q + \bar{Q} + G = 1 .$$

That is to say, the quark model predicts the following structure for hadrons: as their constituents they have the valence quarks, the "sea" quarks, and the gluons.

Hence, exact scaling can be accounted for in the framework of non-interacting free quarks. Experiment shows, however, that scaling is violated. This discrepance indicates that, as a matter of fact, the quarks do interact within a hadron. Let us suppose that the "mediators" of the interactions between the quarks are gluons (playing the role analogous to that of photons in electrodynamics). Scaling violation is then due to gluon effects. These effects can be accounted for by quantum chromodynamics which will be considered in the next chapter.

# 12. Quantum Chromodynamics. Perturbation Theory

In quantum chromodynamics, hadrons are considered to be bound-states of quarks. Interactions with a hadron reduce to interactions with its quark constituents. Strong interactions between quarks are mediated by *gluons* – the gauge fields of the *colour group* $SU_3$. The coloured $SU_3$-symmetry is considered *exact*, and therefore the gluon mass is zero.

At present there exist two main lines in investigating the quantum chromodynamics: (i) *perturbative approach*, based on perturbation theory, and (ii) *non-perturbative approach*, which does not use perturbation theory.

In this chapter we shall construct the covariant perturbation theory for quantum chromodynamics. The applications of this theory will be illustrated by several examples. Finally, the basic methods of quantum chromodynamics will be presented which make it possible to take into account the contribution of the gluon effects to the structure functions.

The non-perturbative version of quantum chromodynamics, based on approximating the continuous space-time by a discrete lattice of finite size, will be dealt with in the next chapter.

## 12.1 Covariant Perturbation Theory for Quantum Chromodynamics

### 12.1.1 The Lagrangian for Quantum Chromodynamics

Consider the $SU_3$-triplet of spinor fields

$$\psi^a(x) = \begin{pmatrix} \psi^1(x) \\ \psi^2(x) \\ \psi^3(x) \end{pmatrix} . \tag{12.1.1}$$

The free Langrangian for such a triplet is written in the form

$$L(\psi^a(x), \partial_\mu \psi^a(x)) = \mathrm{i}\,\bar{\psi}^a \gamma_\mu \partial_\mu \psi^a - M \bar{\psi}^a \psi^a . \tag{12.1.2}$$

It is invariant under the global non-Abelian group of $SU_3$-transformations in a space referred to as the colour space:

$$\psi^a(x) \to \psi'^a(x) = \left[\exp\left(-\mathrm{i}\frac{g}{2}\lambda_m \varepsilon_m\right)\right]_{ab} \psi^b(x)\,,$$
$$\bar{\psi}^a(x) \to \bar{\psi}'^a(x) = \bar{\psi}^b(x)\left[\exp\left(\mathrm{i}\frac{g}{2}\lambda_m \varepsilon_m\right)\right]_{ba}\,. \tag{12.1.3}$$

Here $\varepsilon_m$ are the (constant) parameters of the group $SU_3$, $\lambda_m$ are the Gell-Mann matrices defined by (1.2.13); $g$ is the coupling constant.

According to (12.1.3) we have for the infinitesimal transformations of functions:

$$\delta\psi^a(x) = -\mathrm{i}\frac{g}{2}(\lambda_m)^{ab}\varepsilon_m \psi^b(x)\,,$$
$$\delta\bar{\psi}^a(x) = \mathrm{i}\frac{g}{2}\bar{\psi}^b(x)(\lambda_m)^{ba}\varepsilon_m\,, \tag{12.1.4}$$

or for the generators of the transformations

$$T^k_{ab} = -\mathrm{i}\frac{g}{2}(\lambda_k)_{ab}\,. \tag{12.1.5}$$

Taking into account (1.2.14) we find

$$[T^k, T^l]^{ab}_- = -\frac{g^2}{4}[\lambda_k, \lambda_l]^{ab}_- = g f_{klm} T^m_{ab}\,.$$

Let us now turn to the local group of $SU_3$-transformations. According to (2.1.12), the covariant derivative has the form

$$\nabla_\mu \psi^a(x) = \partial_\mu \psi^a(x) + \mathrm{i}\frac{g}{2}(\lambda_m)^{ab}\psi^b(x)\, V^m_\mu(x)\,. \tag{12.1.6}$$

As can be seen, in this case the octet of the vector fields $V^k_\mu(x)$ is gauged. These fields are called the *gluon fields.*

The Lagrangian for the gluon fields reads, by virtue of (2.2.23, 24),

$$\mathscr{L} = -\tfrac{1}{4}F^k_{\mu\nu}F^k_{\mu\nu}\,, \tag{12.1.7}$$

where

$$F^k_{\mu\nu} = \partial_\mu V^k_\nu - \partial_\nu V^k_\mu - \frac{g}{2}f_{npk}(V^n_\mu V^p_\nu - V^n_\nu V^p_\mu)$$

is the gluon field tensor.

The gluon fields transform, according to (2.2.7), as follows:

$$\delta V^m_\mu(x) = g f_{npm} V^p_\mu(x)\varepsilon_n(x) + \partial_\mu \varepsilon_m(x)\,. \tag{12.1.8}$$

For the total locally invariant Lagrangian we have

$$\mathscr{L} = \mathrm{i}\,\bar{\psi}^a \gamma_\mu \partial_\mu \psi^a - M\,\bar{\psi}^a \psi^a - \frac{g}{2}\,\bar{\psi}^a \gamma_\mu (\lambda_m)^{ab} V_\mu^m \psi^b$$

$$-\frac{1}{4}\left[\partial_\mu V_\nu^m - \partial_\nu V_\mu^m - \frac{g}{2} f_{npm}(V_\mu^n V_\nu^p - V_\nu^n V_\mu^p)\right]^2 . \tag{12.1.9}$$

The constant $g$ plays the role of the coupling constant for the fields. Equation (12.1.9) expresses the *Lagrangian for quantum chromodynamics.*

As follows from (2.2.33), to the Lagrangian (12.1.9) corresponds a conserved current which has four components with respect to the Lorentz group and eight components with respect to $SU_3$:

$$J_\mu^m(x) = \frac{\partial \mathscr{L}(x)}{\partial \nabla_\mu \psi^a}\,\frac{\mathrm{i}g}{2}(\lambda^m)^{ab}\psi^b(x) - 2g\frac{\partial \mathscr{L}(x)}{\partial F_{\mu\nu}^n} f^{mpn} V_\nu^p(x) . \tag{12.1.10}$$

### 12.1.2 Covariant Perturbation Theory

As discussed at length in Chap. 10, the magnitude of the effective (or running) coupling constant $\alpha_s(q^2)$ is small for the case of large momentum transfers $q^2$ (or small distances) in quantum chromodynamics. Thus, perturbation theory is applicable in this region. At small $q^2$, or large distances, the effective coupling constant $\alpha_s(q^2)$ becomes large, and perturbation theory fails to be applicable.

Let us formulate the covariant perturbation theory for quantum chromodynamics.

The Lagrangian for quantum chromodynamics is given by (12.1.9). The part of the Lagrangian associated with the gluon field can be represented, for convenience, as the sum of two Lagrangians (in the $\alpha$-gauge),

$$\mathscr{L}_0 = -\frac{1}{4}(\partial_\mu V_\nu^m - \partial_\nu V_\mu^m)^2 - \frac{1}{2\alpha}(\partial_\mu V_\mu^m)^2 \tag{12.1.11}$$

and

$$\mathscr{L}_{0\mathrm{I}} = g f_{npm}(\partial_\mu V_\nu^m)\, V_\mu^n V_\nu^p - \frac{g^2}{4} f_{jpm} f_{lnm} V_\mu^j V_\nu^p V_\mu^l V_\nu^n . \tag{12.1.12}$$

The first of them corresponds to a non-interacting gluon field, and the second one to the self-interaction of gluon fields.

The generating functional $W$ of quantum chromodynamics in the $\alpha$-gauge, according to (5.1.26), reads as follows:

$$W = R\int \mathscr{D}V_\mu^m(x)\,\mathscr{D}\bar{\psi}^a(x)\,\mathscr{D}\psi^a(x)\,\mathscr{D}\bar{c}^m(x)\,\mathscr{D}c^m(x)$$

$$\times \exp\left\{\mathrm{i}\int dx\left[\frac{1}{2}V_\mu^m(x)\,\delta^{mn}\left(\Box g_{\mu\nu} - \left(1-\frac{1}{\alpha}\right)\partial_\mu\partial_\nu\right)V_\nu^n(x)\right.\right.$$

$$+\,\bar{\psi}^a(x)\,\delta^{ab}(\mathrm{i}\gamma_\mu\partial_\mu - M)\,\psi^b(x) + \bar{c}^m(x)\,\delta_{mn}\Box\, c^n(x)$$

$$+J_\mu^m(x)\,V_\mu^m(x)+\bar{\eta}^a(x)\,\psi^a(x)+\bar{\psi}^a(x)\,\eta^a(x)+\bar{\chi}^m(x)\,c^m(x)$$
$$\left.\left.+\bar{c}^m(x)\,\chi^m(x)\right]\right\}, \tag{12.1.13}$$

where

$$R=\exp\left\{\mathrm{i}\int dx\,\mathcal{L}_\mathrm{I}\left(\frac{1}{\mathrm{i}}\,\frac{\delta}{\delta J_\mu^m(x)},\ldots\right)\right\}.$$

Upon some transformations, the form of (5.1.30) is recovered:

$$W(J_\mu^m,\bar{\eta}^a,\eta^a,\bar{\chi}^m,\chi^m)$$
$$=R\exp\{-\mathrm{i}\int dx\,dy\,[\tfrac{1}{2}J_\mu^m(x)(K_{\mu\nu}^{mn}(x-y))^{-1}J_\nu^n(y)$$
$$+\bar{\eta}^a(x)\,\mathcal{K}_{ab}^{-1}(x-y)\,\eta^b(y)+\bar{\chi}^m(x)\,\mathcal{K}_{mn}^{-1}(x-y)\,\chi^n(y)]\}. \tag{12.1.14}$$

For the generating functional for the matrix elements, by virtue of (5.1.31), we have

$$S(V_{\mu 0}^m,\ldots)=R\exp\{-\mathrm{i}\int dx(V_{\mu 0}^m J_\mu^m+\bar{\psi}_0^a\eta^a+\bar{\eta}^a\psi_0^a)$$
$$-\mathrm{i}\int dx\,dy\,[\tfrac{1}{2}J_\mu^m(x)(K_{\mu\nu}^{mn}(x-y))^{-1}J_\nu^n(y)+\bar{\eta}^a(x)\,\mathcal{K}_{ab}^{-1}(x-y)\,\eta^b(y)$$
$$+\bar{\chi}^m(x)\,\mathcal{K}_{mn}^{-1}(x-y)\,\chi^n(y)]\}|_{J_\mu^m=\ldots=0}\,. \tag{12.1.15}$$

The operators $[K_{\mu\nu}^{mn}(x-y)]^{-1}$, $\mathcal{K}_{ab}^{-1}(x-y)$, $\mathcal{K}_{mn}^{-1}(x-y)$ are given by (5.3.7–9).

A series expansion of the differential operator figuring in (12.1.15) with respect to the coupling constant yields the following perturbation-theory expression for the case of quantum chromodynamics:

$$S(V_{\mu 0}^n,\ldots)=\left[1+\frac{g}{2}\int dx\,\frac{\delta}{\delta\eta^a(x)}\,\gamma_\mu(\lambda_m)^{ab}\,\frac{\delta}{\delta\bar{\eta}^b(x)}\,\frac{\delta}{\delta J_\mu^m(x)}\right.$$
$$-gf_{npm}\int dx\left(\partial_\mu\frac{\delta}{\delta J_\nu^m(x)}\right)\frac{\delta}{\delta J_\mu^n(x)}\,\frac{\delta}{\delta J_\nu^p(x)}$$
$$-\frac{\mathrm{i}g^2}{4}\int dx\,f_{jpm}f_{lnm}\frac{\delta}{\delta J_\mu^j(x)}\,\frac{\delta}{\delta J_\nu^p(x)}\,\frac{\delta}{\delta J_\mu^l(x)}\,\frac{\delta}{\delta J_\nu^n(x)}$$
$$\left.+g\int dx\,f_{pnm}\left(\partial_\mu\frac{\delta}{\delta\chi^m(x)}\right)\frac{\delta}{\delta J_\mu^n(x)}\,\frac{\delta}{\delta\bar{\chi}^p(x)}+\ldots\right]$$
$$\times\exp\{-\mathrm{i}\int dx(V_{\mu 0}^m J_\mu^m+\bar{\psi}_0^a\eta^a+\bar{\eta}_a\psi_0^a)$$
$$-\mathrm{i}\int dx\,dy\,[\tfrac{1}{2}J_\mu^m(K_{\mu\nu}^{mn})^{-1}J_\nu^n$$
$$+\bar{\eta}^a\,\mathcal{K}_{ab}^{-1}\eta^b+\bar{\chi}^m\,\mathcal{K}_{mn}^{-1}\chi^n]\}|_{J_\mu^m=\ldots=0}\,. \tag{12.1.16}$$

The expressions for the amplitudes in the case of quantum chromodynamics involve various combinations of the functions of free particles (quarks, gluons), propagators (for quarks, gluons, ghost fields), interaction vertices

**Table 12.1.** Correspondence rules for the amplitudes in quantum chromodynamics

| Physical state | Mathematical expression | Diagram |
|---|---|---|
| Quark in the initial state | $v_r^{(-)}(p)$ |  |
| Anti-quark in the initial state | $\bar{v}_r^{(-)}(p)$ |  |
| Quark in the final state | $\bar{v}_{r'}^{(+)}(p')$ |  |
| Anti-quark in the final state | $v_{r'}^{(+)}(p')$ |  |
| Gluon in the initial or the final state | $\varepsilon_\mu^k$ | 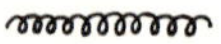 |
| Motion of virtual quark from $a$ to $b$ | $-\dfrac{\delta_{ab}}{\not{p}-m}$ | 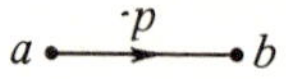 $a$ $p$ $b$ |
| Motion of virtual anti-quark from $a$ to $b$ | $\dfrac{\delta_{ab}}{\not{p}+m}$ | 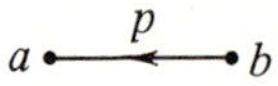 $a$ $p$ $b$ |
| Motion of virtual gluon between the states $\alpha a$ and $\beta b$ | $\left[g_{\alpha\beta}-(1-\alpha)\dfrac{k_\alpha k_\beta}{k^2}\right]\dfrac{\delta_{ab}}{k^2}$ | 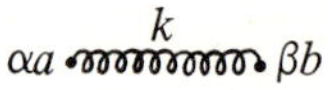 $\alpha a$ $k$ $\beta b$ |
| Motion of virtual ghost | $\delta_{ab}\dfrac{1}{k^2}$ | 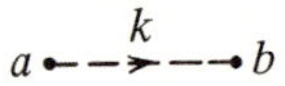 $a$ $k$ $b$ |
| Quark-gluon interaction vertex | $g\gamma_\alpha(\lambda_a)^{cb}$ | 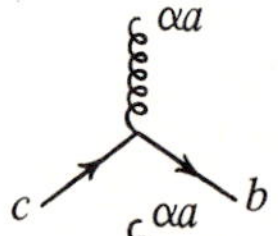 $\alpha a$, $c$, $b$ |
| Ghost-gluon interaction vertex | $\mathrm{i}gf_{abc}q_\alpha$ | 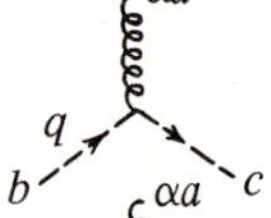 $\alpha a$, $q$, $b$, $c$ |
| Three-gluon interaction vertex | $-\mathrm{i}gf_{abc}[(r-q)_\alpha g_{\beta\gamma}$ $+(p-r)_\beta g_{\alpha\gamma}+(q-p)_\gamma g_{\alpha\beta}]$ | 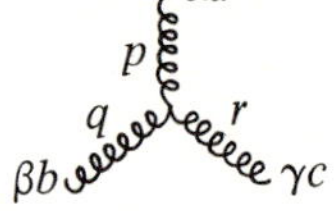 $\alpha a$, $p$, $q$, $r$, $\beta b$, $\gamma c$ |
| Four-gluon interaction vertex | $-\mathrm{i}g^2f_{kac}f_{kbd}(g_{\alpha\beta}g_{\gamma\delta}-g_{\alpha\delta}g_{\beta\gamma})$ $-\mathrm{i}g^2f_{kad}f_{kbc}$ $\times(g_{\alpha\beta}g_{\gamma\delta}-g_{\alpha\gamma}g_{\beta\delta})-\mathrm{i}g^2f_{kab}f_{kcd}$ $\times(g_{\alpha\gamma}g_{\beta\delta}-g_{\alpha\delta}g_{\beta\gamma})$ |  $\alpha a$, $\beta b$, $\gamma c$, $\delta d$ |

(of three gluons, four gluons, one gluon and two ghosts, one gluon and two quarks). The expressions for these propagators and interaction vertices have already been found and listed in Tables 5.3, 4.

The *correspondence rules* for the amplitudes in the momentum representation for quantum chromodynamics are summarized in Table 12.1.

The Lagrangian (12.1.9) for quantum chromodynamics has formal similarity with the Lagrangian (5.1.7) for quantum electrodynamics. They differ, however, in that one photon of quantum electrodynamics is replaced by eight gluons in quantum chromodynamics and, as distinct from photons, the gluons

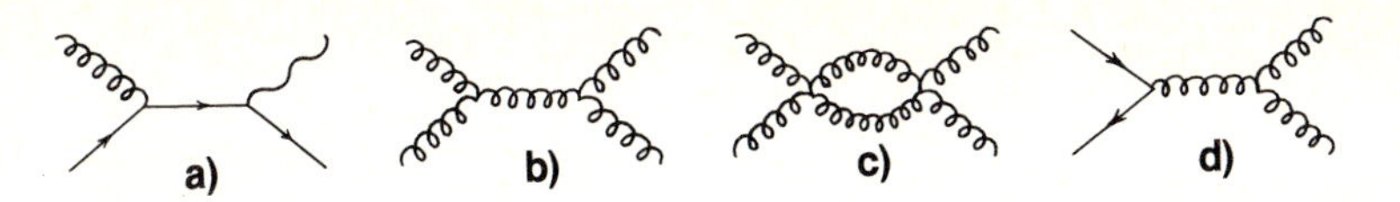


**Fig. 12.1a – d.** Specific diagrams of quantum chromodynamics: (**a**) gluon scattering on a quark with photon production; (**b**), (**c**) gluon-gluon scattering; (**d**) quark pair transformation into two gluons

are self-interacting. A comparison of Table 12.1 with Table 7.2 elucidates the specifics of quantum chromodynamics most clearly. The non-Abelian character of the gluon field brought about the appearance of structure constants and generators of the gauge group in the expressions for quantum chromodynamics; the self-interaction of the gluon field gave rise to additional (three-gluon and four-gluon) interaction vertices. Several Feynman diagrams, characteristic of quantum chromodynamics and having no analogues in quantum electrodynamics, are depicted in Fig. 12.1.

### 12.1.3 Renormalizability of Quantum Chromodynamics

The Lagrangian for quantum chromodynamics is invariant under gauge transformations, and the usual renormalization procedure (Sect. 9.2) is applicable to quantum chromodynamics. The Lagrangian and the currents in quantum chromodynamics do not contain the $\gamma_5$-matrix, i.e., quantum chromodynamics does not exhibit anomalies. Hence, quantum chromodynamics is renormalizable.

## 12.2 Examples of Perturbation-Theory Calculations

### 12.2.1 Basic Processes

One can see that perturbation theory is applicable to processes leading to production, in the final state, of a particle (hadron, photon, lepton) with large momentum transfer $q^2$ or of a system with large virtual or actual mass. Such processes are called *hard*. The following processes are of the hard type.

1) Deep-inelastic lepton scattering on nucleons (Sect. 11.1).
2) Lepton-pair annihilation (e.g., $e^+e^- \to h$; $e^+e^- \to h+X$, where $h$ denotes a hadron and $X$ denotes all other particles produced in the inclusive process).
3) Production, upon collision of hadrons, of
   i) a muon pair with large effective mass ($pp \to \mu^+\mu^- X$) (the Drell-Yan process);
   ii) a pair of charmed particles ($pp \to c\bar{c}X$);
   iii) $\pi$-mesons with large momentum transfer ($pp \to \pi X$).
4) Formation of jets, i.e., of particle fluxes concentrated within small solid angles. (Jets may be initiated both by quarks and by gluons.)

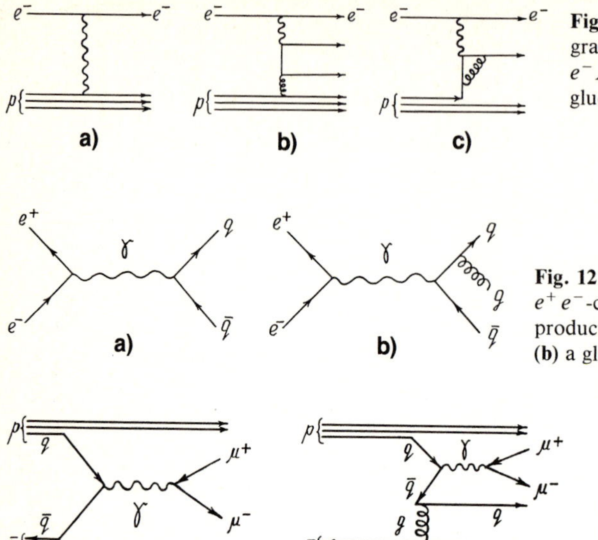


Fig. 12.2a–c. Feynman diagrams for the process $e^- p \to e^- X$: (a) exact scaling; (b), (c) the gluon corrections

Fig. 12.3a, b. Feynman diagrams for $e^+e^-$-collision accompanied by the production of (a) a pair of quarks; (b) a gluon and a pair of quarks

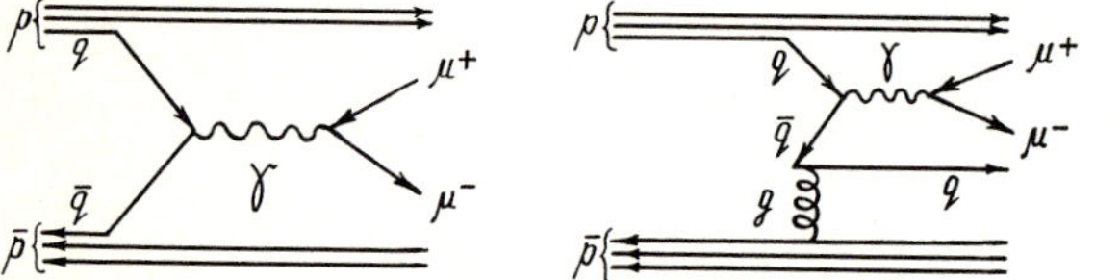


Fig. 12.4. Feynman diagrams for the Drell-Yan process $pp \to \mu^+ \mu^- X$

Studying processes by means of quantum chromodynamics is carried out in two steps: (i) the cross-sections are calculated for the sub-processes involving quarks and gluons; (ii) one goes over from quarks and gluons to hadrons.

Examples of Feynman diagrams for some of the processes mentioned are depicted, within perturbation theory, in Figs. 12.2–4. The corresponding sub-processes are also shown in these figures.

## 12.2.2 Cross-Sections for the Sub-Processes

As an example of calculations in the first non-vanishing order of perturbation theory, we shall derive an expression for the sub-process of conversion of two gluons into a charm quark pair: $gg \to c\bar{c}$ (Fig. 12.5). Making use of Table 12.1 we find the expression for the amplitude of this sub-process in the first non-vanishing order of perturbation theory:

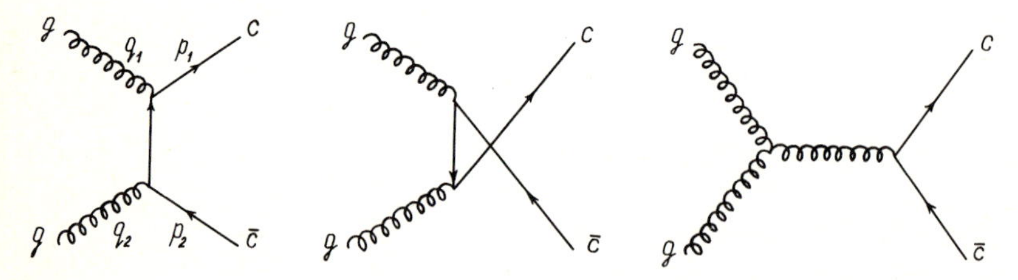


Fig. 12.5. Feynman diagrams in the first non-vanishing order of perturbation theory of the sub-process $gg \to c\bar{c}$

$$M = g^2 \bar{v}^{(+)}_{r_c}(p_1)[O_1 + O_2 + O_3]\, v^{(+)}_{r_{\bar{c}}}(p_2)\,. \tag{12.2.1}$$

Here

$$O_1 = \frac{1}{4}\lambda^a_{ik}\lambda^b_{kj}\frac{\gamma_\alpha[\not{p}_1 - \not{q}_1 + m]\,\gamma_\beta}{m^2 - \hat{t}}\,\varepsilon^a_{1\alpha}\varepsilon^b_{2\beta}\,,$$

$$O_2 = \frac{1}{4}\lambda^b_{ik}\lambda^a_{kj}\frac{\gamma_\beta[\not{p}_1 - \not{q}_2 + m]\,\gamma_\alpha}{m^2 - \hat{u}}\,\varepsilon^a_{1\alpha}\varepsilon^b_{2\beta}\,,$$

$$O_3 = \frac{1}{2\hat{s}}\,\gamma_\delta\lambda^c_{ij}f_{abc}[(r - q_2)_\alpha g_{\beta\delta} + (q_1 - r)_\beta\, g_{\alpha\delta} + (q_2 - q_1)_\delta g_{\alpha\beta}]\,\varepsilon^a_{1\alpha}\varepsilon^b_{2\beta}\,,$$

$$\hat{s} = (q_1 + q_2)^2\,, \quad \hat{t} = (p_1 - q_1)^2\,, \quad \hat{u} = (p_1 - q_2)^2\,, \quad r = q_1 + q_2\,;$$

the $\lambda_i$-matrices are given by (1.2.13), $\varepsilon^a_\mu$ is the four-vector of gluon polarization, and $m$ is the quark mass.

Then we have for the square of the absolute value, $|\bar{M}|^2$, of the matrix element of the sub-process

$$|\bar{M}|^2 = \frac{1}{64}\,\frac{1}{4}\sum_{\text{colour}}\sum_{\text{spin}}|M|^2$$

$$= \frac{g^4}{256}\sum_{\text{colour}}\sum_{\text{spin}}|\bar{v}^{(+)}_{r_c}(p_1)(O_1 + O_2 + O_3)\,v^{(+)}_{r_{\bar{c}}}(p_2)|^2\,. \tag{12.2.2}$$

The summation is carried out over the colour states of the initial gluons and the final quarks as well as over the spin projections of gluons and quarks, while the averaging is performed over the types and the spins of the initial gluons. In summing over the colour states, it should be borne in mind that

$$\lambda^a_{ik}\lambda^b_{kj}\lambda^b_{jn}\lambda^a_{ni} = \mathrm{Tr}\{\lambda^a\lambda^b\lambda^b\lambda^a\} = \frac{256}{3}\,,$$

$$\lambda^b_{ik}\lambda^a_{kj}\lambda^a_{jn}\lambda^b_{ni} = \mathrm{Tr}\{\lambda^b\lambda^a\lambda^a\lambda^b\} = \frac{256}{3}\,,$$

$$f_{abc}\lambda^c_{ij}\lambda^d_{ji}f_{abd} = f_{abc}f_{abd}\mathrm{Tr}\{\lambda^c\lambda^d\} = 48\,,$$

$$\lambda^a_{ik}\lambda^b_{kj}\lambda^a_{jn}\lambda^b_{ni} = \mathrm{Tr}\{\lambda^a\lambda^b\lambda^a\lambda^b\} = -\frac{32}{3}\,,$$

$$-\mathrm{i}\lambda^a_{ik}\lambda^b_{kj}\lambda^c_{ji}f_{abc} = -\mathrm{i}f_{abc}\mathrm{Tr}\{\lambda^a\lambda^b\lambda^c\} = 48\,,$$

$$-\mathrm{i}\lambda^b_{ik}\lambda^a_{kj}\lambda^c_{ji}f_{abc} = -\mathrm{i}f_{abc}\mathrm{Tr}\{\lambda^b\lambda^a\lambda^c\} = -48\,.$$

In summing over the gluon spin projections, it should be taken into account that, due to the gauge chosen, $q^\alpha_1\varepsilon^a_{1\alpha} = 0$ and $q^\beta_2\varepsilon^b_{2\beta} = 0$. Then the divergence of the matrix element (12.2.1) is zero, and the summation in (12.2.2) can be carried out over physical and unphysical gluon polarizations by replacing $\not{\varepsilon}^a_1$

and $\not{\varepsilon}_2^b$ by $\gamma_\alpha$ and $\gamma_\beta$, respectively. All this being considered, we obtain, e.g., for the third term in the formula (12.2.2)

$$\sum_{\text{colour}} \sum_{\text{spin}} |\bar{v}_{r_c}^{(+)} O_3 v_{r_{\bar{c}}}^{(+)}|^2 = \mathrm{Tr}\{(\not{p}_1+m) O_3 (\not{p}_2-m)\bar{O}_3\}$$

$$= \frac{1}{\hat{s}^2} f_{abc} f_{abd} \mathrm{Tr}\{\lambda^c \lambda^d\} \mathrm{Tr}\{(\not{p}_1+m)[g_{\alpha\beta}(\not{q}_2-\not{q}_1)+2\gamma_\alpha q_{1\beta}-2\gamma_\beta q_{2\alpha})]$$

$$\times (\not{p}_2-m)[g_{\alpha\beta}(\not{q}_2-\not{q}_1)+2\gamma_\alpha q_{1\beta}-2\gamma_\beta q_{2\alpha}]\}$$

$$= \frac{48\cdot 16}{\hat{s}^2}(m^2-\hat{u})(m^2-\hat{t}) .$$

Calculating, in a similar way, the other terms of (12.2.2) and subsequently summing them we arrive at the following expression for the differential cross-section of the sub-process $gg\to c\bar{c}$:

$$d\hat{\sigma} = \frac{\alpha_s}{16\hat{s}^2}\left[\frac{12}{\hat{s}^2}(m^2-\hat{t})(m^2-\hat{u}) + \frac{8}{3}\,\frac{(m^2-\hat{t})(m^2-\hat{u})-2m^2(m^2+\hat{t})}{(m^2-\hat{t})^2}\right.$$

$$+\frac{8}{3}\,\frac{(m^2-\hat{t})(m^2-\hat{u})-2m^2(m^2+\hat{u})}{(m^2-\hat{u})^2} - \frac{2m^2}{3}\,\frac{(\hat{s}-4m^2)}{(m^2-\hat{t})(m^2-\hat{u})}$$

$$\left. -6\frac{(m^2-\hat{t})(m^2-\hat{u})+m^2(\hat{u}-\hat{t})}{\hat{s}(m^2-\hat{t})} - 6\frac{(m^2-\hat{t})(m^2-\hat{u})+m^2(\hat{t}-\hat{u})}{\hat{s}(m^2-\hat{u})}\right], \tag{12.2.3}$$

where

$$\alpha_s = \frac{g^2}{4\pi} .$$

To find the total cross-section of the sub-process, (12.2.3) has to be integrated over $\hat{t}$ in the physical range

$$m^2 - \frac{\hat{s}}{2} - \frac{1}{2}\sqrt{\hat{s}(\hat{s}-4m^2)} \leqslant \hat{t} \leqslant m^2 - \frac{\hat{s}}{2} + \frac{1}{2}\sqrt{\hat{s}(\hat{s}-4m^2)} .$$

This yields

$$\hat{\sigma}(\hat{s}) = \frac{\pi\alpha_s^2}{3\hat{s}}\left[-\left(7+\frac{31m^2}{\hat{s}}\right)\frac{1}{4}X + \left(1+\frac{4m^2}{\hat{s}}+\frac{m^4}{\hat{s}^2}\right)\ln\frac{1+X}{1-X}\right], \tag{12.2.4}$$

where

$$X = \sqrt{1-\frac{4m^2}{\hat{s}}} .$$

Likewise, the cross-sections of other sub-processes can be calculated.

### 12.2.3 Radiative Corrections

As an example of calculations within higher orders of perturbation theory for quantum chromodynamics, we shall find the *radiative corrections* to the Green's functions (similar calculations for another model have been made in Chap. 10; that model includes quantum chromodynamics as a special case).

1) Let us first concern ourselves with the Green's function $D_{\mu\nu}^{ab}(k)$ for the Yang-Mills field. The corresponding Feynman diagrams, within the second order of perturbation theory and in the Landau gauge, are shown in Fig. 12.6; the bare Green's function has the form

$$D_{\mu\nu}^{ab}(k) = \mathscr{D}_{\mu\nu}^{ab}(k) + \frac{\mathrm{i}g_0^2}{(2\pi)^4}\, \mathscr{D}_{\mu\lambda}^{ac}(k)\,\Pi_{\lambda\varrho}^{cd}(k)\,\mathscr{D}_{\varrho\nu}^{db}(k)\,, \tag{12.2.5}$$

where $\mathscr{D}_{\mu\nu}^{ab}(k)$ is the renormalized Green's function, $\Pi_{\lambda\varrho}^{cd}(k)$ is the polarization operator, and $g_0$ is the bare coupling constant of the gauge field.

Let us find the expression for the polarization operator in the asymptotic region for the case of $SU_n$-symmetry. By means of calculations analogous to those performed at the beginning of Sect. 10.2, and retaining the terms which contain $\ln(\Lambda/\mu)$ we find

$$\Pi_{\lambda\varrho}^{cd}(k) = -\mathrm{i}\pi^2\left(\frac{13}{3}n - \frac{4}{3}m\right)\ln\frac{\Lambda}{\mu}\,\delta^{cd}k^2\left(g_{\lambda\varrho} - \frac{k_\lambda k_\varrho}{k^2}\right), \tag{12.2.6}$$

where $m$ is the number of fundamental fermion multiplets of the group $SU_n$.

Let us choose the propagator of the Yang-Mills field in the form

$$\mathscr{D}_{\mu\nu}^{ab}(k) = \frac{\delta^{ab}}{k^2}\left(g_{\mu\nu} - \frac{k_\mu k_\nu}{k^2}\right). \tag{12.2.7}$$

By substituting (12.2.6, 7) into (12.2.5) we find the desired expression for the bare Green's function of the Yang-Mills field in the second order or perturbation theory:

$$D_{\mu\nu}^{ab}(k) = Z_2\,\mathscr{D}_{\mu\nu}^{ab}(k)\,. \tag{12.2.8}$$

Here $Z_2$ is the renormalization factor (in the $g^2$-approximation) given by

$$Z_2 = 1 + \frac{g_0^2}{16\pi^2}\left(\frac{13}{3}n - \frac{4}{3}m\right)\ln\frac{\Lambda}{\mu}\,. \tag{12.2.9}$$

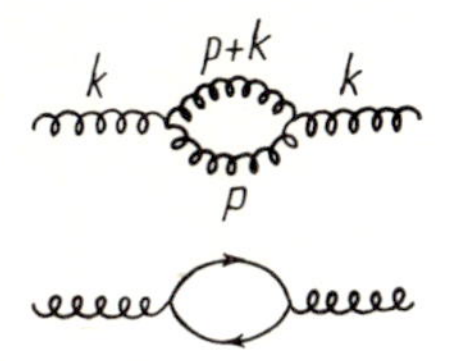


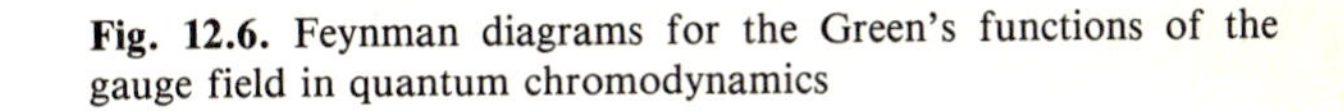

**Fig. 12.6.** Feynman diagrams for the Green's functions of the gauge field in quantum chromodynamics

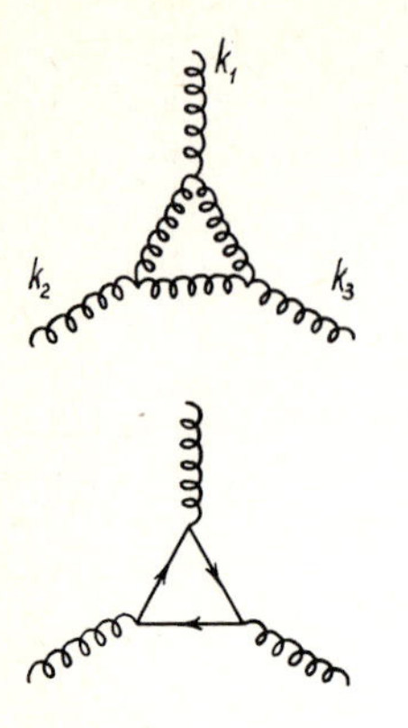


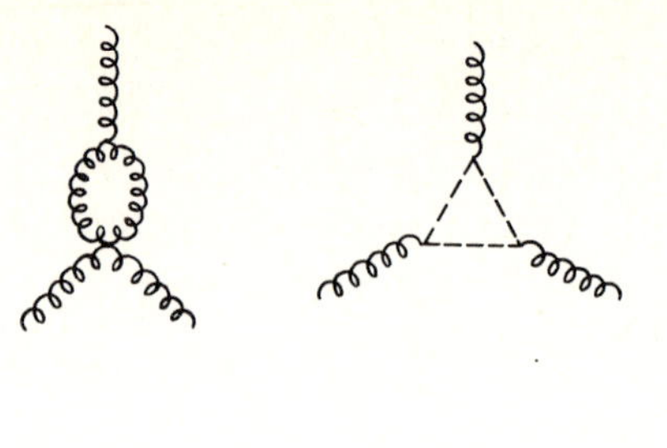

**Fig. 12.7.** Feynman diagrams for the Green's functions with three external gauge fields in quantum chromodynamics

2) By similar calculations, one can find the radiative corrections to the Green's functions $\Gamma^{abc}_{\mu\nu\varrho}(k_1, k_2, k_3)$ involving three external Yang-Mills fields in the third order of perturbation theory in the Landau gauge. The corresponding Feynman diagrams are given in Fig. 12.7. Calculating $\Gamma^{abc}_{\mu\nu\varrho}(k_1, k_2, k_3)$ in the asymptotic region for the case of $SU_n$-symmetry yields

$$\Gamma^{abc}_{B\mu\nu\varrho}(k_1, k_2, k_3) = Z_1^{-1}\Gamma^{abc}_{\mu\nu\varrho}(k_1, k_2, k_3)\,,$$

where

$$Z_1^{-1} = 1 - \frac{g_0^2}{16\pi^2}\left(\frac{17}{6}n - \frac{4}{3}m\right)\ln\frac{\Lambda}{\mu}\,. \tag{12.2.10}$$

### 12.2.4 The Effective (or Running) Coupling Constant

Let us find the expression for the coupling constant in quantum chromodynamics. The effective charge is determined by (10.1.20) and the function $\beta(g)$ by (10.1.16).

The renormalized charge $g$ is related to the bare charge $g_0$ by $g = Z_g g_0 = Z_1^{-1} Z_2^{3/2} g_0$. Therefore, by using (12.2.9, 10), we obtain in the one-loop approximation (which involves the diagrams of Figs. 12.6, 7), in the case of $SU_3$-symmetry

$$\beta(g) = -\frac{g^3}{48\pi^2}(33 - 2n_\mathrm{f})\,. \tag{12.2.11}$$

From this, we have for the effective charge in the one-loop approximation

$$\alpha_\mathrm{s}(k^2) = \frac{g^2(k^2)}{4\pi} = \frac{g^2}{4\pi\left[1 + ag^2\ln\left(\dfrac{k^2}{\mu^2}\right)\right]}\,, \tag{12.2.12}$$

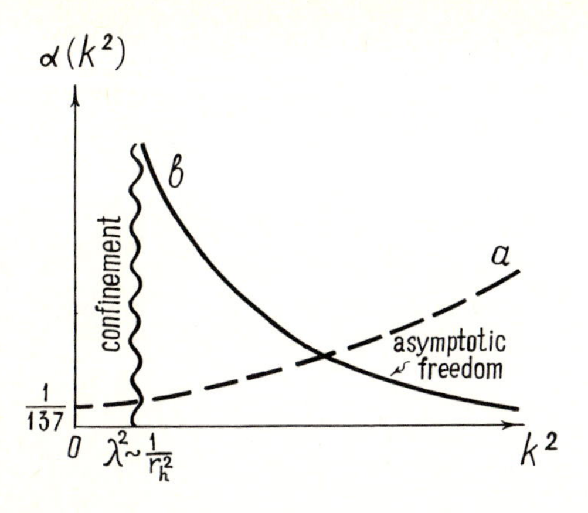


**Fig. 12.8.** $k^2$-dependence of the effective (running) charge: (*a*) for electrodynamics; (*b*) for chromodynamics ($r_h$ is the radius of hadron)

with $a=(33-2n_f)/(48\pi^2)$ and $n_f$ denoting the number of flavours of the quarks $(u, d, s, c, \ldots)$ involved in the theory.

In quantum electrodynamics (Fig. 12.8, curve *a*), the coupling constant $e$ is, by definition, equal to the effective charge $e(k^2)$ at $k^2=0$ (the Thompson limit): $e=e(k^2=0)$.

In quantum chromodynamics (Fig. 12.8, curve *b*), the limiting values of the effective charge $g(k^2)$ are either infinite $(k^2=0)$ or zero $(k^2=\infty)$, and, as a consequence, the coupling constant is determined as the value of the effective charge $g(k^2)$ taken at some arbitrary value of $\mu^2$. At the same time the value of $\alpha_s(k^2)$ must be independent of the choice of the renormalization point $\mu^2$. Therefore, taking $\mu_1^2$ instead of $\mu^2$ will yield for the expression contained in (12.2.12)

$$\frac{1}{g^2(\mu^2)}-a\ln\mu^2=\frac{1}{g^2(\mu_1^2)}-a\ln\mu_1^2\equiv -a\ln\lambda^2 .$$

From this, it follows that the true parameter is not $\mu^2$ or $g(\mu^2)$, but rather the quantity $\lambda$, which has the dimension of mass and is independent of $\mu^2$:

$$\ln\lambda^2=-\frac{1}{ag^2(\mu^2)}+\ln\mu^2 .$$

Substituting the latter expression into (12.2.12) we obtain the desired expression for the coupling constant of quantum chromodynamics in the one-loop approximation:

$$\alpha_s(k^2)=\frac{1}{4\pi a\ln(k^2/\lambda^2)}=\frac{12\pi}{(33-2n_f)\ln(k^2/\lambda^2)} . \tag{12.2.13}$$

Hence, in quantum chromodynamics there appears a constant $\lambda$ with the dimension of mass. This constant has no connection with the arbitrary choice of the renormalization point and is not given by the theory.

In a similar way, one can find the expression for the coupling constant $\alpha_s(k^2)$ in the two-loop approximation.

The calculated value of the constant $\lambda$ depends on the renormalization prescription and the choice of the gauge. This dependence is due to the fact that the expression of the two-loop corrections are different for different renormalization schemes and various choices of the gauge. If we perform the *regularization in momentum space* and choose the *Feynman gauge* (Sect. 5.3), then we obtain an expression for the quantity $\lambda$ which is denoted by $\lambda_{\mathrm{MOM}}$. The value of $\lambda_{\mathrm{MOM}}$ can be found from experimental data. By analyzing the entirety of experimental data pertaining to various processes, it has been found that

$$\lambda_{\mathrm{MOM}}^{\exp} \sim 100\text{—}200\ \mathrm{MeV}\,.$$

### 12.2.5 Asymptotic Freedom of Quantum Chromodynamics

It has been demonstrated in Sect. 10.1 that a theory is asymptotically free provided that $\beta < 0$. Accordingly, it follows from (12.2.11) that quantum chromodynamics is asymptotically free as long as $n_f \leqslant 16$, i.e., as long as it involves not more than sixteen flavours of quarks.

### 12.2.6 Scaling Violation

As shown in Sect. 11.2, the model with non-interacting quarks leads to exact scaling. However, there is experimental evidence (Sect. 11.1) for scaling violation, implying that the structure functions $W_i$ not only depend on $x$ but also on $q^2$. Scaling violation may stem from the interaction between the quarks caused by the exchange of gluons. This opens a possibility of accounting for scaling violation within the framework of quantum chromodynamics.

Quantum chromodynamics can be most effectively applied to analyzing the hard processes. For them, the coupling constant is small, and perturbation theory may be utilized.

A comparison of the results of calculations with the corresponding experimental data provides a means of verifying quantum chromodynamics.

The aim of the theory is to calculate the dependence of various quantities (primarily, structure functions) on the momentum transfer $q^2$. For this purpose, a number of methods have been propounded. We shall schematically expose three of them.

## 12.3 The Method of Operator Product Expansion

This rather formal method of calculating the dependence of the structure functions on the momentum transfer $q^2$ is based on using the Wilson expansion of the operator product for the currents as well as the renormalization group equation. We shall spell the main steps of the method by considering the example of the deep-inelastic electron scattering on protons (Sect. 11.1).

1) We proceed from the expression (11.1.4) for the amplitude $W_{\mu\nu}$ that describes the strong interaction vertex associated with the gluons:

$$W_{\mu\nu}(p,q) = \frac{1}{4\pi} \sum_{\text{spin}} \int d\zeta \, e^{iq\zeta} \langle p | J_\mu(\zeta) J_\nu(0) | p \rangle \,. \tag{12.3.1}$$

2) By virtue of the optical theorem, the amplitude $W_{\mu\nu}$ is related to the amplitude $T_{\mu\nu}$ of the elastic virtual photon scattering on a proton by

$$W_{\mu\nu} = \frac{1}{2\pi} \operatorname{Im}\{T_{\mu\nu}\} \,, \tag{12.3.2}$$

where

$$T_{\mu\nu}(p,q) = i \int d\zeta \, e^{iq\zeta} \langle p | T(J_\mu(\zeta) J_\nu(0)) | p \rangle \,. \tag{12.3.3}$$

Here we have for $T_{\mu\nu}$, by analogy with (11.1.5),

$$T_{\mu\nu}(\nu,q^2) = \left(-g_{\mu\nu} + \frac{q_\mu q_\nu}{q^2}\right) T_1(\nu,q^2) + \frac{1}{M^2}\left(p_\mu - \frac{pq}{q^2} q_\mu\right)\left(p_\nu - \frac{pq}{q^2} q_\nu\right) T_2(\nu,q^2) \,. \tag{12.3.4}$$

It is more convenient to use first $T_{\mu\nu}$ and then return, with the help of (12.3.2), to $W_{\mu\nu}$.

3) We make use of the fact that the main contribution to the integral (12.3.3) in the deep-inelastic limit is due to the region in the vicinity of the light-cone. In fact, in the nucleon rest frame,

$$p = (M,0,0,0) \quad \text{and} \quad q = (\nu,0,0,\sqrt{\nu^2 - q^2}) \,.$$

In the deep-inelastic region ($\nu \to \infty$, $-q^2 \to \infty$, $x = -q^2/(2\nu M)$ and finite) we have

$$(q\zeta) \simeq \nu(\zeta_0 - \zeta_3) - Mx\zeta_3 \,.$$

The exponential factor $\exp(iq\zeta)$ is fastly-oscillating, and the integral will have a significant magnitude only if

$$\zeta_0 - \zeta_3 = O\left(\frac{1}{\nu}\right),$$

i.e., in the vicinity of the light-cone. Therefore, we confine ourselves to dealing with this region.

4) Let us make use of the *Wilson expansion* to treat the $T$-product of the currents entering (12.3.3), in the vicinity of the light-cone. For the product of two local scalar operators, this expansion reads

$$J(x)J(y) \approx \sum_n F_n(z^2) z_{\mu_1} z_{\mu_2} \ldots z_{\mu_n} O_n^{\mu_1 \mu_2 \ldots \mu_n}(R) . \tag{12.3.5}$$

Here $z = x - y$, $R = \frac{1}{2}(x+y)$; $F_n(z^2)$ denotes non-operator functions generally having a singularity at $z^2 \to 0$; $O_n$ are local operators which do not contain singularities.

It should be emphasized that all singularities on the light-cone occurring in the expansion (12.3.5) are contained only in the non-operator factor.

It should further be noted that Wilson's expansion is a generalization of results obtained by analyzing a number of field theoretical models.

5) By analogy with (12.3.5), the Wilson expansion for the $T$-product of vector current operators is written as

$$\begin{aligned} T(J_\mu(x) J_\nu(y)) \sim & \frac{g^{\mu\nu}}{(z^2 - \mathrm{i}\varepsilon)^2} \sum_i \sum_{n=1}^{\infty} \frac{\mathrm{i}^{n-1}}{(n-1)!} C_{ni}(z^2 - \mathrm{i}\varepsilon) z_{\mu_1} \ldots z_{\mu_n} O_{ni}^{\mu_1 \mu_2 \ldots \mu_n}(R) \\ & - \frac{1}{z^2 - \mathrm{i}\varepsilon} \sum_i \sum_{n=3}^{\infty} \frac{\mathrm{i}^{n-1}}{(n-1)!} D_{ni}(z^2 - \mathrm{i}\varepsilon) z_{\mu_3} \ldots z_{\mu_n} O_{ni}^{\mu\nu\mu_3 \ldots \mu_n}(R) , \end{aligned} \tag{12.3.6}$$

where $i$ is the number of terms corresponding to a given value of $n$.

6) Substitute (12.3.6) into (12.3.3). Introduce further the notation

$$\frac{1}{2} \sum_s \langle p, s | O_{ni}^{\mu_1 \ldots \mu_n}(0) | p, s \rangle = 2 A_{ni} p^{\mu_1} \ldots p^{\mu_n} + \ldots$$

and take the limit $-q^2 \to \infty$.

By comparing the result obtained with (12.3.4) we arrive at the Wilson expansion for the functions $T_i$:

$$\begin{aligned} T_1 &\sim \sum_i \sum_{n=1}^{\infty} A_{ni} c_{ni}(q^2) x^{-n} , \\ \nu T_2 &\sim 2 \sum_i \sum_{n=2}^{\infty} A_{ni} d_{ni}(q^2) x^{-n+1} , \end{aligned} \tag{12.3.7}$$

where

$$c_{ni}(q^2) = \frac{\mathrm{i}}{\pi^2} \frac{1}{(n-1)!} (-q^2)^n \left( \frac{\partial}{\partial q^2} \right)^n \int d\zeta \, \mathrm{e}^{\mathrm{i}q\zeta} \frac{C_{ni}(\zeta^2 - \mathrm{i}\varepsilon)}{(\zeta^2 - \mathrm{i}\varepsilon)^2} ,$$

$$d_{ni}(q^2) = -\frac{\mathrm{i}}{4\pi^2} \frac{1}{(n-2)!} (-q^2)^{n-1} \left( \frac{\partial}{\partial q^2} \right)^{n-2} \int d\zeta \, \mathrm{e}^{\mathrm{i}q\zeta} \frac{D_{ni}(\zeta^2 - \mathrm{i}\varepsilon)}{\zeta^2 - \mathrm{i}\varepsilon} .$$

7) Let us now pass from the functions $T_i$ to the functions $W_i$. For that, we make use of the *dispersion relations* for the functions $T_i$. An analysis based on the Regge theory shows that in the limit $\nu \to \infty$,

$$T_1(\nu, q^2) \sim \nu^\alpha, \quad T_2(\nu, q^2) \sim \nu^{\alpha-2}, \quad \alpha \leqslant 1,$$

and the dispersion relations can be therefore written as

$$T_1(\nu, q^2) = T_1(0, q^2) + \nu \int \frac{d\nu'}{\nu'} \frac{\operatorname{Im}\{T_1(\nu', q^2)\}}{\nu' - \nu},$$
$$T_2(\nu, q^2) = \int d\nu' \frac{\operatorname{Im}\{T_2(\nu', q^2)\}}{\nu' - \nu}. \tag{12.3.8}$$

Taking into account (12.3.2), $\operatorname{Im}\{T_i\}$ can be replaced in the expressions (12.3.8) by $W_i$. Multiplying both sides by corresponding power of $x$ and subsequently integrating yields for the moments of the structure functions $W_i$

$$M_n^{(1)} = \int_0^1 dx\, x^{n-1} W_1(x, q^2) \sim \sum_i A_{ni} c_{ni}(q^2),$$
$$M_n^{(2)} = \int_0^1 dx\, x^{n-2} \nu W_2(x, q^2) \sim 2 \sum_i A_{ni} d_{ni}(q^2). \tag{12.3.9}$$

These expressions indicate that the functions $W_1$ and $\nu W_2$ will possess exact scaling properties only if the coefficients $c_{ni}$, $d_{ni}$ are independent of $q^2$.

8) In order to find the dependence of the functions $c_{ni}$, $d_{ni}$ on $q^2$ in an explicit form, we make use of the *renormalization group equation* (10.1.15). The latter can be rewritten, e.g., for the function $c_{ni}$

$$\left[\frac{\partial}{\partial t} - \beta(g) \frac{\partial}{\partial g}\right] c_{ni}\left(\frac{q^2}{\lambda^2}; g\right) = \sum_j \gamma_{ij}^n(g) c_{nj}\left(\frac{q^2}{\lambda^2}; g\right). \tag{12.3.10}$$

Here $t = \ln(-q^2/\lambda^2)/2$, $\lambda$ denoting an unknown parameter; $\beta(g)$ and $\gamma(g)$ are unknown functions, and $j$ is the number of Wilson operators of the type $i$.

The solution of the equation (12.3.10) is written in the form (10.1.19):

$$c_{ni}\left(\frac{q^2}{\lambda^2}; g\right) = \sum_j \left\{T' \exp\left[-\int_0^t \gamma_{ij}^n(g(t'))\, dt'\right]\right\} c_{nj}(1; g(t)), \tag{12.3.11}$$

$$\frac{d}{dt} g(t) = \beta(g(t)), \quad g(0) = g, \tag{12.3.12}$$

where $g(t)$ is the effective charge and $T'$ denotes the ordering operation with respect to the variable $t$. In other words, the dependence of the functions $c_{ni}$ on $q^2$ is determined by the functions $\beta(g(t))$ and $\gamma(g(t))$. The expression for $d_{ni}(q^2/\lambda^2; g)$ can be written in the form similar to (12.3.11).

9) Deep-inelastic processes take place at small distances implying that the effective coupling constant is sufficiently small for *perturbation theory* of quantum chromodynamics to be applicable in calculating the functions $\beta(g)$

and $\gamma(g)$. The calculations can be carried out in different approximations: one-loop approximation, two-loop approximation, etc. In the one-loop approximation, the functions $\beta(g)$ and $\gamma(g)$ have the form

$$\beta(g) = -bg^3 + O(g^5), \tag{12.3.13}$$

$$\gamma_{ij}^n(g) = \lambda_{ij}^n g^2 + O(g^4). \tag{12.3.14}$$

Here $b$ and $\lambda_{ij}^n$ are coefficients calculated within the one-loop approximation.

The solution (12.3.11) corresponding to the values of the functions $\beta(g)$ and $\gamma(g)$ in the one-loop approximation is commonly referred to as the *leading logarithmic approximation.*

10) The solution of (12.3.12) is written, by virtue of (12.3.13), in the form (10.1.24):

$$g^2(t) = \frac{g^2}{1+2bg^2t} \underset{t\to\infty}{\sim} \frac{1}{2bt}. \tag{12.3.15}$$

Let $\lambda_k^n$ be the eigenvalues of the matrices $\lambda_{ij}^n$ entering (12.3.14) (where $k=1,2$ in our case). Substitution of (12.3.14, 15) into (12.3.11) and a subsequent substitution of the resulting expression into (12.3.9) yields

$$M_n^{(1)}(q^2) = \sum_{i,k=1}^{2} a_{ik}^n \left(\ln\frac{-q^2}{\lambda^2}\right)^{-\lambda_k^n/2b}. \tag{12.3.16}$$

Here $a_{ik}^n$ are unknown quantities independent of $q^2$. To determine them, we use the experimental values of the moments at some fixed value of $q^2 (q^2 = q_0^2)$:

$$M_n^{(1)}(q_0^2) = \sum_{i,k=1}^{2} a_{ik}^n \left(\ln\frac{-q_0^2}{\lambda^2}\right)^{-\lambda_k^n/2b}.$$

Substituting the last expression into (12.3.16) we eventually obtain the formula which determines the desired $q^2$-dependence of the moments of the structure functions (for $-q^2 > -q_0^2$):

$$M_n^{(1)}(q^2) = \sum_{k=1}^{2} M_n^{(1)}(q_0^2) \left(\ln\frac{-q_0^2}{\lambda^2} \Big/ \ln\frac{-q^2}{\lambda^2}\right)^{\lambda_k^n/2b}. \tag{12.3.17}$$

11) From the moments of the structure functions, the functions themselves can be reconstructed (the inverse problem).

It should be stressed that in deriving the formula (12.3.17) we have considered only the contribution from the leading singularity on the light-cone and have utilized the expressions for the functions $\beta$ and $\gamma$ obtained in the first order of perturbation theory (with respect to $\alpha_s$). However, as analysis shows, the neglected contributions due to non-leading singularities on the light-cone (higher twists) as well as higher-order approximations for the functions $\beta$ and

$\gamma$ might appear important. This should be borne in mind when comparing the results obtained by using (12.3.17) with experimental data.

Note also that the Wilson operator product expansion for the currents proves to work effectively only in the case of deep-inelastic scattering processes and of $e^+e^-$-annihilation, i.e., the method described has a rather limited range of applicability. Therefore, other, more effective methods have been developed. Below we shall present one of such methods.

## 12.4 Evolution Equations

The method for calculating the $q^2$-dependence of the structure functions we are going to describe is based on using the quark (parton) model, together with evolution equations. Hereafter, for convenience we sometimes use the notation $Q^2 \equiv |q^2|$.

1) Qualitatively, the dependence of the distribution function on the momentum transfer square $q^2$ can be understood as follows. Let a hadron be bombarded with a virtual particle (photon, $W$-, $Z$-boson) having the momentum square $q^2$. Such a particle distinguishes between quarks and gluons of the size $1/Q^2$ (Fig. 12.9). These quarks and gluons are characterized by the distribution functions $q_0(x)$ and $g_0(x)$ in $x$. The particles with a momentum $Q_1^2 > Q_0^2$ will distinguish between quarks and gluons of smaller sizes (Fig. 12.9). In the framework of quantum chromodynamics in the lowest approximation, a quark may appear as a quark accompanied by a gluon (Fig. 12.9a), while a gluon may appear as a quark-anti-quark pair (Fig. 12.9b) or two

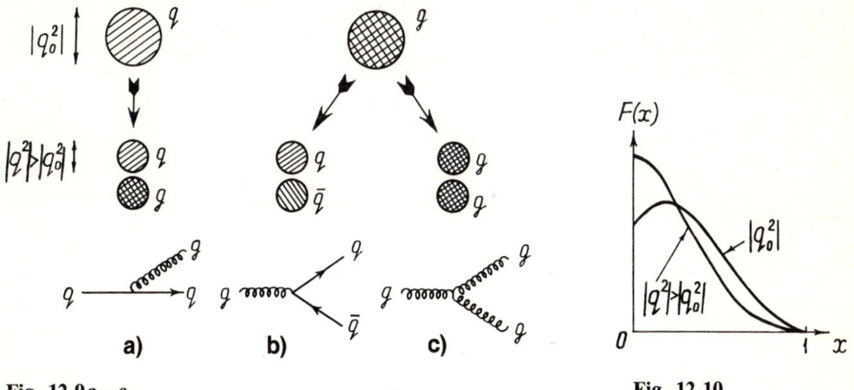


**Fig. 12.9 a–c** **Fig. 12.10**

**Fig. 12.9.** Qualitative picture of the dependence of the distribution functions on the transferred momentum

**Fig. 12.10.** Structure function for two different values of $q^2$

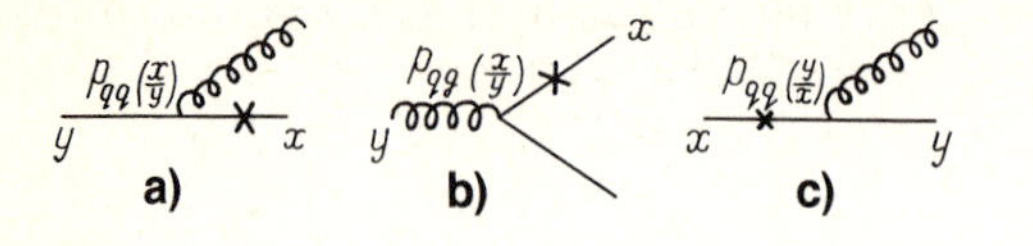

**Fig. 12.11a–c.** To the derivation of the evolution equation for the quark distribution function

gluons (Fig. 12.9c). The momentum $xp$ (where $p$ is the hadron momentum) of each particle is arbitrarily shared between the new particles. Consequently, the distribution functions $q_1(x)$ and $g_1(x)$ of the appeared quarks and gluons will differ from $q_0(x)$ and $g_0(x)$. A similar situation will be encountered in the case of momenta $Q_2^2 > Q_1^2$, $Q_3^2 > Q_2^2$, etc. This leads to the conclusion that the distribution functions of quarks and gluons in a hadron not only depend on $x$ but also on $Q^2$ (Fig. 12.10). It follows for the mechanism described that the particle momenta keep fractioning, so that the number of particles with small momenta grows as $Q^2$ increases (Fig. 12.10).

For the sake of simplicity, we shall assume that the quark masses are zero; the colour interactions for different flavours will be the same in this case.

2) Let us give a quantitative description of the dependence of the distribution functions for quarks and gluons on $Q^2$. To produce the corresponding equations, we observe that the change of the number of quarks and gluons within a certain interval $dx$, upon changing $Q^2$, is equal to the difference between the number of particles arriving at the interval $dx$ and the number of particles leaving it.

In the interval $dx$, there arrive quarks with the initial momentum $yp$ (where $p$ is the hadron momentum) which acquire the momentum $xp$ after emitting a gluon (Fig. 12.11 a) as well as those quarks with the momentum $xp$ which are produced by a gluon of momentum $yp$ (Fig. 12.11 b). Those quarks which, after emitting a gluon, acquire the momentum $yp$ (Fig. 12.11 c) will leave the interval $dx$. The particles for which we perform counting are marked by crosses in Fig. 12.11.

Let us consider, for example, the process depicted in Fig. 12.11 a. Let $q(y,t)$ denote the number of quarks in the interval $dy$, where $t = \ln(Q^2/Q_0^2)$ is a new variable, and let $P_{qq}(x/y)$ denote the probability of producing a quark with the momentum $xp$ by a quark with the momentum $yp$ after emitting a gluon, (or the probability of finding a quark with the momentum $xp$ within a quark with the momentum $yp$). Then the number of quarks coming from the interval $dy$ to the interval $dx$ is

$$P_{qq}(x/y)\,q(y,t)\,. \tag{12.4.1}$$

To obtain the total number of quarks arriving in the interval $dx$, the expression (12.4.1) has to be integrated over $y$ in the form

$$\int_x^1 \frac{dy}{y}\, q(y,t)\, P_{qq}\left(\frac{x}{y}\right).$$

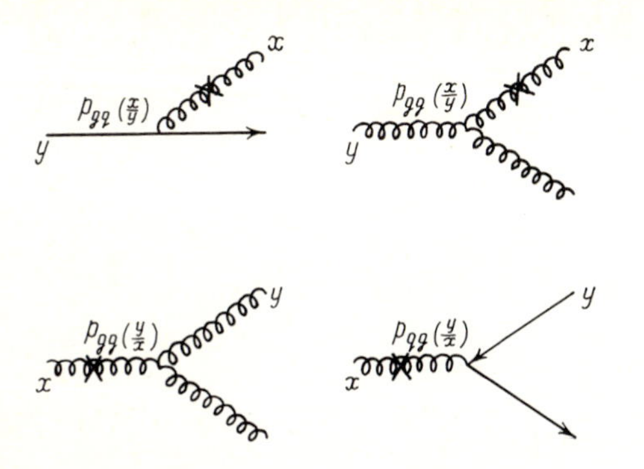


**Fig. 12.12.** To the derivation of the evolution equation for the gluon distribution function

Considering the remaining processes mentioned (Fig. 12.11 a, b) we eventually find that the net change of the number of quarks in the interval $dx$ due to the variation of $Q^2$ (or $t$) is described by

$$\frac{dq(x,t)}{dt} = \int_x^1 \frac{dy}{y}\left[q(y,t)P_{\text{qq}}\left(\frac{x}{y}\right) + 2n_{\text{f}}g(y,t)P_{\text{qg}}\left(\frac{x}{y}\right)\right]$$
$$-\int_0^x \frac{dy}{y}q(x,t)P_{\text{qq}}\left(\frac{y}{x}\right), \tag{12.4.2}$$

where $q(x,t)$ and $g(x,t)$, respectively, are the distribution functions of quarks and gluons in a hadron, and $P_{\text{qg}}(x/y)$ is the probability of formation of a quark with the momentum $xp$ by a gluon with the momentum $yp$ upon creation of a quark pair.

The factor $n_{\text{f}}$ takes care of the existence of different ways of gluon conversions, $g \to q + \bar{q}$, and the factor 2 originates from the fact that the quarks generated by a gluon are of the same type (quarks and anti-quarks contributions being indistinguishable).

A similar expression can be found for the net variation of the number of gluons in the interval $dx$ (cf. Fig. 12.12):

$$\frac{dg(x,t)}{dt} = \int_x^1 \frac{dy}{y}\left[q(y,t)P_{\text{gq}}\left(\frac{x}{y}\right) + 2g(y,t)P_{\text{gg}}\left(\frac{x}{y}\right)\right]$$
$$-\int_0^x \frac{dy}{y}\left[g(x,t)P_{\text{gg}}\left(\frac{y}{x}\right) + n_{\text{f}}g(x,t)P_{\text{qg}}\left(\frac{y}{x}\right)\right]. \tag{12.4.3}$$

Here $P_{\text{gg}}(x/y)$ is the probability that a gluon with an initial momentum $yp$ will generate two other gluons one of which has the momentum $xp$; $P_{\text{gq}}(x/y)$ is the probability of generating a gluon with the momentum $xp$ by a quark of the momentum $yp$ upon emitting another gluon.

Equations (12.4.2, 3) are called the *evolution* or the *Altarelli-Parisi equations* for the quark and gluon distribution functions varying with $t$.

Similar evolution equations may be written for the distribution functions of quarks and anti-quarks, quarks and anti-quarks of each single flavour, valence quarks, etc. For example, the evolution equation for the distribution functions of valence quarks, $q_v(x,t)$, reads

$$\frac{dq_v(x,t)}{dt} = \int_x^1 \frac{dy}{y} q_v(y,t) P_{qq}\left(\frac{x}{y}\right) - \int_0^x \frac{dy}{y} q_v(x,t) P_{qq}\left(\frac{y}{x}\right).$$

It follows from (12.4.2) by neglecting the second term.

3) The functions $P_{ab}(z)$ can be calculated by means of perturbation theory of quantum chromodynamics. As can be shown in the lowest order of perturbation theory they have the form

$$P_{qq}(z) = c_F \frac{1+z^2}{1-z}, \quad P_{qg}(z) = \frac{1}{2}[z^2+(1-z)^2], \quad P_{gq}(z) = c_F \frac{1+(1-z)^2}{z}, \tag{12.4.4}$$

$$P_{gg}(z) = c_v \left[\frac{1-z}{z} + \frac{z}{1-z} + z(1-z)\right],$$

where, for the group $SU_n$, $c_v = n$ and $c_F = (n^2-1)/(2n)$.

It should be emphasized that in calculating these functions only the terms containing $\ln Q^2$ have been retained, i.e., the functions $P_{ab}$ express the coefficients of $\ln Q^2$. The constant terms and the power terms of the type $(1/Q^2)^n$ have been omitted. Likewise, the higher-order contributions of perturbation theory with respect to the constant $\alpha_s$ have been neglected.

4) By multipyling (12.4.2, 3) by $x^{n-1}$ and subsequently integrating from zero to unity we obtain a set of differential equations for the *moments* (of the distribution functions or the structure functions)

$$\frac{dM_n^a(t)}{dt} = A_n^{ab} M_n^b(t). \tag{12.4.5}$$

Here

$$M_n^a(t) = \int_0^1 dx\, x^{n-1} a(x,t)$$

are the moments of the functions $a(x,t)$, $a(x,t)$ denoting the structure functions or the distribution functions;

$$A_n^{ab} = \int_0^1 dz\, z^{n-1} P_{ab}(z)$$

are the moments of the functions $P_{ab}(z)$.

5) Solving (12.4.5) provides an explicit form of the moments of the functions. From them, the functions themselves can be reconstructed. In particular, we have for the moments $M_n^{(1)}(q^2)$ of the structure functions $W_1(x,q^2)$ for the deep-inelastic electron scattering on protons (Sect. 12.3)

$$M_n^{(1)}(q^2) = M_n^{(1)}(q_0^2)\left[\frac{\ln(-q_0^2/\lambda^2)}{\ln(-q^2/\lambda^2)}\right]^{\lambda_k^n/2b}, \qquad (12.4.6)$$

where $\lambda_k^n$ are the eigenvalues of the matrices $A_n^{ab}$.

The coefficients $A_n^{ab}$ coincide with the coefficients $\gamma_{ij}^n$ involved in (12.3.11). Expressions (12.3.17) and (12.4.6) are thus equivalent.

Again, it should be noted that the above results have been obtained by neglecting the constant and the power terms of the type $(1/Q^2)^n$ (which is equivalent to neglecting the non-leading singularities on the light-cone as done in the previous method) as well as the higher orders of perturbation theory in the constant $\alpha_s$.

## 12.5 The Summation Method of Feynman Diagrams

This method of calculating the $q^2$-dependence of the quantities of interest is based on singling out and summing up only those Feynman diagrams whose contribution is of the order of $(\alpha_s \ln -q^2)^n$ and neglecting the diagrams of the type $(\alpha_s \ln -q^2)^n \alpha_s^k$. That is to say, in this method the summation of the Feynman diagrams is performed in the *leading logarithmic approximation*.

1) As analysis shows, it is instructive to choose the *axial gauge* (Sect. 4.2), in order to single out the diagrams which correspond to the leading logarithmic approximation. This gauge effectively considers only the transversal gluons and permits treating the internal lines of the Feynman diagrams as the lines of quark propagation.

2) In the axial gauge, the leading logarithmic approximation involves the summation only over the diagrams depicted in Fig. 12.13 (the ladder diagrams). The diagrams of the type of those in Fig. 12.13a describe the interaction of a valence quark, and of the type of those in Fig. 12.13b represent the contribution of a sea quark.

3) Let us consider, for example, the Drell-Yan process

$$h_1(p_1) + h_2(p_2) \to \gamma(q) + X = \mu^+ + \mu^- + X. \qquad (12.5.1)$$

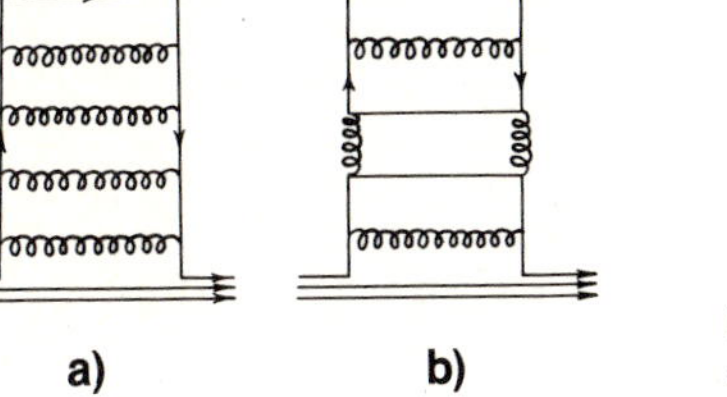


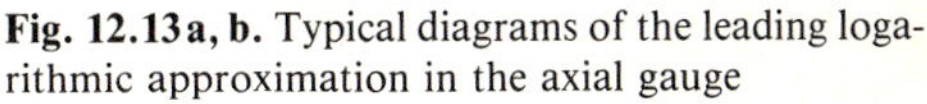

**Fig. 12.13a, b.** Typical diagrams of the leading logarithmic approximation in the axial gauge

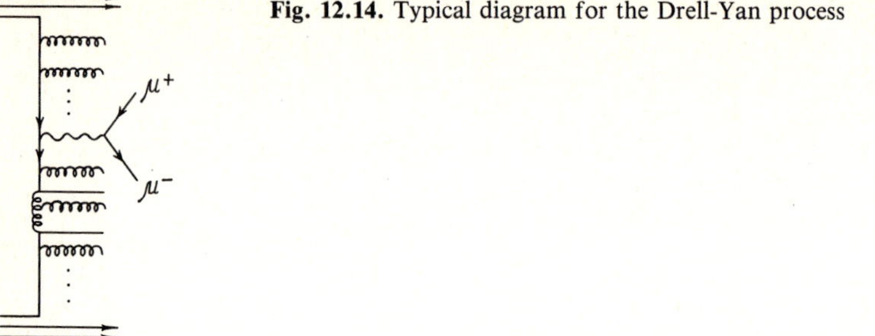


**Fig. 12.14.** Typical diagram for the Drell-Yan process

In this process, a quark and an anti-quark of each of the hadrons $h_1$ and $h_2$ annihilate to produce a photon which generates a pair of $\mu$-mesons.

The leading logarithmic approximation for the process (12.5.1) in the axial gauge is represented by a set of graphs of the type shown in Fig. 12.14. The cross-section of the Drell-Yan process (Fig. 12.15a) is proportional to the absorptive part of the diagram given by Fig. 12.15b. Integrating over the transversal momenta (Fig. 12.15c) we obtain the desired expression for the differential cross-section of the Drell-Yan process in the leading logarithmic approximation:

$$\frac{d\sigma}{dq^2} = \frac{4\pi\alpha_e}{9q^4} \sum_f Q_f^2 \int dx_1 dx_2 \delta\left(\frac{q^2}{s} - x_1 x_2\right) [D_{h_1}^q(x_1, q^2) D_{h_2}^{\bar{q}}(x, q^2) + D_{h_1}^{\bar{q}}(x_1, q^2) D_{h_2}^q(x_2, q^2)] \,. \tag{12.5.2}$$

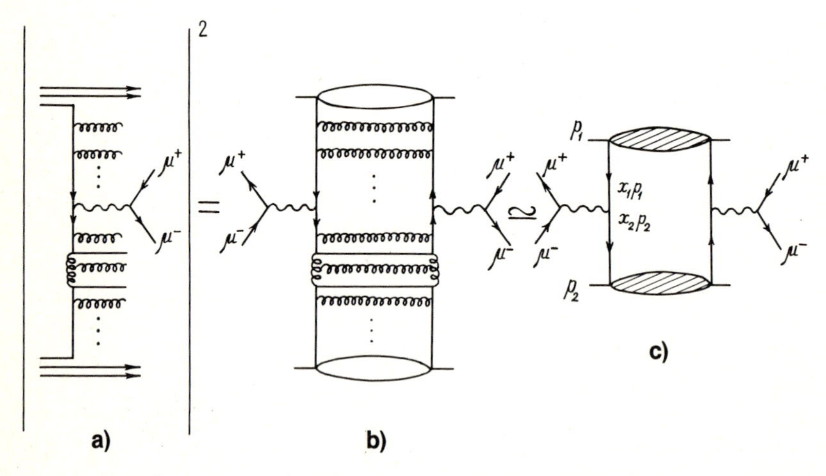


**Fig. 12.15.** Main stages of calculating the cross-section for the Drell-Yan process in the leading logarithmic approximation

Here $f$ is the quark flavour, $\alpha_e$ is the electromagnetic interaction constant, $Q_f$ is the charge of the quark with the flavour $f$, $s = (p_{\mu^+} + p_{\mu^-})^2$; $D_h^q(x, q^2)$ and $D_h^{\bar{q}}(x, q^2)$ are the distribution functions of quarks and anti-quarks in the hadrons $h_1$ and $h_2$.

As can be seen, the distribution functions $D^q(x, q^2)$ and $D^{\bar{q}}(x, q^2)$ depend on $q^2$. Furthermore, this dependence is the same as for the deep-inelastic scattering. In the model with free quarks, a similar expression is obtained for the cross-section of the Drell-Yan process (Fig. 12.14), but with the distribution functions $D^q(x)$ and $D^{\bar{q}}(x)$ which are independent of $q^2$. This can be put in the following way: the method of diagram summation enables one to calculate the $q^2$-dependence of the distribution functions stemming from the gluon effects.

In much the same way one can consider the sets of the Feynman diagrams for other hard processes and calculate the $q^2$-dependence of the corresponding differential cross-sections.

In contrast to the two foregoing methods, the one under consideration makes it possible to calculate also the $q^2$-dependence of more detailed characteristics than the structure functions. The most interesting result pertains to the $q^2$-dependence of the transversal momentum transfer $q_\perp^2$. Calculations show that the transverse momentum increases with $q^2$.

## 12.6 Quark and Gluon Fragmentation into Hadrons

To be in a position to compare the predictions of quantum chromodynamics with experimental data, one has to go over from considering quarks and gluons to considering hadrons, for hadrons, rather than quarks and gluons, are directly observable in experiment. The process of quark or gluon conversion into hadrons is called *hadronic fragmentation* of the quark or the gluon. Unfortunately, we are not able as yet to give a description of the hadronic fragmentation process. Probably, coloured quarks or gluons created at small distances fly apart. During the flight, a "cloud" of quarks, anti-quarks and gluons forms around them, and finally they transform to colourless hadrons.

A basic difficulty is connected with the necessity of describing the behaviour of quarks and gluons at large distances in the fragmentation process. Here perturbation theory fails because of the largeness of the interaction. Since we are not able to overcome this difficulty, a more or less plausible hypothesis ought to be dealt with. Thus, one may calculate the total cross-section of the $e^+e^-$-annihilation to quarks and gluons and then assume that a subsequent transformation of quarks and gluons into hadrons at large energies will not change the result.

An elaboration of the properties of the final hadronic states requires introducing additional assumptions. In particular, a conceivable form of existence of the quarks emitted from within small distances is a hadronic jet with restricted transversal momenta with respect to the direction of the parent quark and with a characteristic distribution over longitudinal momenta. One may anticipate that also gluons emitted at small distances will realize as some kind of hadronic jets whose properties differ from those of the hadronic jets generated by quarks. The state-of-the-art experimental data are in accordance with the notion of jets.

These ideas reflect a development referred to as the naive quantum chromodynamics.

In conclusion, we would like to emphasize that two basic problems are encountered in verifying quantum chromodynamics by comparing its predictions with experiment. The first problem is connected with sub-processes and consists in the necessity of taking into account the higher-order contributions of perturbation theory with respect to the constant $\alpha_s$ and of the contributions from the power terms of the form $(1/q^2)^n$. The second problem pertains to the mechanism of quark and gluon fragmentation into hadrons and concerns the role of interactions at large distances. One will have to wait with a final judgement on the validity of quantum chromodynamics as long as these two problems remain unresolved.

# 13. Lattice Gauge Theories. Quantum Chromodynamics on a Lattice

In this chapter we shall present one of the versions of quantum chromodynamics which provides a means of calculating physical quantities without using perturbation theory. This version is based on replacing the infinite four-dimensional space-time by a discrete space-time in form of a *lattice of finite dimensions*. Introducing a finite-size lattice makes it possible to carry out computer simulations not involving perturbation theory. It turned out that the *Monte Carlo method* is most suited for this purpose.

In addition, an approximate analytical method for calculations on the lattice has been developed which is based on an expansion in the inverse of the coupling constant; this method is referred to as the *strong coupling expansion*.

On passing to a lattice theory, the relativistic invariance gets violated; the gauge invariance may be preserved, however. The lattice spacing plays the role of a cut-off parameter and, therefore, theories on a lattice do not contain ultraviolet divergences, i.e., they are regularized (cf. Sect. 9.2).

We shall first formulate the classical and the quantum gauge-invariant chromodynamics on a lattice. Then we shall present the fundamentals of both the strong coupling expansion and the Monte Carlo simulations and discuss the main physical results.

## 13.1 Classical and Quantum Chromodynamics on a Lattice

### 13.1.1 The Lattice and Its Elements

Let us pass from the pseudo-Euclidean to the Euclidean space-time, i.e., make the change of variables $x_0 \to it$. Let us assume that the Euclidean space-time has a discrete structure with respect to both the spatial and the temporal coordinates. That is to say, the Euclidean space-time can be represented as four-dimensional lattice of finite dimensions[1].

The basic elements of the lattice are as follows (Fig. 13.1):

1) *A lattice site*, or a point of the lattice specified by the coordinates $x_\mu = a n_\mu$, where $n_\mu$ is a vector with the components $n_x$, $n_y$, $n_z$, $n_t$; $a$ is the

[1] A different formulation of the lattice theory, with a discrete structure of the three-dimensional space but with continuous time, has also been considered. We shall not deal with this version, however.

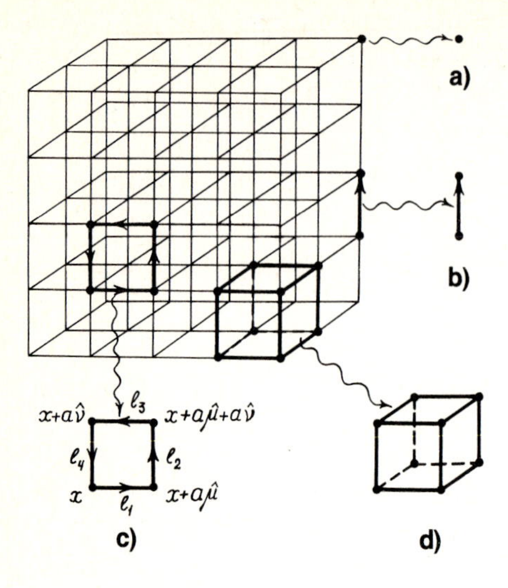


**Fig. 13.1a–d.** Three-dimensional representation of the basic elements of the lattice: **(a)** site; **(b)** link; **(c)** plaquette; **(d)** elementary cube

*lattice spacing*, i.e., the distance between neighbouring lattice sites (Fig. 13.1a); we shall assume that the lattice spacing is the same for all directions.

2) A *link*, or a line connecting two neighbouring sites; a link $l$ is given by the coordinate $x_\mu$ of its origin and by the direction $\mu$ of the corresponding axis in the space ($\mu = 1,2,3,4$): $l = (x, \mu)$; the link connects the sites with the coordinates $x$ and $x + a\hat{\mu}$, where $\hat{\mu}$ denotes a unit vector in the direction $\mu$ (Fig. 13.1b).

3) A *plaquette*, or an elementary square bounded by four links (Fig. 13.1c); a plaquette $p$ is given by the coordinate of the site lying in its corner and by the directions $\mu$ and $\nu$ of the adjacent links forming the sides of the square: $p = p(x, \mu, \nu)$.

4) A *three-dimensional cube* which is formed by six plaquettes joining edge-to-edge (Fig. 13.1d).

### 13.1.2 Gauge Fields on a Lattice

Let us consider gauge fields on a lattice.

1) In the case of continuous space-time (Chap. 2), the gauge field is characterized by the vector potential $A_\mu^k(x)$ and is described by the Lagrangian (2.2.23). After the change

$$A_\mu^k(x) \to \frac{1}{g} A_\mu^k(x) \tag{13.1.1}$$

this Lagrangian takes the form

$$\mathscr{L} = -\frac{1}{4g^2} F_{\mu\nu}^k F_{\mu\nu}^k , \quad F_{\mu\nu}^k = \partial_\mu A_\nu^k - \partial_\nu A_\mu^k - \frac{1}{2} f_{lmk} [A_\mu^l, A_\nu^m]_- , \tag{13.1.2}$$

where $F^k_{\mu\nu}(x)$ is the gauge field strength; furthermore

$$\nabla_\mu u_i = \partial_\mu u_i - T^k_{ij} A^k_\mu u_j . \tag{13.1.3}$$

It is seen that the coupling constant $g$ does not enter the covariant derivative and $F^k_{\mu\nu}$, but rather appears as a common factor in the Lagrangian.

It is convenient to consider, instead of the vector potential $A^k_\mu(x)$, its product with, e.g., a generator $(-\frac{i}{2}\lambda_k)$ of the fundamental representation of the Lie algebra of the gauge group $SU_3$:

$$A_\mu(x) = -\frac{i}{2} \sum_{k=1}^{8} A^k_\mu(x) \lambda_k . \tag{13.1.4}$$

Here $\lambda_k$ are the matrices (1.2.13); the corresponding field strength is

$$F_{\mu\nu}(x) = \partial_\mu A_\nu(x) - \partial_\nu A_\mu(x) - [A_\mu(x), A_\nu(x)]_- . \tag{13.1.5}$$

The quantities (13.1.4), which are anti-Hermitian matrices, are elements of the Lie algebra of the gauge group.

2) In the lattice theory, the gauge field is described by a *matrix* $U_{x,\mu}$ defined on a lattice link. The matrices $U_{x,\mu}$ are elements of the gauge group itself.

The quantities $U_{x,\mu}$ are related in a simple way to the vector potentials $A_\mu(x)$ of the gauge fields of the continuum theory. To establish this relationship, we express $U_{x,\mu}$ as a series expansion in the lattice spacing $a$:

$$U_{x,\mu} = 1 + A_\mu(x) a + O(a^2) . \tag{13.1.6}$$

As the quantities $A_\mu(x)$ are the elements of the Lie algebra of the gauge group, they transform under the gauge transformations as follows:

$$A_\mu(x) \to A^\omega_\mu(x) = \omega^{-1}(x) A_\mu(x) \omega(x) + \omega^{-1}(x) \partial_\mu \omega(x) , \tag{13.1.7}$$

where $\omega(x)$ is the matrix of continuous gauge transformations.

Replacing the derivatives in (13.1.7) by finite differences,

$$\omega_{x+a\hat{\mu}} = \omega(x) + \frac{\partial \omega(x)}{\partial x_\mu} a + O(a^2) ,$$

and substituting the resulting expression into (13.1.6) we find, with the accuracy of $O(a^2)$, the form of the transformation of $U_{x,\mu}$:

$$U_{x,\mu} \to U^\omega_{x,\mu} = \omega_x^{-1} U_{x,\mu} \omega_{x+a\hat{\mu}} . \tag{13.1.8}$$

Here $\omega_x$ denotes the values of $\omega$ at the lattice sites.

We further subdivide the link connecting the points $x$ and $x + a\hat{\mu}$ into small intervals and define $U_{x,\mu}$ as the *matrix product* which is *ordered* along the loop:

$$U_{x,\mu} = P \prod_{j=1}^{n} \left[ 1 + \frac{a}{n} A_\mu \left( x + j \frac{a}{n} \right) \right], \tag{13.1.9}$$

with $P$ denoting the matrix ordering operator.

According to (13.1.8), the matrices $\omega$ cancel out at the internal points of a link; therefore (13.1.9) defines an exact transformation of $U_{x,\mu}$ (in all orders in $a$). In the limit of the intervals in (13.1.9) tending to zero, we obtain an expression that provides the desired relation between the lattice matrix $U_{x,\mu}$ and the vector potential $A_\mu(x)$ of the continuum theory:

$$U_{x,\mu} = P \exp \int_0^{x+a\hat{\mu}} d\xi \, A_\mu(\xi) \, . \tag{13.1.10}$$

The right-hand side of the formula (13.1.10) is called the *ordered loop product.* As can be seen, in the lattice gauge theory it is necessary to specify the *orientation* of a given link. Let us assume that $\mu$ corresponds to the positive and $(-\mu)$ to the negative direction (Fig. 13.1 c), i.e., that the orientations of the links $(x, \mu)$ and $(x + a\hat{\mu}, -\mu)$ are opposite.

It follows from (13.1.10) that the matrices associated with the links with the positive and the negative orientations are related through

$$U_{x+a\hat{\mu},\, -\mu} = U_{x,\mu}^{-1} \, . \tag{13.1.11}$$

Because of the unitary of the matrices $U_{x,\mu}$, the last relation assumes the form

$$U_{x+a\hat{\mu},\, -\mu} = U_{x,\mu}^{\dagger} \, . \tag{13.1.12}$$

In constructing the lattice gauge theory, the *elementary closed loop* which coincides with the perimeter of a plaquette (Fig. 13.1 c) plays the main role. Taking into account (13.1.12), the following ordered product is associated with this loop:

$$U_p = U_{x,\mu} U_{x+a\hat{\mu},\, \nu} U_{x+a\hat{\nu},\, \mu}^{\dagger} U_{x,\nu}^{\dagger} \, . \tag{13.1.13}$$

According to (13.1.8), the product $U_p$ transforms under the gauge transformation as

$$U_p \to U_p^{\omega} = \omega_x^{-1} U_p \omega_x \, , \tag{13.1.14}$$

i.e., *the trace of the matrix $U_p$ is gauge-invariant.* It is this quantity which is used in constructing the gauge-invariant action of the lattice theory.

3) Using the simplest gauge-invariant quantities $\mathrm{Tr}\{U_p\}$ one can express the action $I(U_p)$ for the gauge field on a lattice in the case of $SU_N$-symmetry as

$$I(U_p) = \beta \sum_{\text{n.p.}} \left( 1 - \frac{1}{N} \mathrm{Re}\{\mathrm{Tr}\, U_p\} \right), \tag{13.1.15}$$

where $\beta$ is a constant. Here the summation is carried out over all elementary plaquettes of the lattice, i.e., over all $x$, $\mu$, $\nu$, regardless of their orientation (with "n.p." standing for non-oriented plaquettes). On reversing the orientation of a plaquette, the quantity $\mathrm{Tr}\{U_p\}$ goes over, as follows from (13.1.12), into its complex-conjugate:

$$\mathrm{Tr}\{U_p\} \to (\mathrm{Tr}\{U_p\})^* .$$

Due to this fact the action (13.1.15) can be rewritten in the form

$$I(U_p) = \frac{\beta}{2} \sum_{\text{o.p.}} \left(1 - \frac{1}{N} \mathrm{Tr}\{U_p\}\right), \tag{13.1.16}$$

where the summation is carried out over two possible orientations of each plaquette of the lattice (with "o.p." standing for oriented plaquettes).

4) Let us demonstrate that the lattice action (13.1.15) coincides, in the limit $a \to 0$, with the action of the continuum gauge theory. Let us first consider the theory invariant under the Abelian gauge group $U_1$. Using the Stokes theorem we have

$$U_p = \exp\left[\int_{\partial p} d\xi\, A_\mu(\xi)\right] = \exp\left[\int_p F_{\mu\nu}(z)\, d\sigma_{\mu\nu}(z)\right], \tag{13.1.17}$$

where $\partial p$ is the set of the four links bounding the plaquette $p$ and $F_{\mu\nu}(z)$ is the field strength.

For $a \to 0$, the field strength $F_{\mu\nu}(z)$ may be taken out of the integrand in the surface integral (with the accuracy of the terms of order $a^3$); this yields

$$U_p = \exp[F_{\mu\nu}(x) a^2 + O(a^3)] . \tag{13.1.18}$$

Let us make in this formula a series expansion of the exponential and take into account that in the continuum limit $(a \to 0)$

$$a^4 \sum_p \underset{a \to 0}{\longrightarrow} \frac{1}{2} \int d^4x \sum_p . \tag{13.1.19}$$

Then the term linear in $F_{\mu\nu} a^2 + O(a^3)$ vanishes on summing over the positive and the negative $\mu$ and $\nu$, while the quadratic term leads to the customary expression for the action in the continuum theory, assuming that $\beta = 1/g^2$:

$$I(A_\mu) = -\frac{1}{4g^2} \int dx\, F_{\mu\nu}^2(x) .$$

A similar result also follows in the case of the invariance under the non-Abelian gauge group $SU_N$ assuming that $\beta = 2N/g^2$:

$$I(A_\mu) = -\frac{2N}{g^2} \int dx\, \mathrm{Tr}\{F_{\mu\nu}^2(x)\} . \tag{13.1.20}$$

Hence, at $a \to 0$ and for a fixed $g$ the lattice action (13.1.15) recovers the action of the continuum theory.

### 13.1.3 The Spinor (Quark) Field on a Lattice

Let us formulate a lattice theory of the spinor field describing a quark.

1) In the continuum theory, the gauge transformation of the spinor field reads

$$\psi(x) \to \psi^{\omega}(x) = \omega^{\dagger}(x)\,\psi(x)\,, \qquad \bar{\psi}(x) \to \bar{\psi}^{\omega}(x) = \bar{\psi}(x)\,\omega(x)\,, \tag{13.1.21}$$

and the quantity $\bar{\psi}(x)\,\psi(x)$ is gauge-invariant.

2) The spinor field can be defined on the lattice sites, instead of the links, as distinct from the case of the gauge fields.

In the lattice theory the gauge transformations analogous to (13.1.21) read

$$\psi_x \to \psi_x^{\omega} = \omega_x^{\dagger}\psi_x\,, \qquad \bar{\psi}_x \to \bar{\psi}_x^{\omega} = \bar{\psi}_x\omega_x\,. \tag{13.1.22}$$

Here the spinor field $\psi_x$ and the gauge transformation matrix $\omega_x$ are defined on the lattice sites. Taking into account this fact, the action $I_0(\psi)$ for the free spinor field on the lattice has the form

$$I_0(\psi) = -\frac{\mathrm{i}}{2}\sum_{x,\mu}(\bar{\psi}_x\gamma_\mu\psi_{x+a\hat{\mu}} - \bar{\psi}_{x+a\hat{\mu}}\gamma_\mu\psi_x) - ma\sum_x \bar{\psi}_x\psi_x\,. \tag{13.1.23}$$

Here the summation is carried out over the sites $x$ and the directions $\mu$; $\gamma_\mu$ denotes the Euclidean Dirac matrices, and $m$ is the particle mass.

In the case of $a \to 0$, a slowly varying spinor field $\psi_x$ on the lattice and the spinor field $\psi(x)$ in the continuum theory are related by

$$\begin{aligned}
&\frac{\partial}{\partial x_\mu}\psi(x) = \frac{1}{a}(\psi_{x+a\hat{\mu}} - \psi_x)\,,\\
&\psi_x = a^{3/2}\psi(x)\,, \qquad \bar{\psi}_x = a^{3/2}\bar{\psi}(x)\,,\\
&\psi_{x+a\hat{\mu}} = a^{3/2}\left[\psi(x) + a\frac{\partial\psi}{\partial x_\mu}\right], \qquad \bar{\psi}_{x+a\hat{\mu}} = a^{3/2}\left[\bar{\psi}(x) - a\frac{\partial\bar{\psi}}{\partial x_\mu}\right].
\end{aligned} \tag{13.1.24}$$

The factor $a^{3/2}$ owes to the fact that the function $\psi_x$ is dimensionless whereas $\psi(x)$ has the dimension of (length)$^{-3/2}$. Let us substitute (13.1.24) into (13.1.23), take the limit $a \to 0$ and perform the change

$$a^4\sum_x \to \int d^4x\,.$$

As a result, we obtain the usual action for the spinor field in the continuum theory (Sect. 4.1).

3) However, as will be shown now, there are several different spinor fields, and not only one, to which the action (13.1.23) corresponds in the continuum

limit ($a \to 0$). In fact, on one hand the action (13.1.23) can be rewritten as (Sect. 5.3)

$$I_0(\psi) = \frac{1}{(2\pi)^4} \int_{-\pi/a}^{+\pi/a} dp\, \bar{\psi}_p G^{-1}(p)\, \psi_p\,, \tag{13.1.25}$$

where $G(p)$ is the propagator of the spinor field and $\psi_p$ is the Fourier transform of the field $\psi_x$:

$$\psi_p = a^{5/2} \sum_x \psi_x \mathrm{e}^{-\mathrm{i}px}\,, \qquad \bar{\psi}_p = a^{5/2} \sum_x \bar{\psi}_x \mathrm{e}^{\mathrm{i}px}\,. \tag{13.1.26}$$

On the other hand, substituting the last expression into (13.1.23) yields

$$I_0(\psi) = \frac{1}{(2\pi)^4} \frac{\mathrm{i}}{2a} \sum_\mu \int_{-\pi/a}^{+\pi/a} dp\, \bar{\psi}_p \gamma_\mu (\mathrm{e}^{-\mathrm{i}p_\mu a} - \mathrm{e}^{\mathrm{i}p_\mu a})\, \psi_p - \frac{m}{(2\pi)^4} \int_{-\pi/a}^{+\pi/a} dp\, \bar{\psi}_p \psi_p\,. \tag{13.1.27}$$

By comparing (13.1.25) and (13.1.27) we find the expression for the propagator of the spinor field:

$$G^{-1}(p) = \frac{1}{a}\left[\sum_{\mu=1}^{4} \frac{\mathrm{i}}{2} \gamma_\mu (\mathrm{e}^{-\mathrm{i}p_\mu a} - \mathrm{e}^{\mathrm{i}p_\mu a}) - ma\right] = \frac{1}{a}\left[\sum_{\mu=1}^{4} \gamma_\mu \sin(p_\mu a) - ma\right]. \tag{13.1.28}$$

If $pa \ll 1$, we can expand $\sin(pa)$ into a series in $pa$ which leads to the expression for the propagator of the spinor field in the continuum theory:

$$G^{-1}(p) = \sum_{\mu=1}^{4} \gamma_\mu p_\mu - m\,. \tag{13.1.29}$$

In the general case, the propagator $G(p)$ has the form

$$G(p) = \frac{a \sum_{\mu=1}^{4} \gamma_\mu \sin(p_\mu a)}{-\left[\sum_{\mu=1}^{4} \sin^2(p_\mu a) + (ma)^2\right]}\,. \tag{13.1.30}$$

Let us now turn back from the Euclidean metric to the Minkowski metric (changing $p_0 \to \mathrm{i}E$). The poles of the propagator $G(p)$ are then determined by the zeros of the denominator in (13.1.30), i.e., by the roots of the equation

$$\sinh^2(Ea) = \sum_{\mu=1}^{3} \sin^2(p_\mu a) + (ma)^2\,. \tag{13.1.31}$$

Consider the roots of this equation corresponding to particles (which implies the positiveness of energy, $E > 0$). Let a particle be moving along the $z$-axis, i.e., $p^{(1)} = (E, 0, 0, p_z)$. In this case, according to (13.1.31),

$$\sinh^2(Ea) = \sin^2(p_z a) + (ma)^2 \,. \tag{13.1.32}$$

The sine being a periodic function, another vector, $p^{(2)} = (E, 0, 0, \pi/a - p_z)$, as well as six more vectors obtained from $p^{(1)}$ and $p^{(2)}$ by replacing zeros by $\pi/a$,

$$p^{(3)} = \left(E, \frac{\pi}{a}, 0, p_z\right), \ldots, p^{(8)} = \left(E, \frac{\pi}{a}, \frac{\pi}{a}, \frac{\pi}{a} - p_z\right), \tag{13.1.33}$$

will satisfy the last relation.

The wave functions with different momenta,

$$\psi^{(j)}(x, y, z, t) = \exp(\mathrm{i}Et - \mathrm{i}p_x^{(j)}x - \mathrm{i}p_y^{(j)}y - \mathrm{i}p_z^{(j)}z)\,, \quad j = 1, \ldots, 8\,, \tag{13.1.34}$$

describe different states of particles. Thus, for example, the wave function of the state with the momentum $p^{(3)}$ differs from that of the state with the momentum $p^{(1)}$ by the factor $(-1)^{n_x}$ (where $n_x$ is the coordinate of the origin of the link directed along the $x$-axis). This means that in the continuum limit $(a \to 0)$ the wave function with the momentum $p^{(3)}$ strongly varies within one step in the $x$-direction.

Thus, to the state of the particle with a given energy there belong eight different states with the values of the momentum $p^{(1)}, \ldots, p^{(8)}$, i.e., the state of the system with a given energy is eight-fold degenerate.

4) To remove the above degeneracy, we rewrite the action (13.1.23) in a different form:

$$I(\psi) = k \sum_{x,\mu} [\bar{\psi}_x(1 - \mathrm{i}\gamma_\mu)\psi_{x+a\hat{\mu}} + \bar{\psi}_{x+a\hat{\mu}}(1 + \mathrm{i}\gamma_\mu)\psi(x)] - \sum_x \bar{\psi}_x \psi_x\,, \tag{13.1.35}$$

where $k$ is the *hopping parameter*. The last expression differs from (13.1.23) in that the projection operators $(1 \pm \mathrm{i}\gamma_\mu)$ are now present.

In the continuum limit $(a \to 0)$

$$\psi_x \to \left(\frac{a^3}{2k}\right)^{1/2} \psi(x)\,, \qquad \psi_{x+a\hat{\mu}} \to \left(\frac{a^3}{2k}\right)^{1/2} \left[\psi(x) + \frac{\partial \psi}{\partial x_\mu} a\right], \tag{13.1.36}$$

and (13.1.35) goes over into the expression for the action of the spinor field of the continuum theory,

$$I(\psi) = \int dx \left[-\mathrm{i} \sum_{\mu=1}^{4} \bar{\psi}(x)\gamma_\mu \partial_\mu \psi(x) - m\bar{\psi}(x)\psi(x)\right], \tag{13.1.37}$$

provided that the value of $m = (1 - 8k)/2ka$ is set. The quantity $m$ plays the role of the spinor particle mass in the continuum limit. The mass of a particle has to remain finite as $a \to 0$, i.e., the condition $ma \ll 1$ must be fulfilled; in this case

$$k = \frac{1}{8+2ma} \simeq \frac{1}{8} - \frac{ma}{32} \simeq \frac{1}{8}. \tag{13.1.38}$$

Let us demonstrate that the action (13.1.35) does not possess degenerate states in the continuum limit ($a \to 0$). To the action (13.1.35) corresponds the propagator

$$G^{-1}(p) = -\frac{1}{a}\left\{1 - k \sum_{\mu=1}^{4} [(1+\mathrm{i}\gamma_\mu)\mathrm{e}^{-\mathrm{i}p_\mu a} + (1-\mathrm{i}\gamma_\mu)\mathrm{e}^{\mathrm{i}p_\mu a}]\right\}. \tag{13.1.39}$$

Passing to the Minkowski metric we find that this propagator has the poles at the points where the energy and the momentum are interrelated by

$$\cosh(Ea) = \frac{4k^2 + \left[1 - 2k \sum_{\mu=1}^{3} \cos(p_\mu a)\right]^2 + 4k^2 \sum_{\mu=1}^{3} \sin^2(p_\mu a)}{4k\left[1 - 2k \sum_{\mu=1}^{3} \cos(p_\mu a)\right]}. \tag{13.1.40}$$

As distinct from (13.1.31), the right-hand side of (13.1.40) is a non-periodic function; consequently, (13.1.40) has only one solution in the limit $a \to 0$, $k \to 1/8$. In fact, in the case of small energies $E \sim m$, the left-hand side of (13.1.40) can be replaced by unity and in the right-hand side $k$ can be set equal to 1/8. As a result, we obtain the following equation for determining the admissible solutions for the momentum of the particle:

$$\left[3 - \sum_{\mu=1}^{3} \cos(p_\mu a)\right]^2 + \left[3 - \sum_{\mu=1}^{3} \cos^2(p_\mu a)\right] = 0. \tag{13.1.41}$$

This equation has the only solution $p_1 = p_2 = p_3 = 0$.

In the case of large energies $E > m$, (13.1.40) is fulfilled, for $a \to 0$, only when the usual relation between the energy and the momentum of the particle, $E^2 = p^2 + m^2$, holds, i.e., (13.1.40) has again a single solution.

The states with other momenta (13.1.33) become, in the continuum limit ($a \to 0$), immaterial. For instance, the state with the momentum $p = (p_1 = \pi/a$, $p_2 = p_3 = 0)$ has, according to (13.1.40), the energy

$$E = \frac{1}{a} \operatorname{arc\,cosh}\left(\frac{4k^2 + (1-2k)^2}{4k(1-2k)}\right). \tag{13.1.42}$$

It can be seen that for $a \to 0$ the energy $E \to \infty$, and the contribution of this state, which is proportional to $\exp(-Et)$, becomes negligibly small.

### 13.1.4 Classical Chromodynamics on a Lattice

The simplest gauge invariant expression for the action on a lattice which describes a system of interacting gauge and spinor fields can be written, by virtue of (13.1.16) and (13.1.35), as

$$I(\psi, U_p) = \frac{1}{2}\beta \sum_{\text{o.p.}} \left(1 - \frac{1}{N}\mathrm{Tr}\{U_p\}\right) + k \sum_{x,\mu} [\bar{\psi}_x(1 - \mathrm{i}\gamma_\mu) U_{x,\mu} \psi_{x+a\hat{\mu}}$$
$$+ \bar{\psi}_{x+a\hat{\mu}}(1 + \mathrm{i}\gamma_\mu) U^\dagger_{x,\mu} \psi_x] - \sum_x \bar{\psi}_x \psi_x . \qquad (13.1.43)$$

Making use of (13.1.6) and (13.1.24) we obtain in the continuum limit ($a \to 0$):

$$\bar{\psi}_x \gamma_\mu U_{x,\mu} \psi_{x+a\hat{\mu}} \to a^3 \bar{\psi}(x) \gamma_\mu \psi(x) + a^4 \bar{\psi}(x) \nabla_\mu \gamma_\mu \psi(x) + O(a^5) ,$$
$$\bar{\psi}_{x+a\hat{\mu}} \gamma_\mu U^\dagger_{x,\mu} \psi_x \to a^3 \bar{\psi}(x) \gamma_\mu \psi(x) - a^4 \nabla_\mu \bar{\psi}(x) \gamma_\mu \psi(x) + O(a^5) ,$$

where $\nabla_\mu = \partial/(\partial x_\mu) + A_\mu$ is the covariant derivative.

The expression for the action in classical chromodynamics in the Euclidean space is obtained, for the continuum case, by taking the limit of $a \to 0$ and by making use of the limit expressions for $I(U_p)$ and $I(\psi)$ derived earlier.

### 13.1.5 Quantum Chromodynamics on a Lattice

As already mentioned (Chap. 4), one of the quantization methods is to represent the amplitude as a functional integral. In the lattice theory, the amplitude expressed in terms of the functional integral over all the fields, reads

$$S = \int \prod_{x,\mu} \mathcal{D}U_{x,\mu} \mathcal{D}\bar{\psi}_x \mathcal{D}\psi_x \exp[-I(\psi, U_p)] , \qquad (13.1.44)$$

where $I(\psi, U_p)$ is the action given by (13.1.43), and $\prod_{x,\mu} \mathcal{D}U_{x,\mu} \mathcal{D}\bar{\psi}_x \mathcal{D}\psi_x$ is the invariant measure over the group.

The integral (13.1.44) is defined in the Euclidean space and as such is convergent.

In the formula (13.1.44), the gauge is not fixed; therefore this formula provides an example of non-perturbative quantization with unfixed gauge.

### 13.1.6 The Continuum Limit

We have already seen that in the limiting case of $a \to 0$ the lattice action of quantum chromodynamics goes over into the continuum action. Besides it is important to know the behaviour of *physical quantities* in the continuum limit so that one would be able to judge whether the lattice quantum chromodynamics goes over into the continuum theory of strong interactions of hadrons as $a \to 0$.

Let us first consider the behaviour at $a \to 0$ of the effective charge. Because of the renormalizability of quantum chromodynamics, for the effective charge $g(a)$ at $a \to 0$, one may apply (10.1.20) which in the two-loop approximation is written as

$$a \frac{dg(a)}{da} \equiv \bar{\beta}(g(a)) = \beta_0 g^3 + \beta_1 g^5 + O(g^7), \tag{13.1.45}$$

where

$$\beta_0 = \frac{1}{48\pi^2}(33 - 2n_f), \quad \beta_1 = \frac{1}{256\pi^4}\left(102 - \frac{38}{3}n_f\right).$$

The solution of this equation has the form

$$a\lambda_l = (\beta_0 g^2)^{-\beta_1/2\beta_0} \exp(-1/2\beta_0 g^2), \tag{13.1.46}$$

where $\lambda_l$ is an integration constant. It follows from the last formula that if $a \to 0$, then $g \to 0$ and vice versa. That is to say, in the continuum limit the lattice theory is asymptotically free; moreover, the smallness of $g^2$ implies that the continuum transition set in and the invariance under the Lorentz group may be recovered.

Now let us turn to the behaviour at $a \to 0$ of quantities with physical significance. The lattice spacing $a$ plays the role of a cut-off parameter. The physical quantities $M_i$ do not depend on the cut-off parameter and therefore obey at $a \to 0$ a renormalization group equation similar to (10.1.14):

$$\left[a\frac{\partial}{\partial a} - \beta(g)\frac{\partial}{\partial g}\right] M_i(a,g) = 0. \tag{13.1.47}$$

In the case of $a \to 0$, $g \to 0$, the solution of this equation, e.g., for $M_i$ having the dimension of inverse length, has the form

$$M_i = C_i \lambda_l = \frac{C_i}{a}(\beta_0 g^2)^{-\beta_1/2\beta_0} \exp(-1/2\beta_0 g^2). \tag{13.1.48}$$

Here $C_i$ is a coefficient characterizing the physical quantity $M_i$. In numerical calculations the magnitude of the coefficient $C_i$ is singled out and the dependence of $M_i$ on $g^2$ is analyzed. If, starting from some value of $g^2$, the dependence of $M_i$ on $g^2$ obeys the formula (13.1.48), then the value obtained pertains to the continuum limit.

A comparison of the results of the lattice and the continuum theory requires the knowledge of the relation between the corresponding parameters $\lambda_l$ and $\lambda_{\text{MOM}}$ (Sect. 12.2). In the framework of the $SU_3$-symmetry, this relation can be shown to be

$$\lambda_{\text{MOM}}^{\alpha=1} = 83.5\,\lambda_l.$$

A calculation yields $\lambda_l \sim 2$ MeV. From this we find for $\lambda_{\mathrm{MOM}}^{\alpha=1}$ a value which is close to the experimental one (Sect. 12.2).

There are two basic ways of calculating the functional integrals in the lattice theory: an approximate analytical method (strong coupling expansion) and a numerical method based on computer simulations.

Let us consider first the strong coupling expansion.

## 13.2 Strong Coupling Expansion for the Gauge Fields

### 13.2.1 The Loop Average

Let us consider in some detail the quantum theory of only the gauge fields. The action $I(U_p)$ for these fields is determined by (13.1.16), the functional integral has the form

$$S = \int \prod_{x,\mu} \mathscr{D} U_{x,\mu} \exp[-I(U_p)] . \tag{13.2.1}$$

The group measure is invariant under the left- and the right-hand-side multiplication by an arbitrary element of the gauge group:

$$dU = d(\omega U) = d(U\omega) . \tag{13.2.2}$$

This provides the gauge invariance of the functional integral (13.2.1).

The basic quantity in the lattice theory of gauge fields is the *average over the loop* (or the vacuum expectation value of Wilson loop) $\langle \mathrm{Tr}\{U_{C_1}\} \times \mathrm{Tr}\{U_{C_2}\} \dots \rangle$, which is defined by

$$\langle \mathrm{Tr}\{U_{C_1}\}\mathrm{Tr}\{U_{C_2}\}\dots\rangle = \frac{\int \prod_{x,\mu} \mathscr{D} U_{x,\mu} \exp[-I(U_p)](\mathrm{Tr}\{U_{C_1}\}\mathrm{Tr}\{U_{C_2}\}\dots)}{\int \prod_{x,\mu} \mathscr{D} U_{x,\mu} \exp[-I(U_p)]} . \tag{13.2.3}$$

With the aid of (13.1.6) and (13.1.20), (13.2.3) reduces, in the continuum limit ($a \to 0$), to the loop average of the continuum theory:

$$\langle \mathrm{Tr}\{U(C_1)\}\mathrm{Tr}\{U(C_2)\}\dots\rangle$$

$$= \frac{\int \prod_{x,\mu} \mathscr{D} A_\mu(x) \exp\left[-\frac{1}{4g^2}\int dx\, F_{\mu\nu}(x)\, F_{\mu\nu}(x)\right][\mathrm{Tr}\{U(C_1)\}\mathrm{Tr}\{U(C_2)\}\dots]}{\int \prod_{x,\mu} \mathscr{D} A_\mu(x) \exp\left[-\frac{1}{4g^2}\int dx\, F_{\mu\nu}(x) F_{\mu\nu}(x)\right]} , \tag{13.2.4}$$

if the constant $\beta$ is related to the charge $g$ by

$$\beta = 2N/g^2. \tag{13.2.5}$$

For small $\beta$, i.e., for large $g$, the functional integral (13.2.1) can be calculated using a series expansion in $\beta$. Such an expansion is known as the *strong coupling expansion* (or the *high-temperature expansion*). The series expansion reduces the problem to calculating group integrals of the type

$$\int \mathscr{D}U U^{i_1}_{j_1} \dots U^{i_n}_{j_n} U^{\dagger k_1}_{l_1} \dots U^{\dagger k_m}_{l_m}, \tag{13.2.6}$$

the measure being normalized by the condition $\int \mathscr{D}U = 1$.

Let us calculate the value of the simplest loop average $W(\partial p)$ which corresponds to the loop coinciding with the perimeter of the elementary plaquette (Fig. 13.1 c):

$$W(\partial p) = \left\langle \frac{1}{N} \operatorname{Tr}\{U_p\} \right\rangle, \tag{13.2.7}$$

where $\partial p$ is the set of four links forming the perimeter of the plaquette $p$.

We confine ourselves to the first-order terms in the constant $\beta$; this results in

$$W(\partial p) = \frac{\int \prod\limits_{x,\mu} \mathscr{D}U_{x,\mu} \frac{1}{N} \operatorname{Tr}\{U_p\} \left(1 + \beta \sum\limits_{p'} \frac{1}{N} \operatorname{Re}\{\operatorname{Tr} U_{p'}\}\right)}{\int \prod\limits_{x,\mu} \mathscr{D}U_{x,\mu} \left(1 + \beta \sum\limits_{p'} \frac{1}{N} \operatorname{Re}\{\operatorname{Tr} U_{p'}\}\right)}. \tag{13.2.8}$$

Utilizing the formula

$$\int \mathscr{D}U_l U^i_{lj} U^{\dagger n}_{l'm} = \frac{1}{N} \delta_{l'l} \delta^i_m \delta^n_j, \tag{13.2.9}$$

which follows from the unitarity and the orthogonality of the matrices, we obtain for the case of the $SU_N$-group ($N \geqslant 3$)

$$W(\partial p) = \beta/2N^2, \tag{13.2.10}$$

and for the case of the $SU_2$-group

$$W(\partial p) = \beta/4. \tag{13.2.10'}$$

With the aid of (13.2.9) the loop average $W(C)$ can be calculated, in the first order in $\beta$, for the loops $C$ of more complicated shapes. The presence of the symbol $\delta_{l'l}$ in (13.2.9) implies that the integral is non-zero only if the oriented link $l'$ coincides with the link $l$. Consequently, only such integrals will be non-zero for which the elementary plaquettes fill out the entire surface bounded by the loop (Fig. 13.2). In this case, each link appears twice in the integral, namely in the positive and in the negative direction. As a consequence, all the integrals will differ from zero. The corresponding loop average (for simple loops without self-intersections) in the first order with respect to $\beta$ is

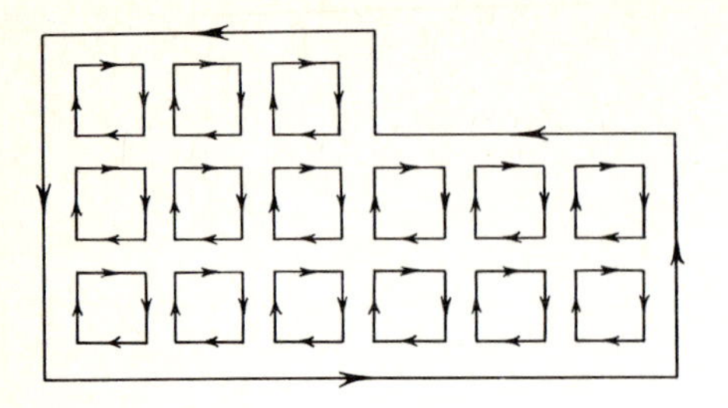

**Fig. 13.2.** Loop filled out with elementary plaquettes

$$W(C)=\left\langle\frac{1}{N}\mathrm{Tr}\{U_C\}\right\rangle=[W(\partial p)]^{S_{\min}(C)}=\exp[-KS_{\min}(C)]\,,\qquad(13.2.11)$$

where $S_{\min}(C)$ is the minimum area bounded by the loop $C$; $K=-\ln W(p)$, and $W(\partial p)$ is given by (13.2.10).

It can be shown that the series in $\beta$ is only convergent if the condition $ma>1$ is fulfilled, i.e., if the size of the lattice is sufficiently large. On the other hand, we are interested in the range of small lattice sizes where $ma\ll 1$. There arises a problem of analytical continuation of the solution from the region $ma>1$ into the region $ma\ll 1$. For this purpose the method of Padé's summation is employed. This requires the knowledge of a large number of the terms in the expansion with respect to $\beta$. Calculating them is connected with great difficulties for which reason the problem of analytical continuation has not been satisfactorily solved to date.

### 13.2.2 The Area Law. The Quark Confinement

The exponential character of the area dependence of the loop average (13.2.11) is referred to as the *area law.*

To get an insight into the physical meaning of the area law, we calculate the loop average over the rectangular contour depicted in Fig. 13.3. Let us choose the gauge $A_0(x,t)=0$. For the loop average $W(R,T)$ we then have

$$W(R,T)=\langle\mathrm{Tr}\{\psi(0)\,\psi^\dagger(T)\}\rangle\,,\qquad(13.2.12)$$

where

$$\psi(0)=P\exp\left\{\int_0^R dx\,A_1(x,0)\right\},\qquad \psi(T)=P\exp\left\{\int_0^R dx\,A_1(x,t)\right\}$$

or, inserting into (13.2.12) the sum over the intermediate states, we get

$$\langle\mathrm{Tr}\{\psi(0)\,\psi^\dagger(T)\}\rangle=\sum_n\langle\psi_{ij}(0)\,|n\rangle\langle n|\,\psi^\dagger_{ji}(T)\rangle\,.\qquad(13.2.13)$$

We further take into account that the states at different moments are connected by

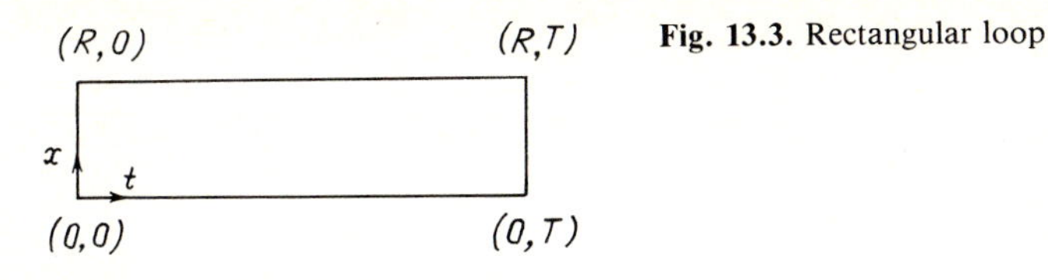


**Fig. 13.3.** Rectangular loop

$$\psi(T) = \mathrm{e}^{-\hat{H}T}\psi(0)\,\mathrm{e}^{\hat{H}T},$$

(where $\hat{H}$ is the lattice Hamiltonian of the system).

Let us substitute the last expression into (13.2.13) and act with the operator $\hat{H}$ on the vacuum state. As a result, we obtain the desired loop average

$$W(R,T) = \sum_n |\langle \psi_{ij}(0)\,|n\rangle|^2 \mathrm{e}^{-E_n T}, \qquad (13.2.14)$$

where $E_n$ is the energy of the system.

In the limit $T \to \infty$ only the ground state with the minimum energy $E_0$ is retained in the sum (13.2.14):

$$W(R,T) \underset{T\to\infty}{\longrightarrow} \mathrm{e}^{-E_0(R)T}. \qquad (13.2.15)$$

The change of variables $T \to \mathrm{i}T$ in this formula leads to the physical energy $E_0(R)$. By definition, the latter quantity characterizes the change of the ground state upon introducing particles (quarks) into the field, i.e., $E_0(R)$ is the energy, or the interaction potential, of the quarks. Hence, the loop average is related to the interaction potential $E_0(R)$ of static quarks.

For a rectangular loop the Eq. (13.2.11) is written as

$$W(R,T) = \mathrm{e}^{-KRT}, \qquad (13.2.16)$$

where $K$ is a constant. A comparison of (13.2.15, 16) gives

$$E_0(R) = KR. \qquad (13.2.17)$$

As can be seen, the strong coupling expansion leads to the area law which, in turn, leads to an interaction potential between the quarks linearly rising with the distance.

The coefficient $K$ is called the *string tension.* This notion originates from the fact that in order to give rise to a linearly increasing potential, the gluon field between the quarks must be constricted into a pipe or a string (Fig. 13.4 also showing for comparison the field in the case of the Coulomb potential). The string does not allow the quarks to spread out over large distances. Thus, if the loop average for large loops tends exponentially to zero, then the potential between the quarks linearly grows with the distance and the quarks cannot leave the hadron. In other words, the area law is a *criterion for the quark confinement.* (This criterion has been formulated by K. Wilson.)

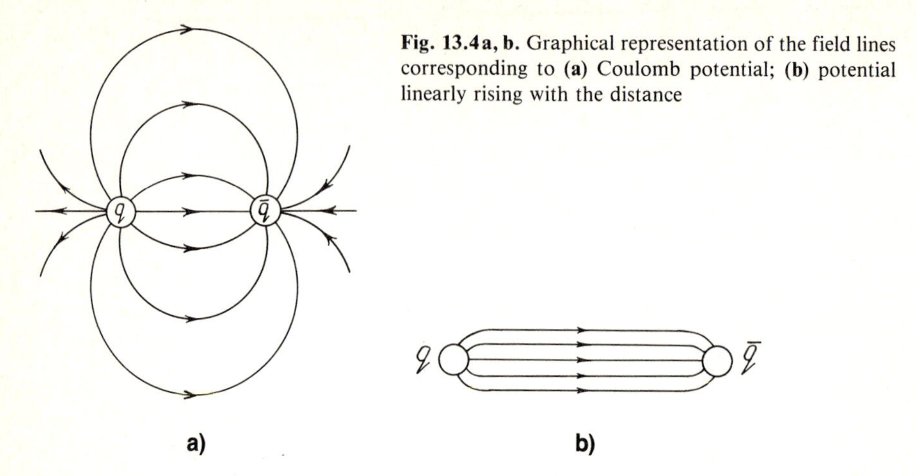


**Fig. 13.4a, b.** Graphical representation of the field lines corresponding to **(a)** Coulomb potential; **(b)** potential linearly rising with the distance

Now we go over to a numerical method of calculating the functional integrals.

## 13.3 Non-Perturbative Calculations in Quantum Chromodynamics by Means of Monte-Carlo Simulations

### 13.3.1 Monte-Carlo Simulations

The lattice formulation of quantum chromodynamics allows the application of computer simulations for numerical calculations of the functional integrals. These calculations can be made directly, for any value of the coupling constant and without using perturbation theory.

As we have seen, the integration in the lattice functional integrals of quantum chromodynamics has to be made in succession over the matrices $U_l$ on each link $l$ and over the spinor fields $\psi_x$ on each site. In actual calculations it is beneficial to replace the multiple integral by the sum over the states of the system and to compute this sum by means of *Monte-Carlo simulations.*

1) First we concern ourselves with integrating over the matrices $U_l$. Let us enumerate all links of the lattice. A state $B$ of the system is then described by the value of the matrix $U_1$ on the first link, the value of $U_2$ on the second link, etc. i.e.,

$$B = \{U_1, U_2, \ldots, U_{N_l}\}. \tag{13.3.1}$$

For the convergent integral of finite multiplicity under consideration, the successive summation over $U_i$ can be expressed as the sum over all states of the system

$$\sum_{U_1} \sum_{U_2} \cdots \sum_{U_{N_l}} = \sum_{\{U_1, U_2, \ldots U_{N_l}\}} . \tag{13.3.2}$$

The average value $\langle Q(B) \rangle$ of a quantity $Q(B)$ is given by the formula

$$\langle Q(B) \rangle = \frac{\sum_B Q(B) \exp[-I(B)]}{\sum_B \exp[-I(B)]}, \tag{13.3.3}$$

where $Q(B)$ and $I(B)$ are the values of the quantity to be averaged and of the action for the given state $B$, respectively.

In the case under consideration the total number of states is large, and considering each of them individually would be impossible. This suggests using for the calculation of the sum (13.3.3) the method of Monte-Carlo simulations.

The method is based on the fact that in case of random processes, it is sufficient to study a small number of realizations (*sampling*) to clarify the regularities in the behaviour for a large number of realizations (*sample states*). Of course, the sampling must be as representative of the sample states as possible. Following this, we choose, at random, a subsequence (sampling) of states

$$A = \{B_0, B_1, \ldots B_N\}. \tag{13.3.4}$$

Let the state $B_n$ occur in the sampling with the probability $P(B_n)$. For the average value (13.3.3) we then have

$$\langle Q(B) \rangle = \frac{\sum_{n=1}^{N} Q(B_n) P^{-1}(B_n) \exp[-I(B_n)]}{\sum_{n=1}^{N} P^{-1}(B_n) \exp[-I(B_n)]}. \tag{13.3.5}$$

In the case of large $\beta$ (small $g$) not all of the states of the sampling will be equiprobable. It is beneficial to assume that the states $B_n$ occur in the sampling with the probability imitating the distribution of the sample states:

$$P(B_n) \simeq \exp[-I(B_n)]. \tag{13.3.6}$$

The average value (13.3.5) will then be written as the arithmetic average

$$\langle Q(B) \rangle \simeq \frac{1}{N} \sum_{n=1}^{N} Q(B_n). \tag{13.3.7}$$

A sequence of states generated with the probability (13.3.6) is called *equilibrium*. To be able to use (13.3.7), one has to choose such a chain of transitions which, starting from any state, would lead in the limit to the equilibrium state. It can be shown that this implies the fulfillment of the condition

$$\exp[-I(B)]\, W(B \to B') = \exp[-I(B')]\, W(B' \to B), \tag{13.3.8}$$

where $W(B \to B')$ is the probability of transition from the state $B$ into the state $B'$. In this case a non-equilibrium sequence of states approaches, in the limit, an equilibrium one, while an equilibrium sequence of states goes over again into an equilibrium one.

In concrete calculations, the probabilities $W(B \to B')$ are chosen in various ways, but so that the relation (13.3.8) be satisfied. The rate of convergence towards the equilibrium state will be different for different $W(B \to B')$.

Monte-Carlo simulations give the average values with a statistical error proportional to $1/\sqrt{N}$.

2) There are also techniques devised for numerical integration over the spinor functions by means of Monte-Carlo simulations. Such an integration can be most easily performed when the contribution of the virtual quark loops, i.e., the contribution due to the sea quarks, is neglected. Otherwise an excessively large computer time is required for the calculations.

The quark loop contribution being neglected, one can express the contribution of the valence quarks in terms of the Green's functions in the external gluon field, i.e., reduce the problem to integrating over the gluon fields. This circumstance is utilized in integrating over the spinor fields.

### 13.3.2 Results of Calculations

Monte-Carlo simulations have been used to compute various physical quantities. We shall briefly consider two typical examples of such computations.

1) *String Tension.* To calculate the string tension $K$, one can make use of the relation between $K$ and the lattice loop average for the gauge fields which is given by the area law (13.2.16), under assumption that this relation holds for any values of $\beta$. More convenient, however, is calculating another quantity $\chi$, defined as a certain combination of the loop averages:

$$\chi(R, T) = -\ln\left(\frac{W(R, T)\, W(R-1, T-1)}{W(R-1, T)\, W(R, T-1)}\right). \tag{13.3.9}$$

Here $W(R, T)$ is the average over a rectangular loop with the dimension $R \times T$ (Fig. 13.3).

For $T \gg R$, the limit (13.2.15) holds. Substituting it into (13.3.9) we find, in the continuum limit ($a \to 0$),

$$\chi = a^2 \frac{dE_0(R)}{dR}. \tag{13.3.10}$$

Hence, the function $\chi$ is proportional to the force between the quarks. It is this function which is usually calculated. Computations in the framework of the $SU_2$- and $SU_3$-symmetry show that, as long as the distance between the quarks is smaller than the confinement radius, the interaction obeys the Coulomb law.

As the distance increases, the Coulomb potential goes over into a linearly rising potential for which the interaction force is independent of the distance. That is to say, the calculations substantiate the quark confinement.

It should be stressed that in the region of small $\beta$ (large $g$), the results of the calculation of the quantity $K$ by means of Monte-Carlo simulations coincide with the corresponding results obtained in the strong coupling approximation. However, the Monte-Carlo method permits calculating $K$ also in the region of large $\beta$ (small $g$) where the strong coupling approximation is inapplicable.

It is furthermore seen from the calculations that for $g^2 < 2$ the dependence of the string tension $K$ on $g^2$ is determined by (13.1.48). This implies that for this region the continuum limit already holds. This suggests that the dependence of $a^2K$ on $g^2$ will not substantially change as $g^2$ is further decreased. The occurrence of the dependence of $a^2K$ on $g^2$, which corresponds to the asymptotic freedom regime, already at $g^2 \sim 2$ is referred to as *early scaling*.

It should be noted that for sufficiently large $R$ and $T$ the loop average obeys the area law:

$$W(R,T) = \exp[-a^2KR\,T]\,.$$

Substituting the last expression into (13.3.9) yields

$$\chi(R,T) = a^2K\,.$$

Thus, in this case the function $\chi$ does not depend on the loop size. The absence of this dependence in computations indicates that the loop average $W(C)$ has started to obey the area law, and the quark confinement regime has set in.

In calculating the string tension, the integration has been carried out only over the gluon states. We shall also give an example of calculations involving the integration over both gluon and spinor states.

2) *Spectrum of Hadronic Mass.* The spectrum of masses of the lower hadronic resonances has been calculated in the case of the $SU_2$- and $SU_3$-symmetry. The magnitudes of these masses are determined by the expectation values of correlation functions for operators $\psi$ carrying the same quantum numbers as hadrons. For the mesons, for example,

$$\Delta(x) = \langle \bar{\psi}(x)\Gamma\psi(x)\,\bar{\psi}(0)\Gamma\psi(0)\rangle_{A,\psi}\,. \tag{13.3.11}$$

Here $\Gamma = 1, \gamma_5, \gamma_\mu, \gamma_\mu\gamma_5$ stand for the scalar, pseudoscalar, vector, and pseudovector mesons, respectively. The averaging is carried out over the gluon as well as the quark states.

In the Minkowski space, to the masses there correspond the poles of the Fourier transform of the correlation function

$$\int dx\, e^{ip_\mu x_\mu}\Delta(x) = \sum_k \frac{C_k}{p^2 - m_k^2}\,.$$

In the Euclidean space, upon the inverse Fourier transformation, the correlation function is written as

$$\Delta(x) = \sum_k C_k \mathrm{e}^{-m_k|x|} .$$

For $x \to \infty$, only the resonance with the minimal mass contributes to this sum:

$$\Delta(x) \sim \mathrm{e}^{-m|x|} .$$

Thus, to determine the particle mass, the correlation functions have to be calculated.

Let us neglect the virtual quark loops. Then, taking into account that the right-hand side of (13.3.11) can be expressed in terms of the Green's function $G(x,0;A_\mu)$ for the quarks in the external gluon field $A_\mu$, we have

$$\langle \bar{\psi} \psi \rangle = \langle \mathrm{Tr}\{G(0,0;A_\mu)\} \rangle_A , \tag{13.3.12}$$

where Tr refers to the trace operation with respect to the colour and spinor indices; furthermore we have set $x = 0$ since the trace of the Green's function is finite at $x \to 0$.

According to (13.3.12), the calculation of $\langle \bar{\psi} \psi \rangle$ requires averaging over the gluon fields only. The Green's function $G(x,0;A_\mu)$ for each configuration of the gluon field was found by numerically solving the lattice analogue for the equation

$$-(\mathrm{i}\gamma_\mu \partial_\mu + m)\, G(x,0;A_\mu) = \delta^4 x .$$

As a result, averaging over the gluon fields reduces to calculating a sum analogous to (13.3.7).

The quark mass is a free parameter. The value of it was chosen by equating the mass of the $\pi$-meson to its experimental value.

The calculated values of the hadronic masses are quite close to the corresponding experimental quantities.

It should be emphasized that perturbation theory on a lattice in each order of the coupling constant $g$ gives zero values for the string tension and the hadronic masses. The non-zero values of these quantities in Monte-Carlo simulations imply that they are determined only by the non-perturbative effects of quantum chromodynamics. Besides, there are physical quantities which differ from zero in the perturbation theory calculations.

Monte-Carlo simulations have been also applied for calculating other physical quantities (the mass of the glueball, the magnitude of the quark and the gluon condensates, critical points for phase transitions, etc.).

Hence, numerical calculations by Monte-Carlo simulations enable one to obtain physical results directly from the Lagrangian of quantum chromodynamics, without making use of perturbation theory.

In assessing the results obtained by Monte-Carlo simulations, one should keep in mind that they contain, besides the statistical error, an error caused by approximations made in the calculations (neglecting the virtual quark loops, using a finite size lattice, etc.). The uncertainties of Monte-Carlo simulations are rather big, and one cannot draw definite conclusions as yet, although the results appears quite promising.

The accuracy of Monte-Carlo simulations is presently limited by two major technical factors: the speed of the computers (not allowing to collect sufficient statistics) and the computer memory (restricting the size of the lattice). There are intensive efforts to improve both of these factors.

# 14. Grand Unification

In Part III, unified gauge theories of the electroweak interaction of particles have been presented. This chapter deals with unified gauge models of strong, electromagnetic, and weak interactions. Constructing such models is based on the fact that, as will be demonstrated below, the effective coupling constants of various groups behave differently with increasing energy: the coupling constants of the groups $SU_3$ and $SU_2$ decrease (the former more rapidly than the latter), while the coupling constant of the group $U_1$ increases (Fig. 14.1). As a consequence, it may be expected that at sufficiently large energies all three constants will coincide. Models with a single coupling constant in the high energy range are thus conceivable. The unification procedure resulting in the coincidence of the coupling constants of strong, electromagnetic, and weak interactions at a certain high energy is known as *grand unification.* The grand-unification group is broken by the Higgs fields in the presently accessible energy range so that it manifests itself as the direct product of the group of the strong, electromagnetic, and weak interactions (e.g., $SU_3 \times SU_2 \times U_1$).

As the gauge group of the grand unification, one can choose either a simple group $G$ (e.g., $SU_5$, $SO_{10}$, $E_6$, $SU_8$) or a semi-simple group $G' = G_1 \times G_2$ with an additional discrete symmetry between $G_1$ and $G_2$ (e.g., $SU_4 \times SU_4$). Models based on the groups $G$ and $G'$ involve a single coupling constant.

A great number of grand-unification models have been studied, namely, $SU_5$, $SU_6$, $SU_7$, $SU_8$, $SU_9$, $(SU_4)^4$, $(SU_8)_L \times (SU_8)_R$, $SO_{10}$, $SO_{11}$, $SO_{12}$, $SO_{14}$, $SO_{15}$, $E_6$, $E_7$, $E_8$, etc.

We shall consider in some detail the models based on the group $SU_n$. The simplest of them is the $SU_5$-model.

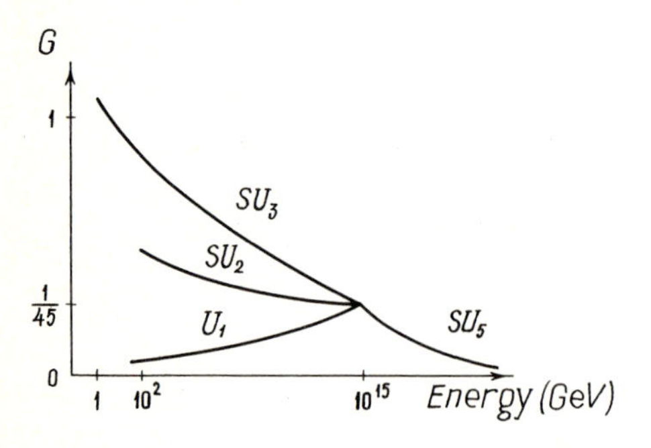


**Fig. 14.1.** Qualitative picture of the behaviour of the coupling constants as a function of energy

## 14.1 The $SU_5$-Model

### 14.1.1 The Lagrangian for the Model

Constructing the model is conducted in the usual way (Chap. 6).

1) Let us demonstrate that, given the group $SU_3 \times SU_2 \times U_1$ as the gauge group in the low energy range, the *minimal group* of the grand-unification is $SU_5$. Indeed, since the sum of the ranks of the groups $SU_3$, $SU_2$, and $U_1$ equals 4, the minimal grand-unification group can only be a group of rank 4. The groups of rank 4 with a single coupling constant are

$$\begin{aligned}&(SU_2)^4, (SO_5)^2, (SU_3)^2, (G_2)^2, \\ &SO_8, SO_9, Sp_8, F_4, SU_5 .\end{aligned} \tag{14.1.1}$$

The first two groups do not include the group $SU_3$ as a sub-group and have to be abandoned for this reason.

The representations of the group $SU_3 \times SU_2$ have the form

$$(3,2)+2(\bar{3},1)+(1,2)+(1,1) . \tag{14.1.2}$$

The numbers in parentheses indicate the dimension of the multiplet with respect to the groups $SU_3$ and $SU_2$, respectively. The representations (14.1.2) are complex, and the representations of the grand-unification group have to be complex as well. Of all the groups (14.1.1), only the groups $(SU_3)^2$ and $SU_5$ have such representations. The group $(SU_3)^2$ is inappropriate from the physical point of view, however. This leads us to conclude that the minimal admissible group of rank 4 for the gauge subgroup $SU_3 \times SU_2 \times U_1$ is $SU_5$. We thus choose $SU_5$ as the gauge group of the model.

2) The initial particles of the model are 6 leptons and 18 quarks

$$\begin{pmatrix} u_1 & u_2 & u_3 & \nu_e \\ d_1 & d_2 & d_3 & e^- \\ c_1 & c_2 & c_3 & \nu_\mu \\ s_1 & s_2 & s_3 & \mu^- \\ t_1 & t_2 & t_3 & \nu_\tau \\ b_1 & b_2 & b_3 & \tau^- \end{pmatrix} . \tag{14.1.3}$$

3) Among the representations of the group $SU_5$, there are those of dimension $\bar{5}$ and 10 whose structure with respect to the subgroup $SU_3 \times SU_2$ is as follows:

$$\begin{aligned}&\bar{5} = (\bar{3},1)+(1,2) ; \\ &10 = (\bar{3},1)+(1,1)+(3,2) .\end{aligned}$$

Let us put the first 8 particles (the *first generation*) contained in (14.1.3), as well as their anti-particles, in these representations:

$$\psi^i = \bar{5} = \begin{pmatrix} d_1^c \\ d_2^c \\ d_3^c \\ e^- \\ \nu_e \end{pmatrix}_L , \qquad \psi^{ij} = 10 = \frac{1}{\sqrt{2}} \begin{pmatrix} 0 & u_3^c & -u_2^c & -u_1 & -d_1 \\ -u_3^c & 0 & u_1^c & -u_2 & -d_2 \\ u_2^c & -u_1^c & 0 & -u_3 & -d_3 \\ u_1 & u_2 & u_3 & 0 & -e^+ \\ d_1 & d_2 & d_3 & e^+ & 0 \end{pmatrix}_L . \tag{14.1.4}$$

Here the index c refers to the charge conjugation; it is sometimes beneficial to deal not with a multiplet but rather with its charge conjugate; then the multiplet chirality ought to be replaced by the opposite one ($L \to R$ or $R \to L$).

The remaining particles of the *second generation* $(c, s, \nu_\mu, \mu^-)$ and of the *third generation* $(t, b, \nu_\tau, \tau^-)$ of (14.1.3) are placed in a similar way. In the $SU_5$-model, all known fermions are thus subdivided into three generations, the particle masses increasing from one generation to the next one. In the representations of the model (14.1.4), there is no place for right-handed neutrino. Like the standard model, the model under consideration contains only left-handed neutrinos.

It should be stressed that in the $SU_5$-model a multiplet comprises both quarks and leptons, i.e., leptons are on *equal footing* with quarks implying that at large energies quarks and leptons exhibit the same interactions.

The charge operator $Q$ is one of the generators of the symmetry group. The eigenvalues $q$ of the charge operator $Q$ are obtained by acting with the operator $Q$ on the representations $\psi$ of the group:

$$Q\psi = q\psi .$$

Because both leptons and quarks are contained in the multiplets of the $SU_5$-model, the values of their charges are multiples (*charge quantization*).

The trace of a generator of the group is zero, and, in particular, $\mathrm{Tr}\{Q\} = 0$. This means that the sum of the charges of the particles of a multiplet must be zero. For example, for the $SU_5$-quintet,

$$\mathrm{Tr}\{Q\} = q_{d_1^c} + q_{d_2^c} + q_{d_3^c} + q_{e^-} + q_{\nu_e} = 0 .$$

From this, it follows for the charge of a $d$-quark: $q_d = -e/3$. Hence, the $SU_5$-symmetry accounts for the fractional charges of quarks. In general, quark charges are non-integer in models based on the simple groups $G$. In models based on a semi-simple group $G' = G_1 \times G_2$ with discrete symmetry, both fractional and integer charges are possible. The latter can be achieved, e.g., in a way described in Sect. 6.2.

4) Spontaneous $SU_5$-symmetry breaking is carried out in two stages:

$$SU_5 \xrightarrow[\text{24-plet}]{} SU_3 \times SU_2 \times U_1 \xrightarrow[\text{5-plet}]{} SU_3 \times U_1 .$$

In the first stage the group $SU_5$ breaks down to $SU_3 \times SU_2 \times U_1$ which, in turn, breaks down to $SU_3 \times U_1$ in the second stage. Realization of the first stage requires introducing a multiplet $\Phi_i^j$ of Higgs scalar fields which transforms according to the adjoint representation of the $SU_5$-group (24-plet). The second stage of spontaneous symmetry-breaking is realized by introducing a multiplet $\Phi^i$ of Higgs scalar fields which transforms according to the fundamental representation of the $SU_5$-group (5-plet).

5) Thus, the model includes three generations of the quark-lepton multiplets $\psi^i$, $\psi^{ij}$ as well as the multiplets $\Phi^i$, $\Phi_i^j$ of Higgs scalar fields.

6) The globally invariant Lagrangian for the model is written as

$$L = \mathrm{i}\,\bar{\psi}^i \gamma_\mu \partial_\mu \psi^i + \mathrm{i}\,\bar{\psi}^{ij} \gamma_\mu \partial_\mu \psi^{ij} + |\partial_\mu \Phi^i|^2 + |\partial_\mu \Phi_i^j|^2 + V(\Phi) + L_{\mathrm{Yuk}} \,.$$

Here $V(\Phi)$ is the Higgs scalar field potential and $L_{\mathrm{Yuk}}$ denotes the Yukawa interaction terms, which make the fermions acquire mass,

$$L_{\mathrm{Yuk}} = h\,\bar{\psi}^{ij}\,\Phi^i \psi^j + \mathrm{h.c.} \tag{14.1.5}$$

7) Let us localize the Lagrangian by using the expression (2.1.12) for the covariant derivative of fields.

8) Let us apply the mechanism of spontaneous symmetry-breaking. The potential $V(\Phi)$ is chosen in the general form

$$\begin{aligned} V(\Phi) = & -\frac{1}{2}\mu_1^2 \Phi_i^j \Phi_j^i + \frac{\varkappa}{4}(\Phi_i^j \Phi_j^i)^2 + \frac{\varrho}{4}\,\Phi_i^j \Phi_j^k \Phi_k^l \Phi_l^i - \mu_2^2 \Phi^{\dagger i} \Phi^i \\ & + \frac{\gamma}{2}(\Phi^{\dagger i} \Phi^i)^2 + \frac{\delta}{2}\,\Phi^{\dagger k} \Phi^k \Phi_i^j \Phi_j^i + \frac{\sigma}{2}\,\Phi^{\dagger k} \Phi_k^j \Phi_j^i \Phi^i \,, \end{aligned} \tag{14.1.6}$$

where $\mu_i$, $\varkappa$, $\varrho$, $\gamma$, $\delta$, and $\sigma$ are the coupling constants.

With the choice of some definite relations between the above coupling constants, this potential leads to the following vacuum expectation values of the functions $\Phi_i^j$ and $\Phi^i$:

$$(\Phi_i^j)_{\mathrm{v}} = \begin{pmatrix} \alpha & & & & \\ & \alpha & & 0 & \\ & & \alpha & & \\ & 0 & & \alpha(-\frac{3}{2}+\varepsilon) & \\ & & & & \alpha(-\frac{3}{2}-\varepsilon) \end{pmatrix}; \quad (\Phi^i)_{\mathrm{v}} = \begin{pmatrix} 0 \\ 0 \\ 0 \\ 0 \\ \beta \end{pmatrix}. \tag{14.1.7}$$

Here $\alpha$, $\beta$, and $\varepsilon$ are some combinations of the coupling constants; they determine the magnitude of particle masses; note that $\alpha \gg \beta$, $\alpha \gg \varepsilon\alpha$.

Let us make a shift in the Higgs fields and diagonalize the free Lagrangian of the model. As in the standard model, one parameter – the Weinberg angle – has to be introduced for that. In the standard model, the Weinberg angle is an arbitrary parameter. In the $SU_5$-model, on the contrary, this angle has a definite value. To find this value, we select, out of 24 gauge bosons of the

$SU_5$-group, those neutral fields $B_\mu^a(x)$ which correspond to the diagonal generators of this group ($a = 1, 2, 3, 4$).

The expressions for the electromagnetic field $B_\mu(x)$ and for the fields $A_\mu^3(x)$, $A_\mu(x)$ which are involved in the Lagrangians (3.4.2) and (3.4.9) for the standard group $SU_2 \times U_1$ assume, in the $SU_5$-model, the form

$$B_\mu(x) = -\sqrt{\frac{2}{3}}\frac{e}{g}B_\mu^3(x),$$

$$A_\mu^3(x) = -\sqrt{\frac{3}{8}}B_\mu^3(x) + \sqrt{\frac{5}{8}}B_\mu^4(x),$$

$$A_\mu(x) = -\sqrt{\frac{5}{8}}B_\mu^3(x) - \sqrt{\frac{3}{8}}B_\mu^4(x).$$

By substituting the last two expressions into (3.4.5) and taking into account the first formula of (6.1.24) we obtain the value of the Weinberg angle searched for:

$$\sin^2\theta_W = 3/8. \qquad (14.1.8)$$

It should be emphasized that this value refers to the energy where the exact $SU_5$-symmetry gets broken. In the low energy range ($\sim 10^2$ GeV), experiment gives another value: $\sin^2\theta_W \sim 0.23$. The relation between these two values of the Weinberg angle will be discussed below.

The $SU_5$-model includes 24 massless gauge fields. Upon spontaneous symmetry-breaking, 9 gauge fields remain massless, while the rest of the gauge fields acquire mass (Table 14.1). Since $\alpha \gg \beta$, $\alpha \gg \varepsilon\alpha$, the masses of $X$- and $Y$-bosons are much larger than those of $W$- and $Z$-bosons.

Gluons are the mediators of strong interactions between quarks (and are associated with the group $SU_3$), while photon, $W$- and $Z$-bosons are the mediators of electroweak interactions between leptons and quarks (associated

**Table 14.1.** Bosons and their masses

| Bosons | | Masses |
|---|---|---|
| Gluons ($SU_3$-octet) | | 0 |
| Photon | | 0 |
| $W^\pm$ | | $g\sqrt{4\alpha^2\varepsilon^2 + \beta^2/2}$ |
| $Z$ | | $\frac{1}{\sqrt{2}} g\beta$ |
| $X^i_{\pm 4/3}$, | $i = 1, 2, 3$ | $g\alpha(5/2 - \varepsilon)$ |
| $Y^i_{\pm 1/3}$, | $i = 1, 2, 3$ | $g\sqrt{\alpha^2(5/2+\varepsilon)^2 + \beta^2/2}$ |

with the standard group $SU_2 \times U_1$). In the $SU_5$-model, there emerge *additional intermediate bosons*, not involved in the standard model, namely, the charged triplets $X^i_{+4/3}$, $X^i_{-4/3}$, $Y^i_{+1/3}$, $Y^i_{-1/3}$, where $i = 1, 2, 3$ is the colour index; the subscripts indicate the value of the electric charge. These bosons take care of the *quark-lepton* and *lepton-quark conversions* accompanied by transfer of the electric charge and colour.

The quarks $d$, $s$, and $b$ as well as the charged leptons $e^-$, $\mu^-$, and $\tau^-$ acquire mass due to the Yukawa interaction (14.1.5). To provide the quarks $u$, $c$, and $t$ with mass, an additional term has to be introduced to describe the Yukawa interaction of fermions with the symmetrical second-rank tensor of Higgs fields. One cannot use the Yukawa term for the interaction between fermions and the Higgs multiplet $\Phi^j_i$, since it leads to much too large fermion masses.

The Lagrangian for the $SU_5$-model includes, besides the free Lagrangian, the interaction Lagrangians of the fields mentioned. Among others, it involves the Lagrangian for the strong interaction between quarks (mediated by gluons; this pertains to quantum chromodynamics); the Lagrangian for the electromagnetic and the weak interactions of quarks and leptons (mediated by photon, $W^\pm$-bosons, and $Z$-boson; this pertains to the standard model); and the Lagrangian describing the quark-lepton and lepton-quark conversions (mediated by the $X$- and $Y$-bosons).

### 14.1.2 Energy Dependence of the Coupling Constants

Let us investigate the character of the variation with energy of the coupling constants $g_3$, $g_2$, and $g_1$ of the subgroups $SU_3$ and $SU_2 \times U_1$. To that end, we make use of (10.1.20),

$$\mu \frac{\partial g_i}{\partial \mu} = \beta_i(g_i) , \qquad t = \ln \mu , \qquad i = 1, 2, 3 , \tag{14.1.9}$$

which in the one-loop approximation is rewritten as

$$\mu \frac{\partial g_i}{\partial \mu} = \frac{1}{4\pi} B_i g_i^3 , \qquad \text{or} \tag{14.1.10}$$

$$\frac{dg_i}{g_i^3} = \frac{1}{4\pi} B_i \frac{d\mu}{\mu} . \tag{14.1.11}$$

Integrating the last equation over the energy from $\mu$ to $M$ yields

$$\frac{1}{g_i^2(\mu)} = \frac{1}{g_i^2(M)} + \frac{1}{4\pi} B_i \ln \left( \frac{M}{\mu} \right)^2 . \tag{14.1.12}$$

With the aid of calculations similar to those of Sect. 10.2 we find

$$B_3 = -\frac{1}{12\pi}(33-2n)\,, \quad B_2 = -\frac{1}{12\pi}(22-m)\,, \quad B_1 = \frac{1}{12\pi}\,\frac{6}{5}\sum_i y_i^2\,. \tag{14.1.13}$$

Here $n$ is the number of quark flavours of the subgroup $SU_3$ and $m$ is the number of doublets of the subgroup $SU_2$; $y_i$ denotes the eigenvalues of the hypercharge operator $Y$ which is given by

$$Y = \begin{pmatrix} -\frac{1}{3} & & & & \\ & -\frac{1}{3} & & 0 & \\ & & -\frac{1}{3} & & \\ & 0 & & \frac{1}{2} & \\ & & & & \frac{1}{2} \end{pmatrix}.$$

The first generation of $SU_5$-model includes two quark flavours $(u, d)$ and four $SU_{2L}$-doublets; furthermore, $\sum y_i^2 = 10/3$. Consequently,

$$B_3 = -\frac{33}{12\pi} + B_1\,, \quad B_2 = -\frac{22}{12\pi} + B_1\,. \tag{14.1.14}$$

The quantities $B_3$ and $B_2$ being negative, the corresponding effective coupling constants $g_3(\mu)$ and $g_2(\mu)$ decrease with increasing energy (Sect. 10.1). Since $B_3 < B_2$, $g_3$ decreases more rapidly than $g_2$ (Fig. 14.1). The quantity $B_1$ is positive, and the effective coupling constant $g_1(\mu)$ increases with the energy (Fig. 14.1).

For the point $M$ at which the $SU_5$-symmetry becomes exact, we have

$$g_3(M) = g_2(M) = g_1(M) = g(M)\,. \tag{14.1.15}$$

From (14.1.12) and (14.1.15) it then follows that

$$\frac{x}{g_3^2(\mu)} + \frac{y}{g_2^2(\mu)} + \frac{z}{g_1^2(\mu)} = \frac{x+y+z}{g^2(M)} + \frac{1}{4\pi}(xB_3 + yB_2 + zB_1)\ln\left(\frac{M}{\mu}\right)^2, \tag{14.1.16}$$

where $x, y, z$ are arbitrary parameters.

To eliminate $B_1$ and the unknown constant $g(M)$ from (14.1.16), we set $x+y+z=0$. Taking into account (14.1.14), Eq. (14.1.16) is then rewritten as

$$\frac{x}{g_3^2(\mu)} + \frac{y}{g_2^2(\mu)} - \frac{x+y}{g_1^2(\mu)} = -\frac{1}{24\pi^2}(33x + 22y)\ln\frac{M}{\mu}\,, \tag{14.1.17}$$

or, using (14.1.8) and (6.1.19, 24), as

$$\frac{x}{\alpha_s(\mu)} - \frac{3/5(x+y) - (3/5x + 8/5y)\sin^2\theta_W(\mu)}{\alpha_e(\mu)} = -\frac{1}{6\pi}(33x + 22y)\ln\frac{M}{\mu}\,, \tag{14.1.18}$$

where $\alpha_s = g_3^2/4\pi$ and $\alpha_e = e^2/4\pi$.

This expression contains two unknown quantities: $M$ and $\sin^2\theta_W(\mu)$. To obtain a formula for each of them, we set in (14.1.18) first $y = -3x/8$ and then $y = -3x/2$. As a result, we find

$$\ln\frac{M}{\mu} = -\frac{8\pi}{33}\left(\frac{1}{\alpha_s(\mu)} - \frac{3/8}{\alpha_e(\mu)}\right), \tag{14.1.19}$$

$$\sin^2\theta_W(\mu) = \frac{5}{9}\frac{\alpha_e(\mu)}{\alpha_s(\mu)} + \frac{1}{6}. \tag{14.1.20}$$

These relations enable us to calculate the values of $M$ and $\sin^2\theta_W(\mu)$ at the energy $\mu \sim 10^2$ GeV. Substituting into (14.1.19, 20) the experimental values of $\alpha_e$ and $\alpha_s$ we obtain

$$\sin^2\theta_W(\mu) \approx 0.20\,, \quad M \sim 10^{15}\,\text{GeV}\,. \tag{14.1.21}$$

As already seen, in the high energy region where the $SU_5$-symmetry is valid, $\sin^2\theta_W = 3/8$. As distinct from this, in the energy range $\mu \sim 10^2$ GeV the relation (14.1.20) leads to the value of $\sin^2\theta_W$ which is close to the experimental one.

The region of symmetry-breaking is determined by the mass of the intermediate boson. As follws from (14.1.21), the $SU_5$-symmetry is exact for energies $\mu > 10^{15}$ GeV. In the energy range $\mu < 10^{15}$ GeV, the $SU_5$-symmetry is broken. Then the corresponding masses of $X$-bosons and $Y$-bosons will be $\sim 10^{15}$ GeV. According to (6.1.26), the $SU_2 \times U_1$-symmetry is broken at energies $\sim 10^2$ GeV.

### 14.1.3 Proton Decay

The most interesting feature of the Lagrangian for the $SU_5$-model is connected with the interaction Lagrangian of $X$- and $Y$-bosons with fermions:

$$\mathscr{L}_X = \frac{g}{\sqrt{2}}\sum_{i=1}^{3} X_\mu^{i-}(-\bar{e}_R\gamma_\mu d_R^{ic} - \varepsilon_{ijk}\bar{u}_L^{jc}\gamma_\mu u_L^k - \bar{e}_L\gamma_\mu d_L^{ic}) + \text{h.c.}$$

$$\mathscr{L}_Y = \frac{g}{\sqrt{2}}\sum_{i=1}^{3} Y_\mu^{i-}(-\bar{\nu}_{eL}\gamma_\mu d_L^{ic} + \bar{e}_R\gamma_\mu u_R^{ic} - \varepsilon_{ijk}\bar{u}_L^{jc}\gamma_\mu d_L^k) + \text{h.c.}$$

This permits realization of the elementary processes of the type

$$ud \underset{X}{\rightarrow} e^+\bar{u}\,, \quad ud \underset{Y}{\rightarrow} \bar{d}\bar{\nu}_e\,, \quad ud \underset{Y}{\rightarrow} e^+\bar{u}\,,$$

which lead to proton decay (Fig. 14.2). The origin of it is that in the $SU_5$-model leptons and quarks belong to one and the same multiplet and can

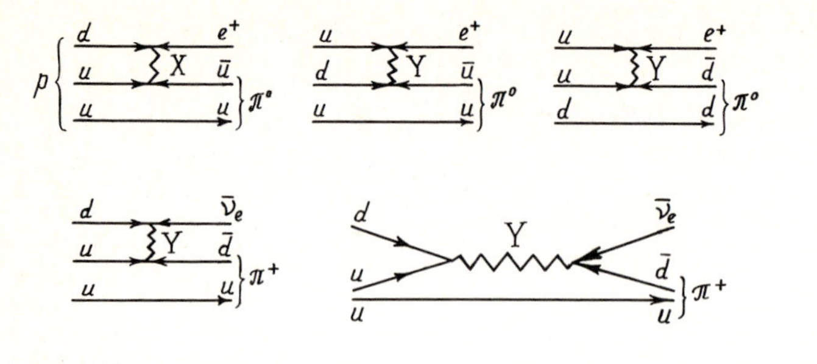


**Fig. 14.2.** Quark diagrams for the proton decay

therefore transform into each others. The lepton and baryon numbers are not conserved in these processes. Other baryons can decay in a similar way.

Elaborate calculations give for the lifetime $\tau_\mathrm{p}$ of a proton the range

$$\tau_\mathrm{p} = (0.6-25)\times 10^{30}\left(\frac{m_X}{5\times 10^{14}\,\mathrm{GeV}}\right)^4 \text{years}\,, \tag{14.1.22}$$

where $m_X$ denotes the $X$-boson mass. Since, according to (14.1.21), $m_X \sim 10^{15}$ GeV, the proton life time is evaluated as

$$\tau_\mathrm{p} = 10^{30\pm 3}\,\text{years}\,.$$

Experiment gives, as a preliminary lower limit, $\tau_\mathrm{p} > 2\times 10^{30}$ years. A number of experimental units are being devised presently, aimed at measuring $\tau_\mathrm{p}$ more precisely.

As can be seen, the $SU_5$-model makes it possible to unite the strong and the electroweak interactions, to describe, at high energies, all interactions in terms of a single coupling constant, and to combine quarks and leptons in one multiplet. It permits accounting for particle charge quantization, calculating the Weinberg angle, explaining the masslessness of the neutrino as well as the massiveness of quarks and leptons, and it predicts the proton decay.

However, the $SU_5$-model has a number of drawbacks: it describes the quark-lepton families separately and does not account for the number of the families; it does not account for fermion mixing and for the origin of *CP* violation; it leads to relations between particle masses which are not all in accordance with available experimental data; it gives a value for the renormalized Weinberg angle which is somewhat smaller than the experimental one; and, finally, it does not explain the tremendous difference of masses of *W*- and *Z*-bosons on one side and *X*- and *Y*-bosons on the other side (the so-called *problem of gauge hierarchies*). That is why, besides the $SU_5$-group, grand-unification models based on other groups, e.g., $SU_n$ have been subjected to analysis. Below we shall consider such models.

## 14.2 Structure of the Fermion Multiplets in the $SU_n$-Model

Let us assume that in the high-energy region there holds the $SU_n$-symmetry which is severely broken by Higgs fields in the presently accessible energy range and therefore manifests itself as the direct product of the strong interaction group $SU_3$ and the electroweak interaction group $SU_2 \times U_1$, i.e., as $SU_3 \times SU_2 \times U_1$.

Let us further assume that

i) quarks and leptons are put in triplet and singlet representations of the colour group $SU_3$, respectively. (This is in accord with quantum chromodynamics);
ii) quarks and leptons have ordinary charges $\mp 1/3$, $\mp 2/3$, $\mp 1, 0$;
iii) electroweak interactions are described by the standard group $SU_2 \times U_1$ which is a subgroup of the group $SU_n$.

All the assumptions mentioned give rise to a definite structure of the multiplets of the group $SU_n$ with respect to the subgroups $SU_3$, $SU_2$ and $U_1$.

### 14.2.1 Structure of $SU_n$-Multiplets with Respect to the Subgroup $SU_3$

The fact that the model involves only quark triplets and lepton singlets implies that the multiplets of the group $SU_n$ must contain no other representations of the colour subgroup $SU_3$ than 3, $\bar{3}$, and 1. It can be demonstrated that this property is characteristic of only the fundamental representation $\psi_1^i$ and the totally anti-symmetric tensors of arbitrary rank, $\psi_m^{i_1 \dots i_m} (m = 2, \dots, n-1)$, of the group $SU_n$. The indices $i_1, \dots, i_m$ run from 1 to $n$.

The interaction Lagrangian of spinor fields $\psi_m^{i_1 \dots i_m}$ with gauge fields $A_\mu^a$ $(a = 1, \dots, n^2 - 1)$ which is invariant under the local group $SU_n$ is written as

$$\mathscr{L}_{\mathrm{I}} = -\frac{g}{2} \bar{\psi}_m^{ii_2 \dots i_m} A_\mu^a \gamma_\mu (\lambda_a)_{ij} \psi_m^{ji_2 \dots i_m}, \tag{14.2.1}$$

where $\lambda_a (a = 1, \dots, n^2 - 1)$ denotes the $n$-dimensional generators of the group $SU_n$, and $g$ is the coupling constant.

Let us consider the multiplets $\psi_m$ as left-handed. We wish to establish a correspondence between the components of the multiplets $\psi_m^{i_1 \dots i_m}$, which are characterized by the indices $i_1, \dots, i_m$, and the fermion type (i.e., quarks or leptons). To be more specific, we choose as the generators of the subgroup $SU_3$ eight generators of the group $SU_n$, with three-row Gell-Mann matrices $\lambda_b$ $(b = 1, \dots, 8)$ placed in the upper left corner and the rest of the elements of the generators being zero. Retaining in the interaction Lagrangian (14.2.1) only the terms associated with the chosen generators $\lambda_b$ of the subgroup $SU_3$ we arrive at the interaction Lagrangian of quantum chromodynamics:

**Table 14.2.** The structure of $SU_n$-multiplets with respect to the subgroup $SU_3$

| Components of the multiplet | Dimension of representation | Number of representations |
|---|---|---|
| $\psi_m^{li_2\ldots i_m}$ | 3 | $C_{n-3}^{m-1}$ |
| $\psi_m^{lki_3\ldots i_m}$ | $\bar{3}$ | $C_{n-3}^{m-2}$ |
| $\psi_m^{i_1\ldots i_m}$ | 1 | $C_{n-3}^{m}$ |
| $\psi_m^{123i_4\ldots i_m}$ | 1 | $C_{n-3}^{m-3}$ |

$$\mathcal{L}_{\mathrm{I}}^{\mathrm{QCD}} = -\frac{g}{2}\bar{\psi}_m^{li_2\ldots i_m} A_\mu^b \gamma_\mu (\lambda_b)_{lk} \psi_m^{ki_2\ldots i_m}$$

$$+\frac{g}{2}\bar{\psi}_m^{iji_3\ldots i_m} \varepsilon_{lij} A_\mu^b \gamma_\mu (\lambda_b)_{kl} \varepsilon_{kpr} \psi_m^{pri_3\ldots i_m} . \qquad (14.2.1')$$

Here $A_\mu^b$ ($b = 1, \ldots, 8$) stands for gluons; $i, j, k, l, p, r = 1, 2, 3$; $i_1, \ldots, i_m = 4, \ldots, n$; $\varepsilon_{ijl}$ is the totally anti-symmetric unit tensor. It follows from the last expression that the multiplet $\psi_m$, whose dimension is $C_n^m$, where $C_n^m = n!/m!(n-m)!$ is the binomial coefficient, decomposes into the multiplets $\psi_m^{li_2\ldots i_m}$, $\psi_m^{lki_3\ldots i_m}$, $\psi_m^{i_1\ldots i_m}$, $\psi_m^{123i_4\ldots i_m}$. A comparison of (14.2.1′) with the Lagrangian of quantum chromodynamics shows that $\psi_m^{li_2\ldots i_m}$ is a triplet representation, $\psi_m^{lki_3\ldots i_m}$ is an anti-triplet representation, and $\psi_m^{i_1\ldots i_m}$, $\psi_m^{123i_4\ldots i_m}$ are singlet representations of the colour group $SU_3$. (Here $l, k = 1, 2, 3$; $i_1, \ldots, i_m = 4, \ldots, n$.) Hence, the multiplet $\psi_m$ of the group $SU_n$ includes a certain number of multiplets 3, $\bar{3}$, and 1 of the subgroup $SU_3$, each of them being specified by the values of indices $i_1, \ldots, i_m$ (Table 14.2). In the table, $l, k = 1, 2, 3$; $4 \leqslant i_1, \ldots, i_m \leqslant n$.

### 14.2.2 Structure of $SU_n$-Multiplets with Respect to the Electric Charge

In the low energy region, the group $SU_n$ is broken down to the subgroup of electromagnetic interaction $U_1$. The generator of this group is the electric charge operator. The gauge field $B_\mu$ corresponding to this operator is the electromagnetic field. The field $B_\mu$ can be expressed as a linear combination of neutral gauge fields $A_\mu^{a_3}, \ldots, A_\mu^{a_{n-1}}$, associated with the diagonal operators $\lambda_{a_p}$ ($p = 1, \ldots, n-1$) of the $SU_n$ group

$$B_\mu = \sum_{p=3}^{n-1} \alpha_{a_p} A_\mu^{a_p} , \qquad (14.2.2)$$

were $\alpha_{a_p}$ are unknown coefficients.

The gluon fields $A_\mu^{a_1} \equiv A_\mu^3$ and $A_\mu^{a_2} \equiv A_\mu^8$ have to be excluded from the expression (14.2.2). Equation (14.2.2) can be reexpressed as

$$A_\mu^{a_p} = \alpha_{a_p} B_\mu + \ldots , p = 3, \ldots, n-1 . \qquad (14.2.2')$$

By substituting (14.2.2′) into (14.2.1) we obtain the Lagrangian for the electromagnetic interaction

$$\mathscr{L}_{\mathrm{I}}^{\mathrm{em}} = -e\,\bar{\psi}_m^{ii_2\ldots i_m} B_\mu \gamma_\mu k_i \psi_m^{ii_2\ldots i_m}, \tag{14.2.3}$$

where $e$ is the electromagnetic interaction constant and

$$k_i = \frac{g}{2e} \sum_{p=3}^{n-1} \alpha_{a_p} (\lambda_{a_p})_{ii}.$$

The trace of the electric charge operator is equal to zero which results in $k_1 = k_2 = k_3 = -\frac{1}{3}K$, where

$$K = \sum_{i=4}^{n} k_i.$$

In accordance with Table 14.2, the electromagnetic interaction of quarks and leptons reads

$$\begin{aligned}
&-e\,\bar{\psi}_m^{li_2\ldots i_m}(k_{i_2}+\ldots+k_{i_m}-\tfrac{1}{3}K)B_\mu\gamma_\mu\psi_m^{li_2\ldots i_m}, && 1\leqslant m\leqslant n-2,\\
&-e\,\bar{\psi}_m^{lki_3\ldots i_m}(k_{i_3}+\ldots+k_{i_m}-\tfrac{2}{3}K)B_\mu\gamma_\mu\psi_m^{lki_3\ldots i_m}, && 2\leqslant m\leqslant n-1,\\
&-e\,\bar{\psi}_m^{123i_4\ldots i_m}(k_{i_4}+\ldots+k_{i_m}-K)B_\mu\gamma_\mu\psi_m^{123i_4\ldots i_m}, && 3\leqslant m\leqslant n-1,\\
&-e\,\bar{\psi}_m^{i_1\ldots i_m}(k_{i_1}+\ldots+k_{i_m})B_\mu\gamma_\mu\psi_m^{i_1\ldots i_m}, && 1\leqslant m\leqslant n-3.
\end{aligned} \tag{14.2.4}$$

Because the model involves only quarks with charges $\mp 1/3$, $\pm 2/3$ and leptons with charges $\pm 1, 0$, the expressions in (14.2.4) lead to the following set of linear algebraic equations for the unknown coefficients $k_i$:

$$k_{j_1}+\ldots+k_{j_{m-1}}-\tfrac{1}{3}K = \begin{cases}\pm\frac{1}{3}\\ \pm\frac{2}{3}\end{cases}, \quad 1\leqslant m\leqslant n-2,$$

$$k_{j_1}+\ldots+k_{j_{m-2}}-\tfrac{2}{3}K = \begin{cases}\pm\frac{1}{3}\\ \pm\frac{2}{3}\end{cases}, \quad 2\leqslant m\leqslant n-1,$$

$$k_{j_1}+\ldots+k_{j_{m-3}}-K = \begin{cases}0\\ \pm 1\end{cases}, \quad 3\leqslant m\leqslant n-1,$$

$$k_{j_1}+\ldots+k_{j_m} = \begin{cases}0\\ \pm 1\end{cases}, \quad 1\leqslant m\leqslant n-3,$$

where $k_{j_1}+\ldots+k_{j_s}$ are all possible sums of $s$ coefficients $k_i\,(i=4,\ldots,n)$. This set of equations has one general solution for the multiplets of arbitrary rank:

$$k_4 = \pm 1,\; k_i = 0, \tag{14.2.5}$$

where $i = 5,\ldots,n$. Besides, there are 4 more solutions to this set, but these have no physical relevance.

**Table 14.3.** The structure of $SU_n$-multiplets with respect to the electric charge

| Components of the multiplet | | Charge | Number of quark triplets or leptons |
|---|---|---|---|
| $\psi_{m_L}^{l4i_3\ldots i_m}$, | $2 \leqslant m \leqslant n-2$ | $\pm 2/3$ | $C_{n-4}^{m-2}$ |
| $\psi_{m_L}^{li_2\ldots i_m}$, | $1 \leqslant m \leqslant n-3$ | $\mp 1/3$ | $C_{n-4}^{m-1}$ |
| $(\psi_m^{lk4i_4\ldots i_m})_R^c$, | $3 \leqslant m \leqslant n-1$ | $\mp 1/3$ | $C_{n-4}^{m-3}$ |
| $(\psi_m^{lki_3\ldots i_m})_R^c$, | $2 \leqslant m \leqslant n-2$ | $\pm 2/3$ | $C_{n-4}^{m-2}$ |
| $\psi_{m_L}^{1234i_5\ldots i_m}$, | $4 \leqslant m \leqslant n-1$ | 0 | $C_{n-4}^{m-4}$ |
| $\psi_{m_L}^{123i_4\ldots i_m}$, | $3 \leqslant m \leqslant n-1$ | $\mp 1$ | $C_{n-4}^{m-3}$ |
| $(\psi_m^{4i_2\ldots i_m})_R^c$, | $1 \leqslant m \leqslant n-3$ | $\mp 1$ | $C_{n-4}^{m-1}$ |
| $(\psi_m^{i_1\ldots i_m})_R^c$, | $1 \leqslant m \leqslant n-4$ | 0 | $C_{n-4}^{m}$ |

Substituting (14.2.5) into (14.2.4) we find the correspondence between the components of the multiplet $\psi_m$ and the values of the charge of leptons and quarks (Table 14.3). In the last column of Table 14.3, the number of particles with a given electric charge contained in the multiplet $\psi_m$ is indicated. Note that $C_n^m = n!/m!(n-m)!$ Furthermore, in the table, $l, k = 1, 2, 3$; $i_1, \ldots, i_m = 5, \ldots, n$, and $(\psi_m^{i_1\ldots i_m})_R^c$ is equivalent to $\psi_{m_L}^{i_1\ldots i_m}$, where the index c denotes the charge conjugation operation. By means of Table 14.3, one can find the structure of any multiplet of the group $SU_n$ with respect to the subgroups $SU_3$ and $U_1$. In particular, for the multiplets $\bar{5}$ and 10 of the group $SU_5$ we arrive at (14.1.4).

### 14.2.3 Structure of $SU_n$-Multiplets with Respect to the Subgroup $SU_2 \times U_1$

With the use of the results obtained, we can find the structure of an $SU_n$-multiplet with respect to the subgroup of electroweak interactions $SU_2 \times U_1$. For this purpose, we first obtain the Lagrangian of electroweak interaction. Following Table 14.3, we choose the expressions for the gauge fields, $W_\mu^i$, ($i = 1, 2, 3$) and $A_\mu$, of the group $SU_2 \times U_1$ in the form

$$
\begin{aligned}
W_\mu^+ &= \frac{1}{\sqrt{2}}(W_\mu^1 + \mathrm{i}\,W_\mu^2) = \frac{1}{\sqrt{2}} A_\mu^a (\lambda_a)_{54}\,, \\
W_\mu^- &= \frac{1}{\sqrt{2}}(W_\mu^1 - \mathrm{i}\,W_\mu^2) = \frac{1}{\sqrt{2}} A_\mu^a (\lambda_a)_{45}\,, \qquad (14.2.6) \\
W_\mu^3 &= -\sqrt{\frac{3}{8}} A_\mu^{a_3} + \sqrt{\frac{5}{8}} A_\mu^{a_4}\,,
\end{aligned}
$$

$$A_\mu = -\sqrt{\frac{5}{8}}A_\mu^{a_3} - \sqrt{\frac{3}{8}}A_\mu^{a_4}. \tag{14.2.6}$$

Using (14.2.2, 5) we arrive at the expression for the electromagnetic field:

$$B_\mu = -\sqrt{\frac{8}{3}}\frac{e}{g}A_\mu^{a_3}. \tag{14.2.7}$$

The fields $W_\mu^3$ and $A_\mu$ are related to the fields $B_\mu$ and $Z_\mu$ through (3.4.5):

$$B_\mu = \cos\theta_W A_\mu + \sin\theta_W W_\mu^3, \qquad Z_\mu = \cos\theta_W W_\mu^3 - \sin\theta_W A_\mu, \tag{14.2.8}$$

where $\theta_W$ is the Weinberg angle.

Determining from (14.2.6 – 8) the expressions for the gauge fields $A_\mu^a$ of the group $SU_n$ in terms of the fields $W_\mu^\pm$ and $Z_\mu$ and substituting the result into (14.2.1) we obtain the Lagrangian of the weak interaction

$$\mathscr{L}_I^W = -\frac{g}{\sqrt{2}}\bar{\psi}_m^{4i_2\ldots i_m} W_\mu^- \gamma_\mu \psi_m^{5i_2\ldots i_m} - \frac{g}{\sqrt{2}}\bar{\psi}_m^{5i_2\ldots i_m} W_\mu^+ \gamma_\mu \psi_m^{4i_2\ldots i_m}$$

$$-\frac{g}{2\cos\theta_W}\bar{\psi}_m^{ii_2\ldots i_m}(Y^W)_{ij} Z_\mu \gamma_\mu \psi_m^{ji_2\ldots i_m}, \tag{14.2.9}$$

where

$$Y^W = \begin{pmatrix} \frac{1}{4} & & & & & & & \\ & \frac{1}{4} & & & & & & \\ & & \frac{1}{4} & & & & & \\ & & & \frac{1}{4} & & & & \\ & & & & -1 & & & \\ & & & & & 0 & & \\ & & & & & & \ddots & \\ & & & & & & & 0 \end{pmatrix}.$$

A comparison of (14.2.9) with the Lagrangian for the standard model of the electroweak interaction shows that the components containing the indices 4 or 5 (i.e., $\psi_m^{\cdots 4\cdots}, \psi_m^{\cdots 5\cdots}$) form doublets with respect to the group $SU_2$. The number of such doublets included in an $SU_n$-multiplet $\psi_m$ is given in Table 14.4, where $l, k = 1, 2, 3$ and $i_2, \ldots, i_m = 6, \ldots, n$. It is seen from Tables 14.3, 4 that the $SU_n$-model includes, in the case of $n \geqslant 6$, both the $V-A$ generations (the first two lines containing the left-handed fermion doublets) and the $V+A$ generations (the third and the fourth lines containing the right-handed fermion doublets).

Notice that from the formulae (14.2.6 – 8) the asymptotic value of the Weinberg angle follows:

$$e = g\sin\theta_W, \qquad \sin^2\theta_W = \frac{3}{8}. \tag{14.2.10}$$

**Table 14.4.** The structure of $SU_n$-multiplets with respect to the doublets of the group $SU_2$

| Doublets of the group $SU_2$ | Number of doublets in the multiplet $\psi_m$ |
|---|---|
| $(\psi_m^{l4i_3\ldots i_m}, \psi_m^{l5i_3\ldots i_m})_L$ | $C_{n-5}^{m-2}$, left-handed quark doublets |
| $(\psi_m^{1234i_5\ldots i_m}, \psi_m^{1235i_5\ldots i_m})_L$ | $C_{n-5}^{m-4}$, left-handed lepton doublets |
| $(\psi_m^{lk4i_4\ldots i_m}, \psi_m^{lk5i_4\ldots i_m})_R^c$ | $C_{n-5}^{m-3}$, right-handed quark doublets |
| $(\psi_m^{4i_2\ldots i_m}, \psi_m^{5i_2\ldots i_m})_R^c$ | $C_{n-5}^{m-1}$, right-handed lepton doublets |

The renormalized value of the Weinberg angle is determined by the identities (14.1.12). If at some stage of $SU_n$-symmetry breaking the symmetry with respect to the subgroup $SU_5$ is preserved and the model has no anomalies, then the quantities $B_i$ are interrelated *via* (14.1.14). The renormalized value of the Weinberg angle in the $SU_n$-model then coincides with that resulting from the $SU_5$-model.

## 14.3 General Requirements and the Choice of Model

Any $SU_n$-model is represented by a set of certain multiplets $\psi_m$ of the group $SU_n$,

$$\sum_{m=1}^{n-1} a_m \psi_{m_L}^{i_1\ldots i_m}, \qquad (14.3.1)$$

where $a_m$ is the number of the multiplets $\psi_m$ involved in the model.

To construct a specific $SU_n$-model, it is necessary to fix the numbers $a_m$ and the rank $n$ of the group. In solving this problem, it is instructive to take into account at once all the general requirements to be fulfilled by the model. A unique choice of the requirements appears not to be possible. Therefore, various authors have been using different sets of requirements. Those used in our treatment are summarized as follows.

1) *Choosing the grand-unification group.* A number of grand-unification models have been subjected to analysis. We have confined ourselves to the group $SU_n$.
2) *Composition of the model with respect to the colour subgroup $SU_3$.* We have assumed that the model includes only the representations 1, 3, and $\bar{3}$ of the colour subgroup $SU_3$.
3) *Fixing the electric charge of the particles involved in the model.* We consider the models including only particles with the charges 0, $\pm 1$, $\pm 2/3$, $\pm 1/3$.

4) *Absence of repeated occurrence of representations of the group $SU_n$.* That is to say, different generations of particles have to be contained in one representation of the group $SU_n$.

5) *Absence of the $\gamma_5$-anomalies.* Absence of anomalies means the renormalizability of the model. Anomalies are absent if the relationship

$$\sum_{m=1}^{n-1} a_m \frac{(n-3)!\,(n-2m)}{(n-m-1)!\,(m-1)!} = 0\,, \tag{14.3.2}$$

holds, i.e., the requirement of the absence of anomalies imposes restrictions on the numbers $a_m$ and on the rank $n$ of the group.

6) *Asymptotic freedom.* The model ought to be asymptotically free with respect to the coupling constant $g$ of the group $SU_n$ as well as with respect to the coupling constant $g_s$ of the subgroup $SU_3$. According to Sect. 10.2, these requirements lead, respectively, to the relations

$$\sum_{m=1}^{n-1} a_m \frac{(n-2)!}{(n-m-1)!\,(m-1)!} < 11\,n\,, \tag{14.3.3}$$

$$\sum_{m=1}^{n-1} a_m \frac{(n-2)!}{(n-m-1)!\,(m-1)!} < 33\,. \tag{14.3.4}$$

7) *Upon $SU_n$-symmetry breaking, the exact $SU_5$-symmetry may either be preserved or not.* If the $SU_5$-symmetry is preserved the $SU_n$-model recovers, in particular, the results of the $SU_5$-model.

8) *Satisfying the condition $N(5)+N(10) = N(\bar{5})+N(\overline{10})$*, where $N(5)$, $N(10)$, $N(\bar{5})$, and $N(\overline{10})$, respectively, are the numbers of representations 5, 10, $\bar{5}$, and $\overline{10}$ of the $SU_5$ subgroup included in the multiplets of the group $SU_n$. For the particles of the observed $SU_5$-families this condition leads to mass formulae that coincide with the corresponding mass formulae of the $SU_5$-model.

9) *Complexity of the representations of the subgroup $SU_3 \times SU_2 \times U_1$.* The representations of the group $SU_5$ do possess this property (14.1.2).

10) *Chirality of the model;* that means that the number of multiplets $\psi_m$ and of multiplets $\psi_{n-m}$ (the latter transforming according to the conjugate representation of the same dimension), which appear in (14.3.1), do not coincide.

11) *The condition for the representations* 3 *and* $\bar{3}$ *of the colour group $SU_3$ to be real* implying the fulfillment of

$$\sum q_L^{2/3} = \sum q_R^{2/3}\,, \quad \sum q_L^{-1/3} = \sum q_R^{-1/3}\,,$$

where $\sum q_L^{2/3}$ and $\sum q_L^{-1/3}$ denote the number of left-handed quark triplets with the charge 2/3 and $-1/3$, respectively, etc. The first of these requirements is automatically fulfilled, even in the absence of the formula (14.3.2). The second requirement leads to the relation

$$\sum_{m=1}^{n-3} a_m \frac{(n-4)!}{(n-m-3)!\,(m-1)!} = \sum_{m=3}^{n-1} a_m \frac{(n-4)!}{(n-m-1)!\,(m-3)!},$$

which coincides with (14.3.2).

12) *Quark-lepton symmetry.* Quarks and leptons appear in the model in a symmetric way, i.e.,

$$\sum l_L^{-1} = \sum q_L^{-1/3}, \qquad \sum l_R^{-1} = \sum q_R^{-1/3}, \tag{14.3.5}$$

$$\sum l_L^0 = \sum q_L^{2/3}, \qquad \sum l_R^0 = \sum q_R^{2/3}, \tag{14.3.6}$$

where $\sum l_L^{-1,0}$ is the number of left-handed leptons with the charge $-1, 0$, etc. The requirements (14.3.5) lead to the relation (14.3.2) and the requirements (14.3.6) to the relations

$$\sum_{m=2}^{n-1} a_m \frac{(n-3)!\,(n-2m+2)}{(n-m)!\,(m-2)!} = 0, \quad \text{and} \tag{14.3.7}$$

$$\sum_{m=1}^{n-2} a_m \frac{(n-3)!\,(n-2m-2)}{(n-m-2)!\,m!} = 0, \tag{14.3.8}$$

respectively.

13) *The condition for the singlet representations of the colour group $SU_3$ to be real* implying the equality of the numbers of left-handed and right-handed leptons,

$$\sum l_L^{-1} = \sum l_R^{-1}, \qquad \sum l_L^0 = \sum l_R^0.$$

This requirement leads to the relations (14.3.2), (14.3.7), and (14.3.8).

14) *The equality of the numbers of generations of quarks and leptons*

$$\sum G_{q_L} = \sum G_{l_L}, \qquad \sum G_{q_R} = \sum G_{l_R},$$

where $\sum G_{q_L}$ and $\sum G_{q_R}$, respectively, are the numbers of the $V-A$ and $V+A$ generations of quarks, etc. This requirement leads to the relation

$$\sum_{m=2}^{n-1} \frac{(n-4)!\,(n-2m+1)}{(n-m-1)!\,(m-2)!} = 0. \tag{14.3.9}$$

15) *The equality of the numbers of generations of right-handed and left-handed fermions*, i.e.,

$$\sum G_{q_L} = \sum G_{q_R}, \qquad \sum G_{l_L} = \sum G_{l_R}.$$

This requirement yields the relation

$$\sum_{m=2}^{n-2} a_m \frac{(n-5)!\,(n-2m)}{(n-m-2)!\,(m-2)!} = 0 . \tag{14.3.10}$$

The relations (14.3.2), (14.3.7 – 10) are linearly independent. In particular, in the $SU_5$-model, there hold the relations (14.3.2), (14.3.7), and (14.3.9).

Some of the listed requirements are mutually contradictory, which is the case, e.g., for requirements (10) and (11 – 15). Various combinations of the above requirements give rise to different models with a definite composition of multiplets. For example, let the requirements (1 – 6), and (11 – 15) be satisfied. Let us consider the $SU_8$-model. The group $SU_8$ has the following anti-symmetric representations:

$$8, 28, 56, 70, \overline{56}, \overline{28}, \bar{8} ;$$

accordingly, the $SU_8$-model is expressed in the general form

$$a_1 \times 8 + a_2 \times 28 + a_3 \times 56 + a_4 \times 70 + a_5 \times \overline{56} + a_6 \times \overline{28} + a_7 \times \bar{8} ,$$

where $a_m$ are unknown integer coefficients. Making use of (14.3.2 – 4) and (14.3.7 – 10) we come to the following set of algebraic relations for the coefficients $a_m$:

$$a_1 + 4a_2 + 5a_3 - 5a_5 - 4a_6 - a_7 = 0 , \tag{14.3.2$'$}$$

$$a_1 + 6a_2 + 15a_3 + 20a_4 + 15a_5 + 6a_6 + a_7 < 88 , \tag{14.3.3$'$}$$

$$a_1 + 6a_2 + 15a_3 + 20a_4 + 15a_5 + 6a_6 + a_7 < 33 , \tag{14.3.4$'$}$$

$$a_2 + 4a_3 + 5a_4 - 5a_6 - 4a_7 = 0 , \tag{14.3.7$'$}$$

$$4a_1 + 5a_2 - 5a_4 - 4a_5 - a_6 = 0 , \tag{14.3.8$'$}$$

$$a_2 + 3a_3 + 2a_4 - 2a_5 - 3a_6 - a_7 = 0 , \tag{14.3.9$'$}$$

$$a_2 + 2a_3 - 2a_5 - a_6 = 0 . \tag{14.3.10$'$}$$

It follows from these relations that

$$a_1 = a_3 = a_5 = a_7 = 1 \quad \text{and} \quad a_2 = a_4 = a_6 = 0 .$$

Thus, the set of requirements used makes it possible to uniquely fix the composition of the $SU_8$-model:

$$8_{\mathrm{L}} + 56_{\mathrm{L}} + \overline{56}_{\mathrm{L}} + \bar{8}_{\mathrm{L}} \quad \text{or} \quad 8_{\mathrm{L}} + 56_{\mathrm{L}} + 56_{\mathrm{R}} + 8_{\mathrm{R}} .$$

## 14.4 The $SU_8$-Model

### 14.4.1 Composition of the Model

Let us consider in some detail the model

$$8+56+\overline{56}+\bar{8}\,.$$

It contains left-handed and right-handed leptons and quarks in equal numbers and describes an equal number (four) of both $V-A$ and $V+A$ generations of quarks and leptons:

| | | |
|---|---|---|
| quarks and leptons of $V-A$ generations: | $u_i,\, d_i;\, \nu_i,\, e_i;$ | $i=1,2,3,4$ |
| quarks and leptons of $V+A$ generations (the so-called mirror fermions): | $U_i,\, D_i;\, N_i,\, E_i;$ | $i=1,2,3,4\,.$ |

The charge of the fermions $u_i$, $U_i$; $d_i$, $D_i$; $\nu_i$, $N_i$; $e_i$, $E_i$ is respectively equal to 2/3, $-1/3, 0$, and $-1$. The particles $u_i$, $d_i$; $\nu_i$, $e_i$ with $i=1,2,3$ are the ordinary quarks and leptons of the three known families $(u,d;\nu_e,e)$, $(c,s;\nu_\mu,\mu)$, and $(t,b;\nu_\tau,\tau)$. These particles are placed in the multiplets of the subgroup $SU_2$ as follows

$$8\colon\ \psi_1^l=D_{4\mathrm{L}};\quad \begin{pmatrix}\psi_1^4\\ \psi_1^5\end{pmatrix}=\begin{pmatrix}E_4^{\mathrm{c}}\\ N_4^{\mathrm{c}}\end{pmatrix}_{\mathrm{L}};\quad \psi_1^a=\nu_{i\mathrm{L}}^{\mathrm{c}};$$

$$56\colon\ \begin{pmatrix}\psi_3^{lk4}\\ \psi_3^{lk5}\end{pmatrix}=\begin{pmatrix}D_4^{\mathrm{c}}\\ U_4^{\mathrm{c}}\end{pmatrix}_{\mathrm{L}};\ \begin{pmatrix}\psi_3^{l4a}\\ \psi_3^{l5a}\end{pmatrix}=\begin{pmatrix}u_i\\ d_i\end{pmatrix}_{\mathrm{L}};\ \begin{pmatrix}\psi_3^{4ab}\\ \psi_3^{5ab}\end{pmatrix}=\begin{pmatrix}E_i^{\mathrm{c}}\\ N_i^{\mathrm{c}}\end{pmatrix}_{\mathrm{L}};$$

$$\psi_3^{123}=E_{4\mathrm{L}};\ \psi_3^{lka}=u_{i\mathrm{L}}^{\mathrm{c}};\ \psi_3^{l45}=U_{4\mathrm{L}};\ \psi_3^{lab}=D_{i\mathrm{L}};\ \psi_3^{45a}=e_{i\mathrm{L}}^{\mathrm{c}};\ \psi_3^{678}=\nu_{4\mathrm{L}}^{\mathrm{c}};$$

$$\overline{56}\colon\ \begin{pmatrix}\chi_3^{lk5}\\ \chi_3^{lk4}\end{pmatrix}=\begin{pmatrix}u_4\\ d_4\end{pmatrix}_{\mathrm{L}};\ \begin{pmatrix}\chi_3^{l5a}\\ \chi_3^{l4a}\end{pmatrix}=\begin{pmatrix}D_i^{\mathrm{c}}\\ U_i^{\mathrm{c}}\end{pmatrix}_{\mathrm{L}};\ \begin{pmatrix}\chi_3^{5ab}\\ \chi_3^{4ab}\end{pmatrix}=\begin{pmatrix}\nu_i\\ e_i\end{pmatrix}_{\mathrm{L}};$$

$$\chi_3^{123}=e_{4\mathrm{L}}^{\mathrm{c}};\ \chi_3^{lka}=U_{i\mathrm{L}};\ \chi_3^{l45}=u_{4\mathrm{L}}^{\mathrm{c}};\ \chi_3^{lab}=d_{i\mathrm{L}}^{\mathrm{c}};\ \chi_3^{45a}=E_{i\mathrm{L}};\ \chi_3^{678}=N_{4\mathrm{L}};$$

$$\bar{8}\colon\ \chi_1^l=d_{4\mathrm{L}}^{\mathrm{c}};\quad \begin{pmatrix}\chi_1^5\\ \chi_1^4\end{pmatrix}=\begin{pmatrix}\nu_4\\ e_4\end{pmatrix}_{\mathrm{L}};\quad \chi_1^a=N_{i\mathrm{L}}\,,$$

where $l,k=1,2,3$; $a,b=6,7,8$; $i=1,2,3$;

$$\chi_{n-m}^{i_{m+1}\dots i_n}=\frac{1}{\sqrt{m!\,(n-m)!}}\,\varepsilon^{i_1\dots i_n}\psi_m^{i_1\dots i_m}\,, \tag{14.4.1}$$

with $\varepsilon^{i_1\dots i_n}$ denoting the totally anti-symmetric unit tensor of rank $n$.

### 14.4.2 The Yukawa Terms and the Fermion Masses

In order to obtain the correct expressions (of the type $m_u \bar{u} u$, $m_e \bar{e} e$) for the mass terms for fermions, we choose the Yukawa terms in the form

$$\bar{\psi}_{mL}^{i_1 \ldots i_k i_{k+1} \ldots i_m} \tilde{\chi}_{pR}^{i_1 \ldots i_k j_{k+1} \ldots j_p} \phi_{j_{k+1} \ldots j_p}^{i_{k+1} \ldots i_m} + \text{h.c.} \tag{14.4.2}$$

Here the right-handed multiplets $\tilde{\chi}_{pR}$, according to (14.4.1), read

$$\tilde{\chi}_{pR}^{i_1 \ldots i_p} = (\chi_p^{i_1 \ldots i_p})_R^c ,$$

with the index $c$ referring to the charge conjugation.

The tensor $\tilde{\chi}_p$ transforms as an anti-symmetric tensor of rank $p$ of the group $SU_n$; its components, like the components of the multiplets $\psi_m$ describe definite particles in accordance with Table 14.3.

The form of the Yukawa interaction terms are given in Table 14.5, together with the non-zero vacuum expectation-values of the components of Higgs multiplets $\phi_{j_1 \ldots j_p}^{i_1 \ldots i_m}$ providing the fermion masses (the indices assuming the values $i = 1,2,3$; $a,b,c = 6,7,8$; and $\alpha, \beta = 1,2,3,4$).

The Higgs field multiplets are anti-symmetric with respect to both the upper and the lower indices and transform according to the irreducible representations of the group $SU_n$. The masses of fermions of the $V-A$ and $V+A$ generations are determined by the vacuum expectation-values of different Higgs multiplets. It can therefore be achieved that the masses of fermions of the $V+A$ generations are much larger than the masses of fermions of the $V-A$ generations. Furthermore it should be noted that the chosen form of the Yukawa interactions (Table 14.5) brings about the fermion mixing.

**Table 14.5.** The form of the Yukawa interaction terms and non-zero vacuum expectation-values

| The Yukawa interaction | Non-zero vacuum expectation-values | Fermions acquiring mass |
|---|---|---|
| $\bar{\psi}_{3L}^{i_1 i_2 j_1} \tilde{\chi}_{5R}^{i_1 i_2 k_1 k_2 k_3} \phi_{k_1 k_2 k_3}^{j_1} + \text{h.c.}$ | $\langle \phi_{5bc}^{a} \rangle, \langle \phi_{678}^{5} \rangle$ | $u_i$, $\nu_i$, $U_4$, $N_4$ |
| $\bar{\psi}_{3L}^{i_1 j_1 j_2} \tilde{\chi}_{3R}^{i_1 k_1 k_2} \phi_{k_1 k_2}^{j_1 j_2} + \text{h.c.}$ | $\langle \phi_{bc}^{5a} \rangle$ | $d_i$, $e_i$ |
| $\bar{\psi}_{5L}^{i_1 i_2 j_1 j_2 j_3} \tilde{\chi}_{3R}^{i_1 i_2 k_1} \tilde{\phi}_{k_1}^{j_1 j_2 j_3} + \text{h.c.}$ | $\langle \tilde{\phi}_{c}^{5ab} \rangle, \langle \tilde{\phi}_{5}^{678} \rangle$ | $U_i$, $N_i$, $u_4$, $\nu_4$ |
| $\bar{\psi}_{3L}^{i_1 j_1 j_2} \tilde{\chi}_{3R}^{i_1 k_1 k_2} \tilde{\phi}_{k_1 k_2}^{j_1 j_2} + \text{h.c.}$ | $\langle \tilde{\phi}_{5c}^{ab} \rangle$ | $D_i$, $E_i$ |
| $\bar{\psi}_{5L}^{i_1 j_1 \ldots j_4} \tilde{\chi}_{1R}^{i_1} \phi^{j_1 \ldots j_4} + \text{h.c.}$ | $\langle \phi^{5678} \rangle$ | $d_4$, $e_4$ |
| $\bar{\psi}_{1L}^{i_1} \tilde{\chi}_{5R}^{i_1 j_1 \ldots j_4} \tilde{\phi}_{j_1 \ldots j_4} + \text{h.c.}$ | $\langle \tilde{\phi}_{5678} \rangle$ | $D_4$, $E_4$ |
| $\bar{\psi}_{3L}^{i_1 i_2 i_3} \tilde{\chi}_{5R}^{j_1 \ldots j_5} \phi_{j_1 \ldots j_5}^{i_1 \ldots i_3} + \text{h.c.}$ | $\langle \phi_{\alpha\beta 5bc}^{\alpha\beta a} \rangle, \langle \phi_{\alpha\beta 678}^{\alpha\beta 5} \rangle$ | $u_i$, $U_4$ |
| $\bar{\psi}_{5L}^{i_1 \ldots i_5} \tilde{\chi}_{3R}^{j_1 j_2 j_3} \tilde{\phi}_{j_1 \ldots j_3}^{i_1 \ldots i_5} + \text{h.c.}$ | $\langle \tilde{\phi}_{\alpha\beta c}^{\alpha\beta 5ab} \rangle, \langle \tilde{\phi}_{\alpha\beta 5}^{\alpha\beta 678} \rangle$ | $u_4$, $U_i$ |
| $\bar{\psi}_{3L}^{i_1 i_2 i_3} \tilde{\chi}_{3R}^{j_1 j_2 j_3} \phi_{j_1 j_2 j_3}^{i_1 i_2 i_3} + \text{h.c.}$ | $\langle \phi_{\alpha bc}^{\alpha 5a} \rangle, \langle \phi_{\alpha 5c}^{\alpha ab} \rangle$ | $d_i$, $e_i$, $D_i$, $E_i$ |

### 14.4.3 Spontaneous Breaking of $SU_8$-Symmetry

Spontaneous breaking of the $SU_8$-symmetry down to $SU_3 \times SU_2 \times U_1$ may be carried out in various ways. Each of them involves several stages; for example,

$$SU_8 \to SU_4 \times SU_4 \times U_1 \to SU_3 \times SU_2 \times SU_2 \times U_1 \to SU_3 \times SU_2 \times U_1 .$$

In every stage, symmetry-breaking is performed with the aid of some Higgs multiplet. Unfortunately, we cannot attribute in a unique way a certain Higgs multiplet to each stage. Usually, in the first stage the symmetry-breaking is carried out by means of the Higgs multiplet transforming according to the adjoint representation of the group $SU_8$, while in the last stage (i.e., in symmetry breaking which reduces $SU_3 \times SU_2 \times U_1$ to $SU_3 \times U_1$), the Higgs multiplet is used that transforms according to the fundamental representation of the group $SU_8$. Let us assume that the Higgs fields are contained in those multiplets of the $SU_8$ group which are obtained by decomposing into irreducible representations the product of the representations underlying the model. As can be seen from Table 14.5, the same multiplets of Higgs fields provide the mass of fermions. These multiplets suffice for carrying out spontaneous symmetry-breaking at all stages.

Among the possible chains of symmetry-breaking, there are some which lead, at some stage, to the exact $SU_5$-symmetry, for instance,

$$SU_8 \to SU_5 \times \ldots \to SU_3 \times SU_2 \times U_1 \times \ldots .$$

In this case, the results of the $SU_5$-model are recovered by the $SU_8$-model. Besides, there exist such chains of symmetry-breaking which do not involve the exact $SU_5$-symmetry. These two possibilities lead to different results. The value of the proton lifetime calculated with (14.1.22) as a function of $\alpha_s(\mu) = g_s^2/4\pi$ and of $\sin\theta_W(\mu)$, where $\mu \simeq 10^2$ GeV, is presented in Fig. 14.3. Curve I in Fig. 14.3 corresponds to those chains of symmetry-breaking which

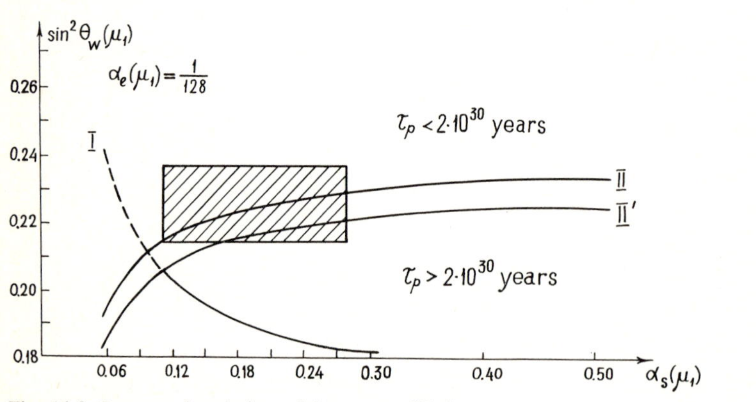


**Fig. 14.3.** Ranges of variation of the proton lifetime

do involve, at some stage, the exact $SU_5$-symmetry (the solid line referring to the portion of the curve where $\tau_p > 2\times 10^{30}$ years); Curves II and II′ (for coefficients 25 and 0.6 in (14.1.22), respectively, correspond to the rest of the chains. As mentioned above, the experiment gives, as the lower boundary, $\tau_p \geqslant 2\times 10^{30}$ years. The dashed area refers to the experimental range of $\alpha_s$ and $\sin^2\theta_W$. One can recognize that the chains of symmetry-breaking which do not involve the exact $SU_5$-symmetry at an intermediate stage are preferable.

To accomplish constructing the $SU_8$-model, it is necessary to specify the Higgs field multiplets and to attribute them to the stages of a given chain of symmetry-breaking. Furthermore, an explicit form for the fermion mixing matrix has to be found and the parameters of mixing have to be expressed in terms of the vacuum expectation values of Higgs multiplets. Eventually, expressions for the fermion masses in the low-energy range which are in agreement with experiment have to be obtained. Obviously, grand-unification models based on symmetry groups of higher rank than $SU_5$ are more powerful than the $SU_5$-model.

## 14.5 The Pati-Salam Model

As an example of a grand-unification model based on the symmetry group $G' = G_1 \times G_2$, we consider a model due to Pati and Salam.

1) As the gauge group for the model, we take $SU_4 \times SU_4$. Upon spontaneous breaking, the initial symmetry goes over into $SU_4 \times (SU_2)_L \times (SU_2)_R$. The gauge fields of the group $SU_4$ provide the strong interaction, whereas the gauge fields of the group $(SU_2)_L \times (SU_2)_R$ provide the electromagnetic and the weak interactions between leptons and quarks.

2) The basic particles of the model comprise four leptons and twelve quarks contained in (6.2.4′).

The quarks may have both fractional and integer charge. Let us consider the model with *integer* quark charges by appropriately choosing the charge operator (Sect. 6.2).

3) Let us combine all the particles into eight left-handed and eight right-handed doublets (the latter may manifest themselves at high energies):

$$L^s_{1a} = \begin{pmatrix} u_s \\ d_s' \end{pmatrix}_L, \quad L^s_{2a} = \begin{pmatrix} c_s \\ s_s' \end{pmatrix}_L, \quad L^4_{1a} = \begin{pmatrix} \nu_e \\ e^- \end{pmatrix}_L, \quad L^4_{2a} = \begin{pmatrix} \nu_\mu \\ \mu^- \end{pmatrix}_L,$$

$$R^s_{1a} = \begin{pmatrix} u_s \\ d_s' \end{pmatrix}_R, \quad R^s_{2a} = \begin{pmatrix} c_s \\ s_s' \end{pmatrix}_R, \quad R^4_{1a} = \begin{pmatrix} \nu_e \\ e^- \end{pmatrix}_R, \quad R^4_{2a} = \begin{pmatrix} \nu_\mu \\ \mu^- \end{pmatrix}_R,$$

$$a = 1, 2; \quad s = 1, 2, 3 .$$

4) In the model under consideration, a complicated combination of scalar field multiplets has to be introduced as Higgs fields.

5) The global Lagrangian for the model reads

$$L = \mathrm{i}\bar{L}^{\alpha}_{ab}\gamma_{\mu}\partial_{\mu}L^{\alpha}_{ab} + \mathrm{i}\bar{R}^{\alpha}_{ab}\gamma_{\mu}\partial_{\mu}R^{\alpha}_{ab} + V(\Phi) + L_{\mathrm{Yuk}}(L, R, \Phi), \tag{14.4.3}$$

where $\alpha = 1,2,3,4$; $a, b = 1,2$; $V(\Phi)$ is the potential of scalar fields giving rise to spontaneous symmetry-breaking, and $L_{\mathrm{Yuk}}(L, R, \Phi)$ stands for the interaction terms of spinor and scalar fields.

The fermion multiplets transform as follows:

i) with respect to the group $SU_4$,

$$L^{\alpha'}_{ab}(x) = \left[\exp\left(-\mathrm{i}\frac{g}{2}\lambda_m\varepsilon_m\right)\right]_{\alpha\beta} L^{\beta}_{ab}(x);$$

$$R^{\alpha'}_{ab}(x) = \left[\exp\left(-\mathrm{i}\frac{g}{2}\lambda_m\varepsilon_m\right)\right]_{\alpha\beta} R^{\beta}_{ab}(x); \quad m = 1, \dots, 15;$$

ii) with respect to the group $(SU_2)_{\mathrm{L}}$,

$$L^{\alpha'}_{ab}(x) = \left[\exp\left(-\mathrm{i}\frac{g'}{2}\tau_k\varepsilon_k\right)\right]_{bc} L^{\alpha}_{ac}(x); \quad R^{\alpha'}_{ab}(x) = R^{\alpha}_{ab}(x); \quad k = 1,2,3;$$

iii) with respect to the group $(SU_2)_{\mathrm{R}}$,

$$R^{\alpha'}_{ab}(x) = \left[\exp\left(-\mathrm{i}\frac{g''}{2}\tau_k\varepsilon_k\right)\right]_{bc} R^{\alpha}_{ac}(x);$$

$$L^{\alpha'}_{ab}(x) = L^{\alpha}_{ab}(x); \quad a, b, c = 1, 2; \quad \alpha, \beta = 1, 2, 3, 4.$$

From the Lagrangian (14.4.3), the local Lagrangian for the model can be derived in a usual way.

# 15. Topological Solitons and Instantons

In some cases the gauge fields permit introducing new objects into the theory – the *solitons*. The solitons are localized states which do not spread out with time and which have a finite energy.

Regarding the solitons, there are two problems to be solved. The first problem is to prove the existence of soliton solutions and to *determine the number of admissible solutions*; the second problem is to find the *explicit form of soliton solutions* of the field equations. Solving the first problem is feasible either by directly investigating the field equations or by using the theory of homotopic mapping and the corresponding homotopy groups, since the number of admissible soliton solutions is determined by the topological properties of the symmetry group of the model. We shall discuss the method of homotopic groups in detail.

First we shall consider the *classical* theory of solitons. One-dimensional (one coordinate and time), two-dimensional (two coordinates and time), three-dimensional (three coordinates and time), and four-dimensional (four coordinates, Euclidean space) models admitting of soliton solutions will be treated. Also, the homotopy groups as applied to calculating the number of the admissible soliton solutions of the field equations will be considered. Finally, the *quantum* theory of solitons will be exposed, exemplified by a one-dimensional model.

## 15.1 One-Dimensional and Two-Dimensional Models

### 15.1.1 One-Dimensional Solitons

We consider two models in one-dimensional space (one coordinate and time: $1+1$), namely, the $\phi^4$-interaction model and the sine-Gordon model.

1) *The $\phi^4$-interaction model* is described by the Lagrangian

$$L=\frac{1}{2}\left[\left(\frac{\partial\phi}{\partial t}\right)^2-\left(\frac{\partial\phi}{\partial x}\right)^2\right]+\frac{1}{2}m^2\phi^2-\frac{1}{4}f\phi^4\,,\quad (m^2,f>0)\,,\quad (15.1.1)$$

where $\phi(x)$ is a real scalar field and $f$ is the self-interaction constant.

The Lagrangian (15.1.1) is invariant under the group of Lorentz transformations and under the group $Z_2$ of discrete transformations consisting of two elements

$$\phi \to \phi , \quad \phi \to -\phi . \tag{15.1.2}$$

By virtue of (15.1.1) we find the following form for the field equations:

$$\frac{\partial^2 \phi}{\partial t^2} - \frac{\partial^2 \phi}{\partial x^2} = m^2 \phi - f \phi^3 . \tag{15.1.3}$$

Equation (15.1.3) has the solutions

$$\phi_{\mathrm{v}} = 0 , \quad \phi_{\mathrm{v}} = \pm m/\sqrt{f} .$$

By calculating the second derivative with respect to $\phi$ of the potential energy entering (15.1.3) and substituting the extremum values into the result of this differentiation we find

$$V''(0) = -m^2 , \quad V''(\pm m/\sqrt{f}) = 2m^2 > 0 ,$$

i.e., the solutions $\phi = \pm m/\sqrt{f}$ are associated with the minimum of energy. Hence, the model under consideration has a doubly degenerate vacuum state (Fig. 15.1 a). From (15.1.1) it further follows for the energy functional that

$$\begin{aligned} E(\phi) &= \int_{-\infty}^{\infty} dx \left[ \frac{1}{2} \left( \frac{\partial \phi}{\partial t} \right)^2 + \frac{1}{2} \left( \frac{\partial \phi}{\partial x} \right)^2 + V(\phi) \right] , \\ V(\phi) &= \frac{1}{4} f \phi^4 - \frac{1}{2} m^2 \phi^2 \end{aligned} \tag{15.1.4}$$

and for the value of the energy relative to the vacuum value that

$$E(\phi) - E(\phi_{\mathrm{v}}) = \int_{-\infty}^{\infty} dx \left[ \frac{1}{2} \left( \frac{\partial \phi}{\partial t} \right)^2 + \frac{1}{2} \left( \frac{\partial \phi}{\partial x} \right)^2 + \frac{f}{4} \left( \phi^2 - \frac{m^2}{f} \right)^2 \right] . \tag{15.1.4'}$$

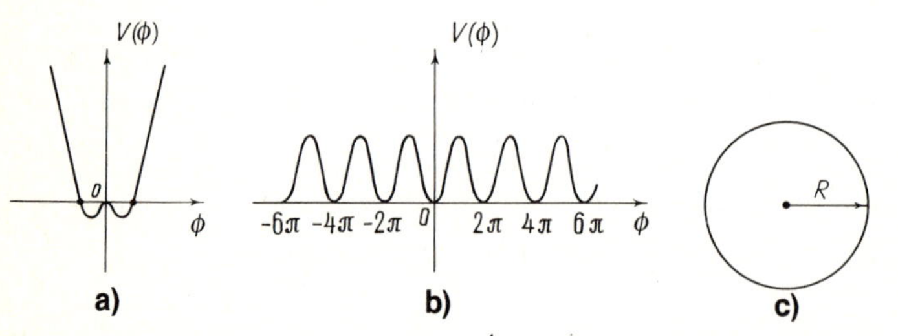


**Fig. 15.1 a – c.** Vacuum states in (**a**) the $\phi^4$-interaction model; (**b**) the sine-Gordon model; (**c**) the Nielsen-Olesen model

Let us find the stationary solutions ($\partial\phi/\partial t = 0$) of the field equations (15.1.3) corresponding to a finite energy. Because all terms in (15.1.4′) are positive, and in view of the infinite limits of integration, the finiteness of the energy implies that at $x \to \pm\infty$

$$\frac{\partial\phi}{\partial x} = 0\,, \tag{15.1.5}$$

$$\frac{f}{4}\left(\phi^2 - \frac{m^2}{f}\right)^2 = 0\,. \tag{15.1.6}$$

From this, it follows that at $x \to \pm\infty$ the function $\phi(x)$ tends to its vacuum value:

$$\phi \to \pm m/\sqrt{f}\,. \tag{15.1.7}$$

In order to find the explicit form of the solutions of (15.1.3), we multiply it by $\partial\phi/\partial x$ and subsequently integrate over $x$; this yields

$$\frac{1}{2}\left(\frac{\partial\phi}{\partial x}\right)^2 = -\frac{m^2\phi^2}{2} + \frac{f\phi^4}{4} + \frac{1}{2}C\,, \tag{15.1.8}$$

where $C$ is an integration constant. From this we have ($X$ denoting another integration constant)

$$x - X = \int_0^{\phi} \frac{d\phi}{\sqrt{\dfrac{f\phi^4}{2} - m^2\phi^2 + C}}\,. \tag{15.1.9}$$

To this general solution corresponds, for arbitrary $C$, an infinite value of the energy $E(\phi)$. To obtain the solution with a finite energy, we make use of the boundary conditions (15.1.5, 7). From (15.1.8) it then follows that $C = m^4/2f$. Putting this value of $C$ into (15.1.9) we come to the solution $\phi_{\rm k}(x)$ of the field equation (15.1.3) with a finite energy:

$$\phi_{\rm k}(x) = \phi(x - X) = \pm\frac{m}{\sqrt{f}}\tanh\left[\frac{m}{\sqrt{2}}(x - X)\right]. \tag{15.1.10}$$

This is called the *kink*. It is represented graphically in Fig. 15.2.

Combining (15.1.4′) with the expression $\phi_{\rm v} = m/\sqrt{f}$ and the expression for $\phi_{\rm k}$ we obtain the energy of the kink relative to the vacuum:

$$E(\phi_{\rm k}) - E(\phi_{\rm v}) = \frac{2\sqrt{2}}{3}\,m^3/f\,.$$

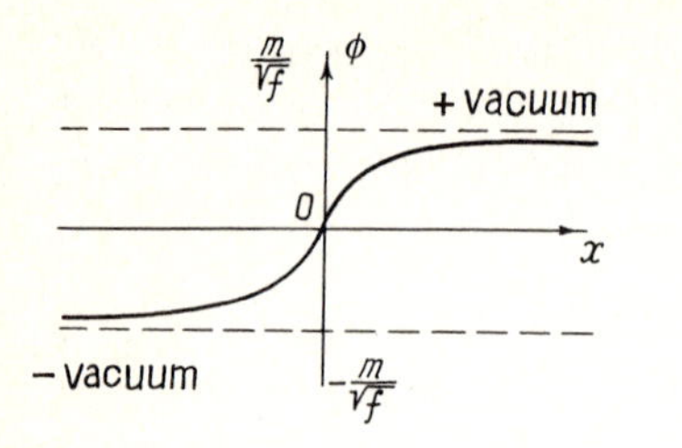


**Fig. 15.2.** Graphical representation of the kink solution

The self-interaction constant $f$ figures in the denominator of the solution (15.1.10). Consequently, this solution cannot be expressed as a series in the coupling constant.

Thus, the model given by the Lagrangian (15.1.1) gives rise to a solution, expressed by (15.1.10), which (i) corresponds to the finite energy $E(\phi_k)$, (ii) asymptotically tends to the vacuum value, when $x \to \pm\infty$, and (iii) does not vary with time, i.e., is a stationary solution $(\partial\phi/\partial t = 0)$.

2) *The sine-Gordon model* is described by the Lagrangian

$$L = \left[\frac{1}{2}\left(\frac{\partial\bar{\phi}}{\partial t'}\right)^2 - \frac{1}{2}\left(\frac{\partial\bar{\phi}}{\partial x'}\right)^2 + \frac{m^4}{f}\left(\cos\frac{\sqrt{f}}{m}\bar{\phi} - 1\right)\right].$$

Upon the change of variables $\phi = (\sqrt{f}/m)\bar{\phi}$, $x = mx'$, $t = mt'$, it is rewritten as

$$L = \frac{m^4}{f}\left\{\frac{1}{2}\left[\left(\frac{\partial\phi}{\partial t}\right)^2 - \left(\frac{\partial\phi}{\partial x}\right)^2\right] + (\cos\phi - 1)\right\}. \tag{15.1.11}$$

This Lagrangian is invariant under the Lorentz group as well as under the following two groups of discrete transformations:

i) $\phi \to \phi + 2\pi n$ (infinite group $Z$), $n = 0, \pm 1, \pm 2, \ldots,$

ii) $\phi \to -\phi$ (finite group $Z_2$).

The field equations corresponding to the Lagrangian (15.1.11) read

$$\frac{\partial^2\phi}{\partial t^2} - \frac{\partial^2\phi}{\partial x^2} = -\sin\phi. \tag{15.1.12}$$

This equation has the solutions $\phi = \pi n$, $n = 0, \pm 1, \pm 2, \ldots$, and the solutions corresponding to the vacuum ($V''(\phi) > 0$) are $\phi = 2\pi n$, $n = 0, \pm 1, \pm 2, \ldots$.

It is recognized that the model has an infinite number of discrete vacua (Fig. 15.1 b).

The general stationary solution $(\partial\phi/\partial t = 0)$ of (15.1.12) has the form (where $x_0$ and $C$ are constants)

$$x - x_0 = \pm\int_{\pi}^{\phi}\frac{d\phi}{\sqrt{2(C - \cos\phi)}}. \tag{15.1.13}$$

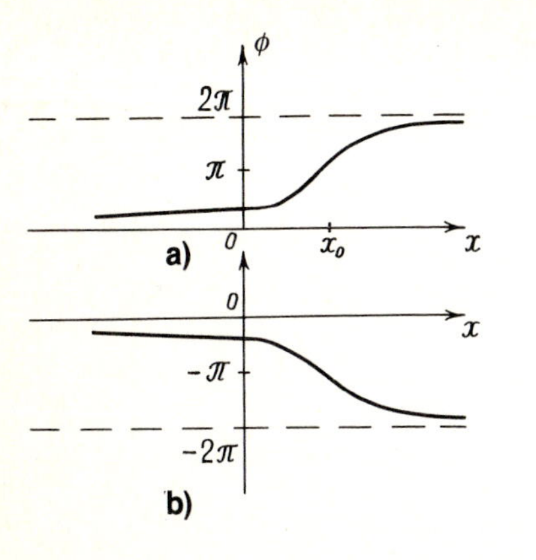


**Fig. 15.3a, b.** Graphical representation of the solution in the sine-Gordon model: (**a**) soliton; (**b**) anti-soliton

From this we find the soliton solution $\phi_s$ corresponding to a finite energy

$$\phi_s(x-x_0) = 4 \arctan \exp(x-x_0) . \tag{15.1.14}$$

It results from (15.1.13) by setting $C=1$ and taking the plus sign on the right-hand side. The soliton solution is graphically represented in Fig. 15.3a.

The energy of the soliton relative to the vacuum is finite and equals

$$E(\phi_s)-E(\phi_v) = \frac{m^4}{f} \int_{-\infty}^{\infty} dx \left[\frac{1}{2}\left(\frac{\partial \phi_s}{\partial x}\right)^2 + 1 - \cos\phi_s\right] = 8\frac{m^4}{f} . \tag{15.1.15}$$

Besides the solution (15.1.14), there is another solution, $\phi_{as}$, to (15.1.12):

$$\phi_{as}(x-x_0) = -4 \arctan \exp(x-x_0) . \tag{15.1.16}$$

It derives from (15.1.13) where now the minus sign is taken and $C=1$. This solution is attributed to the anti-soliton (Fig. 15.3b).

Thus, the sine-Gordon model has the soliton solutions (15.1.14) and (15.1.16) which correspond to a finite energy and are stationary solutions asymptotically tending to the vacuum values. It should be emphasized that in the sine-Gordon model there are no other time-independent solutions. The physical reason for that is as follows: being described by non-linear equations, the solitons are always interacting with one another implying that the behaviour of many-soliton systems is time-dependent.

### 15.1.2 Vortices

Now we pass to the two-dimensional space (two spatial coordinates and time: 2+1). Here we consider the model due to Nielsen and Olesen which is described by the Lagrangian

$$\mathscr{L} = -\tfrac{1}{4}F_{\mu\nu}F_{\mu\nu} + \tfrac{1}{2}(\nabla_\mu\phi)^*(\nabla_\mu\phi) - \tfrac{1}{4}f(\phi\phi^* - m^2/f)^2 . \tag{15.1.17}$$

Here $F_{\mu\nu} = \partial_\mu A_\nu - \partial_\nu A_\mu$ is the tensor of the electromagnetic field, $\nabla_\mu \phi = (\partial_\mu - \mathrm{i}eA_\mu)\phi$ is the covariant derivative, and $\phi(x)$ is a complex scalar field.

This Lagrangian is invariant under the Lorentz group as well as under the local Abelian group $U_1$.

According to (15.1.17), the field equations for the model are written as

$$\begin{aligned} &\partial_\nu F_{\mu\nu} = j_\mu = \tfrac{1}{2}\mathrm{i}e(\phi^*\partial_\mu\phi - \phi\partial_\mu\phi^*) + e^2 A_\mu \phi\phi^* , \\ &\nabla_\mu \nabla_\mu \phi = -f\phi(\phi\phi^* - m^2/f) . \end{aligned} \tag{15.1.18}$$

These equations have the solution

$$F_{\mu\nu} = 0 , \qquad \nabla_\mu \phi = 0 , \qquad |\phi| = m/\sqrt{f} \tag{15.1.19}$$

which corresponds to the absolute minimum of the energy functional. In other terms, the model gives rise (Sect. 3.3) to a continuous degenerate vacuum which is graphically represented as a circle of radius $R = m/\sqrt{f}$ in the complex plane $\phi$ (Fig. 15.1c). It is noted that all particles are massive, the mass of the vector field being $em/\sqrt{f}$ and the mass of the scalar field $m\sqrt{2}$. There is no residual symmetry in the model.

One is not in a position to find the solutions of the field equations (15.1.18) analytically. We shall look for a solution in a symmetric form:

$$A_0 = 0 , \qquad \boldsymbol{A} = \hat{\boldsymbol{r}}_\perp A(r) , \qquad \phi = g(r)\exp(\mathrm{i}n\theta) , \tag{15.1.20}$$

where $r^2 = x^2 + y^2$, $n$ is an integer number, $\theta$ is the rotation angle, and $\hat{\boldsymbol{r}}_\perp$ is a unit vector normal to the radius-vector. Then the set of equations (15.1.18) is reduced to a simpler one:

$$\begin{aligned} &-\frac{1}{r}\frac{d}{dr}\left(r\frac{d}{dr}g\right) + \left[\left(\frac{n}{r} - eA\right)^2 + f\left(g^2 - \frac{m^2}{f}\right)\right] g = 0 , \\ &-\frac{d}{dr}\left[\frac{1}{r}\frac{d}{dr}(rA)\right] + \left(Ae^2 - \frac{ne}{r}\right) g^2 = 0 . \end{aligned} \tag{15.1.21}$$

The $r$-dependence of the function $\phi(r)$ and of the magnetic field $B(r) = \varepsilon_{3ij}\partial_i A_j$ obtained by computer calculations is depicted in Fig. 15.4 for the case $n = 1$. The two-dimensional soliton solutions are called *vortices*.

Let us also calculate the magnetic flux $\Phi$, through the plane $xy$, due to a vortex. This flux is given by

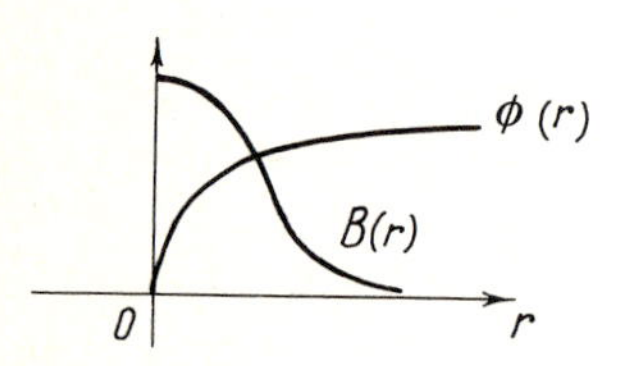


**Fig. 15.4.** $r$-dependence of the functions $\phi(r)$ and $B(r)$

$$\Phi = \int dx\, dy\, B(x,y) = \int_{S^1} A\, dl\,, \tag{15.1.22}$$

where $S^1$ is the circumference of the circle of radius $R$, with $R \to \infty$. According to (15.1.19), $\nabla_\mu \phi = 0$. Furthermore, at $r \to \infty$ the vacuum value $\phi_v$ can be expressed, by virtue of (15.1.20), as

$$\phi_v = |\phi_v| \exp(\mathrm{i} n\theta)\,. \tag{15.1.23}$$

Combining (15.1.19, 20) we find that at $r \to \infty$

$$A \to n/\mathrm{e}r\,, \quad g(r) = m/\sqrt{f}\,. \tag{15.1.24}$$

Substituting (15.1.24) into (15.1.22) we eventually find

$$\Phi = \frac{1}{\mathrm{e}} \int_0^{2\pi} n\, d\theta = 2\pi n/e\,, \tag{15.1.25}$$

i.e., the magnetic flux admits of an infinite number of discrete values.

Hence, the Nielsen-Olesen model has an infinite discrete set of soliton solutions.

### 15.1.3 The Idea of Topological Analysis

By analyzing several models we have demonstrated that they have soliton solutions. Likewise, one may look for soliton solutions of other, more general models associated with spaces of larger dimension and invariant under groups of higher rank. The question emerges of whether every model admits of soliton solutions and, if any, what is their number. An answer to this question may be found by directly investigating the field equations. Besides, a topological method may be applied, because the admissible number of solitons is determined by the topological properties of the symmetry group of the model.

We shall give the *topological method* a more detailed consideration. Its basic idea is as follows. The solitons have a finite energy. This means that the soliton solutions $\Phi_i(r\boldsymbol{k})$ of the field equations tend at infinity to the vacuum values $\lambda_i(\boldsymbol{k})$:

$$\Phi_i(r\boldsymbol{k})_{r\to\infty} \to \lambda_i(\boldsymbol{k})\,, \tag{15.1.26}$$

where $\boldsymbol{k}$ is a unit vector. The asymptotic values of functions of a field depend on the direction of the spatial vector $\boldsymbol{k}$. For example, in the one-dimensional case, the vector $\boldsymbol{k}$ is directed along the axis or has the opposite direction, and the kink solutions assume different values at $x = +\infty$ and $x = -\infty$ (Fig. 15.2). Similarly, in the two-dimensional Nielsen-Olesen model $\boldsymbol{k}$ varies on a one-dimensional spatial sphere $S^1$ (a circle), and the functions $\lambda_i(\boldsymbol{k})$ may be regarded as a mapping of this sphere, taken at infinity, i.e., at $R \to \infty$, onto the vacuum manifold, which in the model under consideration is a sphere $S^1$ as well (Fig. 15.1c).

In the general case, the soliton solutions may be thought of as a *mapping* between an $n$-dimensional spatial sphere taken at infinity and the manifold of the vacuum values of the model.

To each soliton solution there corresponds its own mapping. Thus, the problem is to find all mappings leading to different soliton solutions.

## 15.2 Homotopy Groups

We take the Nielsen-Olesen model to illustrate the basic notions of the topological analysis by way of example. As already mentioned, with this model is associated a mapping of the spatial $S^1$-sphere (circle) taken at infinity on the $S^1$-sphere of vacuum values.

### 15.2.1 Homotopy Classes

Two mappings of the space $X$ on the space $Y$ are called *homotopic* if one of them can be continuously deformed into the other. The set of homotopic mappings form a *homotopy class.* For example, the mapping of the circle $S^1$ on a circle $S^1$ brings the "South Pole" $x_0$ of the circle $S^1$ into a fixed point $y_0$ of the space $S^1$. Let us specify a point on the circle by the polar angle $\phi$. Then each mapping of the circle on a circle can be described by a function $f(\phi)$ satisfying the condition

$$f(2\pi) = f(0) + 2\pi n ,$$

where $n$ is an integer number. This number indicates how many times the circle is wound around itself and is referred to as the winding number. Let us consider the mappings with the same number, for instance, $n = 1$. All such mappings can be continuously deformed into one another, i.e., they are homotopic. They form the homotopy class corresponding to $n = 1$. In a similar way the homotopy classes with other winding numbers ($n = 2, 3, \ldots$) are produced.

Mappings with different winding numbers (e.g., with $n = 1$ and $n = 2$) belong to different classes. They cannot be continuously deformed into one another and are thus homotopically non-equivalent.

Hence, there is an infinite number of homotopically non-equivalent classes ($n = 1, 2, \ldots$) associated with a homotopic mapping of a circle on a circle. This set of homotopy classes is commonly denoted as $\pi_1(S^1)$. The subscript indicates the dimension of the spatial sphere subjected to the mapping; the vacuum manifold onto which the mapping is performed is indicated in the parentheses.

### 15.2.2 Homotopy Group

Introducing a group operation for the obtained set $\pi_1(S^1)$ of homotopy classes makes it a group which is naturally referred to as the *homotopy group.*

The homotopy group has the homotopy classes as its elements. Let us consider two homotopic mappings $f_1(\phi)$ and $f_2(\phi)$ (where $0 \leqslant \phi \leqslant 2\pi$) that start and end at the point $y_0$. We define the product of two homotopic mappings $f_1 * f_2$ as a mapping for which the ultimate point of the curve $f_1(\phi)$ coincides with the starting point of the curve $f_2(\phi)$:

$$f(\phi) = f_1(\phi) * f_2(\phi) = \begin{cases} f_1(2\phi), 0 \leqslant \phi \leqslant \pi \\ f_2(2\phi - 2\pi), \pi \leqslant \phi \leqslant 2\pi , \end{cases}$$

and $n = n_1 + n_2$. The product of two classes of homotopic mappings is then a class of homotopic mappings as well. This group operation has the property of associativity.

The role of the unit element $\{I\}$ of the group plays the class of mappings onto a point ($n = 0$) and that of the inverse element the class of mappings with the opposite direction of loops, i.e., the class of the mappings $f(2\pi - \phi) = f^{-1}(\phi)$.

For example, the Nielsen-Olesen model gives rise to a discrete homotopy group $\pi_1(S^1)$ with an infinite number of elements.

Let us establish a connection between the soliton solutions and the elements of the topological group. As mentioned above, the soliton solutions may be considered as mappings of a spatial sphere taken at infinity onto the vacuum manifold. That is to say, a soliton solution specifies an element of the topological group and, vice versa, to each element of the topological group may correspond a soliton solution. Thus, there is a one-to-one correspondence between the soliton solutions and the elements of the group, and the number of the soliton solutions is equal to the number of the elements of the topological group.

The homotopy group of the Nielsen-Olesen model considered has an infinite number of discrete elements so that this model admits of an infinite number of different soliton solutions. As we have seen, all of them do realize.

### 15.2.3 Determining the Homotopy Groups

In the general case, the homotopy group under consideration is written

$$\pi_k(G/G_0) . \tag{15.2.1}$$

The index $k$ specifies the *dimension* of the sphere which is equal to the dimension of the space: to the one-dimensional space ($S^0$-sphere) there corresponds $k = 0$, to the two-dimensional space ($S^1$-sphere) $k = 1$, to the three-dimensional space ($S^2$-sphere) $k = 2$, and to the four-dimensional space ($S^3$-sphere) $k = 3$. $G$ denotes the group of internal symmetry under which the

**Table 15.1.** The homotopy group $\pi_0(G)$

| $G$ | $\pi_0(G)$ |
|---|---|
| $U_n$, $SU_n$, $SO_n$, $S^n (n \geqslant 1)$, $ISO_n$ | $O$ |
| $GL_n(R)$, $S^0$, $Z_2$ | $Z_2$ |
| $Z$ | $Z$ |

model is invariant and $G_0$ denotes the group of residual symmetry for the model.

The number of admissible soliton solutions for a model is equal to the number of the elements of its homotopy group. This reduces the problem of determining the number of admissible soliton solutions to finding the homotopy groups, i.e., to calculating the number of the elements of these groups. There is a general technique for solving this problem (based on the concept of fibre bundle). We shall not dwell on it, however, and will only give the final results. Some of them are quite obvious and can be obtained heuristically (as we have done above). In the general case this is not possible, however. Tables 15.1 and 15.2 summarize the results of calculating the homotopy groups $\pi_0(G)$ and $\pi_1(G)$, respectively. (The latter group is also referred to as fundamental.) Table 15.3 summarizes the results for the groups $\pi_2(G)$ and $\pi_3(G)$.

Let us give an analysis of the models by means of topological groups. We consider the generalized Nielsen-Olesen model assuming that the Lagrangian (15.1.17) is invariant under the unitary group $\mathrm{SU_n}$, $n \geqslant 2$. In this case (Table 15.2), $\pi_1(SU_n) = 0$, $n \geqslant 2$. This means that the models with the Lagrangian (15.1.17) invariant under the group $SU_n$, $n \geqslant 2$, do not lead to soliton solutions.

**Table 15.2.** The fundamental group $\pi_1(G)$

| | $G$ | $\pi_1(G)$ | Notes |
|---|---|---|---|
| Spheres | $S^n$ | $0 (n > 1)$ | $\pi_1(S^1) = Z$ |
| Rotation groups | $SO_n$ | $Z_2 (n > 2)$ | $\pi_1(SO_2) = Z$ |
| Unitary groups | $U_n$ | $Z (n \geqslant 1)$ | |
| Unitary unimodular groups | $SU_n$<br>$SU_n/Z_n$ | $0 (n \geqslant 1)$<br>$Z_n$ | e.g., $SO_3 = SU_2/Z_2$<br>$SO_6 = SU_4/Z_2$ |
| Symplectic groups | $Sp_{2n}$ | $0$ | |
| Tori | $T^n = (S^1)^n$ | $Z \oplus \ldots \oplus Z$ | |
| Projective spaces | $RP^n$<br>$CP^n$<br>$HP^n$ | $Z_2 (n > 1)$<br>$0 (n \geqslant 1)$<br>$0 (n \geqslant 1)$ | $\pi_1(RP^1) = Z$ |

**Table 15.3.** The homotopy groups $\pi_2(G)$ and $\pi_3(G)$

| | $G$ | $\pi_2(G)$ | $\pi_3(G)$ | Notes |
|---|---|---|---|---|
| Simple Lie groups | $G_0$ | 0 | $Z$ | |
| Compact Lie groups | $G$ | 0 | $Z \oplus \ldots \oplus Z$ | |
| Spheres | $S^n$ | $0 (n \neq 2)$ | $0 (n \neq 2, 3)$ | $\pi_2(S^2) = \pi_3(S^2) = \pi_3(S^3) = Z$ |
| Projective spaces | $RP^n$ | $\pi_2(RP^n) = \pi_2(S^n) = 0 (n > 2)$ | | |
| | $CP^n$ | $\pi_2(CP^n) = \pi_2(S^{2n+1}/S^1) = Z$ | $\pi_3(RP^n) = \pi_3(S^n) = 0$ $(n > 3)$ | |
| | $HP^n$ | $\pi_2(HP^n) = 0$ from $HP^n = S^{4n+1}/S^3$ | | |

Let us now consider the sine-Gordon model. This gives rise to the mapping (Fig. 15.5a) of the zero-sphere ($S^0$) (i.e., of two points) onto the vacuum manifold corresponding to the group of internal symmetry $Z \times Z_2$ (Fig. 15.1b). Unlike the Nielsen-Olesen model, in the model under consideration there is residual symmetry: after fixing a definite vacuum value (the point $y_0$ which we choose as $\phi_v = 2\pi n$), the symmetry with respect to vacuum shifting disappears whereas there remains the symmetry with respect to the change $\phi \to -\phi$ with a subsequent shift by $2n \cdot 2\pi$, i.e., with respect to the group $Z_2$. Since the residual symmetry leads only to equivalent homotopy classes, it has to be excluded. Therefore, the homotopy group for the sine-Gordon model is (Table 15.1)

$$\pi_0(Z \times Z_2/Z_2) = \pi_0(Z) = Z .$$

This is a discrete homotopy group of infinite dimension. Thus, the sine-Gordon model admits of an infinite discrete set of soliton solutions, although not all of the admissible solutions do realize, as has been shown above. The

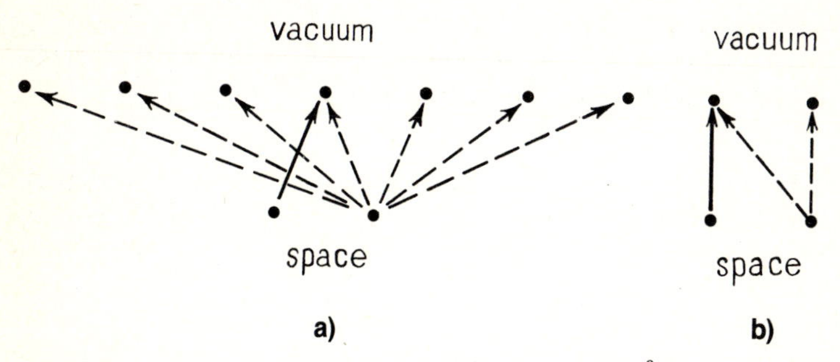


**Fig. 15.5a, b.** Homotopic mappings: (**a**) of the zero-sphere $S^0$ onto the vacuum manifold in the sine-Gordon model; (**b**) of the zero-sphere onto the vacuum manifold in the $\phi^4$-interaction model

reason is that the homotopy analysis has a general character and does not take into account the specifies of the problem which may lead to a reduction in the number of admissible soliton solutions.

Finally, in the case of the $\phi^4$-interaction model, we deal with the mapping of the spatial zero-sphere $S^0$, taken at infinity, onto the vacuum manifold corresponding to the group $Z_2$ (Fig. 15.5b). The homotopy group for this model is (Table 15.1)

$$\pi_0(Z_2) = Z_2 ,$$

i.e., the model gives rise to two solutions: a constant solution $\phi_v$ and a soliton solution $\phi$. The explicit form of these solutions has been found in Sect. 15.1.

### 15.2.4 Topological Charge

According to the Noether theorem, the invariance of the Lagrangian under the continuous group of internal symmetry results in a conserved current associated with a conserved charge. Models with soliton solutions have an additional conserved current and a corresponding charge which is not due to the Noether theorem, though. The occurrence of this charge originates from the topological properties of the model and it is therefore known as the *topological charge.* The number of topological charges is equal to the number of admissible soliton solutions, i.e., to each soliton solution a topological charge is attributed. The topological charge is a conserved quantity.

For the $\phi^4$-interaction model, the conserved topological current is determined by the expression $j_\mu = \frac{1}{2}\varepsilon_{\mu\nu}\partial_\nu\phi$, $\varepsilon_{01} = -\varepsilon_{10} = 1$, $\varepsilon_{00} = \varepsilon_{11} = 0$. The associated conserved topological charge is

$$K = \frac{1}{2}\int_{-\infty}^{\infty} dx\, j_0(x,t) = \frac{1}{2}\int_{-\infty}^{\infty} dx \frac{\partial\phi}{\partial x} = \frac{1}{2}[\phi(+\infty) - \phi(-\infty)] .$$

The vacuum and the kink can be characterized by attributing to them $K=0$ and $K=1$, respectively.

Similar considerations lead to the value of the conserved topological charge for the sine-Gordon model:

$$N = \frac{1}{2\pi}[\phi(+\infty) - \phi(-\infty)] , \qquad N = 1, 2, \ldots .$$

In the Nielsen-Olesen model, the conserved topological current is given by $j_\mu = \partial_\nu F_{\mu\nu}$, and the topological charge turns out to be identical with the magnetic flux,

$$\Phi = \int dx\, dy\, F_{12} = \frac{2\pi}{e} n .$$

Topological charges may be regarded as quantum numbers.

### 15.2.5 Dimension of the Space and Composition of the Model

The composition of a model admitting of soliton solutions depends on the dimensionality of the space. Let us consider a class of models with the local Lagrangian which includes scalar fields $\phi(x)$ and gauge fields $A_\mu^k(x)$. Assume that the energy $E$ of the model can be expressed as

$$E = E_\phi + E_A + V\,, \tag{15.2.2}$$

where

$$\begin{aligned} E_\phi(\phi, A) &= \int d^{\mathscr{D}}x F(\phi)(\nabla_\mu \phi)^* \nabla_\mu \phi\,, \\ E_A(A) &= \tfrac{1}{4}\int d^{\mathscr{D}}x F_{\mu\nu}^k F_{\mu\nu}^k\,, \\ V(\phi) &= \int d^{\mathscr{D}}x\, U(\phi)\,, \end{aligned} \tag{15.2.3}$$

$\mathscr{D}$ denoting the dimension of the space. The functions $F(\phi)$ and $U(\phi)$ are assumed to be positive and not to include any derivatives.

Since the integrands in (15.2.3) are positive, each term in (15.2.2) has to be finite for soliton solutions to exist.

Let us carry out the transformations of the functions

$$\begin{aligned} \phi(x) &\to \phi_\lambda(x) = \phi(\lambda x)\,, \\ A(x) &\to A_\lambda(x) = \lambda A(\lambda x)\,, \\ \nabla_\mu \phi(x) &\to \lambda\, \nabla_\mu \phi(\lambda x)\,, \\ F_{\mu\nu}(x) &\to \lambda^2 F_{\mu\nu}(\lambda x)\,. \end{aligned} \tag{15.2.4}$$

Then

$$\begin{aligned} E_\phi(\phi_\lambda, A_\lambda) &= \lambda^{2-\mathscr{D}} E_\phi(\phi, A)\,, \\ E_A(A_\lambda) &= \lambda^{4-\mathscr{D}} E_A(A)\,, \\ V(\phi_\lambda) &= \lambda^{-\mathscr{D}} V(\phi)\,. \end{aligned}$$

By definition of the soliton, in a static case the energy $E$ must be stationary under arbitrary changes of the field, in particular, under the transformations (15.2.4). This is certainly not the case when the terms in (15.2.2) are either all increasing or all decreasing. The energy may remain constant only if some of the terms in (15.2.2) are decreasing while the others are increasing or, else, if these terms are independent of $\lambda$. The behaviour of the terms in (15.2.2) with the dimension of the space is summarized in Table 15.4. The following notation is adopted: $\uparrow$ means an increase with $\lambda$, $\downarrow$ a decrease with $\lambda$, and 0 indicates the independence of $\lambda$. Table 15.4 clearly demonstrates that one-dimensional models having soliton solutions must necessarily contain an interaction term. The $\phi^4$-interaction model and the sine-Gordon model considered above are of this type. Two-dimensional models have to contain either all three terms or only the term $E_\phi$. An example of the first option is provided by

**Table 15.4.** The behaviour of the terms in Eq. (15.2.2) with the dimension of the space

| $\mathscr{D}$ | $E_\phi$ | $E_A$ | $V$ |
|---|---|---|---|
| 1 | ↑ | ↑ | ↓ |
| 2 | 0 | ↑ | ↓ |
| 3 | ↓ | ↑ | ↓ |
| 4 | ↓ | 0 | ↓ |
| $\mathscr{D} \geqslant 5$ | ↓ | ↓ | ↓ |

the Nielsen-Olesen model. Models of the alternative type will not be considered here. Three-dimensional models necessarily have to contain all three terms. Finally, in four-dimensional models, only gauge fields can be involved. In the class of theories under consideration there are no models of dimension more than 4 which would give rise to soliton solutions.

## 15.3 Monopole

### 15.3.1 The Dirac Monopole

As is well known, the Maxwell equations for the electromagnetic field have the form

$$\partial_\nu F_{\mu\nu} = -j_\mu\,, \tag{15.3.1}$$

$$\partial_\nu \tilde{F}_{\mu\nu} = 0\,. \tag{15.3.2}$$

Here $F_{\mu\nu}$ is the electromagnetic field tensor, i.e., $F_{0i} = -E_i$, $F_{ij} = -\varepsilon_{ijk}B_k$; $j_\mu$ is the current 4-vector $(\varrho, \boldsymbol{j})$; $\boldsymbol{E}$ and $\boldsymbol{B}$ are the electric and the magnetic fields, respectively; and $\tilde{F}_{\mu\nu}$ is the tensor dual to $F_{\mu\nu}$:

$$\tilde{F}_{\mu\nu} = \tfrac{1}{2}\varepsilon_{\mu\nu\varrho\sigma}F_{\varrho\sigma}\,. \tag{15.3.3}$$

The tensor $\tilde{F}_{\mu\nu}$ is obtained from $F_{\mu\nu}$ by the interchange $\boldsymbol{E}\to\boldsymbol{B}$, $\boldsymbol{B}\to-\boldsymbol{E}$.

In the empty space ($j_\mu = 0$), the Maxwell equations are symmetric with respect to the transformations

$$F_{\mu\nu}\to\tilde{F}_{\mu\nu}\,, \quad \tilde{F}_{\mu\nu}\to -F_{\mu\nu} \tag{15.3.4}$$

or, which is equivalent, with respect to the transformations $\boldsymbol{E}\to\boldsymbol{B}$, $\boldsymbol{B}\to-\boldsymbol{E}$.

In order to preserve the symmetry of the Maxwell equations with respect to these transformations in the presence of matter, we introduce, besides the electric current $j_\mu$, the magnetic current $\tilde{j}_\mu$ with the components $(Q_{\mathrm{m}}, \tilde{j})$, where $Q_{\mathrm{m}}$ is the magnetic charge and $\tilde{j}$ is the three-dimensional magnetic current. Then we obtain, instead of (15.3.1, 2) new, generalized equations:

$$\partial_\nu F_{\mu\nu} = -j_\mu, \qquad \partial_\nu \tilde{F}_{\mu\nu} = -\tilde{j}_\mu. \tag{15.3.5}$$

These equations are symmetric with respect to the transformations

$$F_{\mu\nu} \to \tilde{F}_{\mu\nu}, \qquad \tilde{F}_{\mu\nu} \to -F_{\mu\nu}, \qquad j_\mu \to \tilde{j}_\mu, \qquad \tilde{j}_\mu \to -j_\mu. \tag{15.3.6}$$

The electric and the magnetic charges enter the generalized equations (15.3.5) symmetrically. A point-like magnetic charge is usually called the Dirac magnetic monopole or, briefly, the *monopole*. In contradistinction to the electric charge, the monopole has not been experimentally detected to date; a search for it is going on.

It has also been shown by Dirac that equations (15.3.5) can be quantized, provided that the following relation between the electric and the magnetic charges is fulfilled:

$$eQ_{\mathrm{m}} = n, \tag{15.3.7}$$

where $n$ is an integer.

The magnetic field $B_k$ from a monopole with the charge $Q_{\mathrm{m}}$ is given by

$$B_k = x_k \frac{1}{er^3}, \qquad e = \frac{1}{Q_{\mathrm{m}}}. \tag{15.3.8}$$

The monopole hypothesis has been propounded by Dirac relatively long ago. The recent developments of the gauge theory shed a new light on it.

### 15.3.2 Gauge Theory of the Monopole

Let us demonstrate that three-dimensional models (3 coordinates and time: $3+1$) may yield soliton solutions which correspond to the magnetic monopole. We consider the model that is described by the Lagrangian invariant under the local non-Abelian $SO_3$-group (the *'t Hooft-Polyakov* model):

$$\mathscr{L} = -\tfrac{1}{4}F^k_{\mu\nu}F^k_{\mu\nu} + \tfrac{1}{2}(\nabla_\mu\phi^a)(\nabla_\mu\phi^a) - \tfrac{1}{4}f(\phi^a\phi^a - m^2/f)^2, \tag{15.3.9}$$

where $\phi^a$ is an isovector of scalar fields ($a = 1, 2, 3$), and

$$F^k_{\mu\nu} = \partial_\mu A^k_\nu - \partial_\nu A^k_\mu + e\varepsilon_{klm}A^l_\mu A^m_\nu; \qquad \nabla_\mu\phi^k = \partial_\mu\phi^k + e\varepsilon_{klm}A^l_\mu\phi^m.$$

By (15.3.9), the field equations for the model read

$$\nabla_\mu F^k_{\mu\nu} = -e\varepsilon_{klm}\phi^l\nabla_\nu\phi^m, \qquad \nabla_\mu\nabla_\mu\phi^a = -f\phi^a(\phi^{a2} - m^2/f). \tag{15.3.10}$$

These equations have a solution

$$F^k_{\mu\nu} = 0\,, \qquad \nabla_\mu \phi^a = 0\,, \qquad |\phi^a| = m/\sqrt{f}\,, \tag{15.3.11}$$

which corresponds to the absolute minimum of the energy functional, i.e., the model has a continuous degenerate vacuum (the $S^2$-sphere of radius $R = m/\sqrt{f}$). The model has a residual $SO_2$-symmetry (although $\phi_v \neq$ const). It therefore contains not only massive particles (one scalar and two vector mesons) but also a massless particle (photon).

In the case of the three-dimensional model under consideration, the vector $\boldsymbol{k}$ varies on a two-dimensional $S^2$-sphere. From the topological viewpoint the problem thus reduces to mapping the spatial two-dimensional sphere $S^2$, taken at infinity, onto the two-dimensional $S^2$-sphere or vacuum values. According to (15.2.1) and Table 15.3, the number of soliton solutions (or topological charges of the model) equals

$$\pi_2(SO_3/SO_2) = \pi_2(S^2) = Z\,,$$

i.e., the model admits of an infinite discrete set of soliton solutions or topological charges.

Let us look for the solutions of (15.3.10) in the form

$$\begin{aligned} &A^k_0 = 0\,, \qquad A^k_i = \varepsilon_{kij} x_j [1 - g(r)] \frac{1}{er^2}\,, \\ &\phi^a = -x^a h(r) \frac{1}{er^2}\,, \qquad r^2 = x_1^2 + x_2^2 + x_3^2\,. \end{aligned} \tag{15.3.12}$$

Substitution of (15.3.12) into (15.3.10) leads to a simpler set of equations

$$\begin{aligned} r^2 g'' &= g(g^2 - 1) + g h^2\,, \\ r h'' &= 2 h g^2 + (f/e^2)(h^3 - C^2 r^2 h)\,, \end{aligned} \tag{15.3.13}$$

where $C = me/\sqrt{f}$. It follows from the requirement that the energy be finite, that $h \to Cr$ as $r \to \infty$, $g \to 0$.

It is only possible to find the soliton solutions for this system by means of computer calculations.

The mass $M$ of the monopole is given by

$$M = 4\pi \frac{m}{ef} N\left(\frac{f}{e^2}\right) = \frac{M_W}{\alpha} N\left(\frac{f}{e^2}\right),$$

where $N$ is a slowly increasing monotonous function that depends on the ratio of the constants $f$ and $e^2$; $N(0) = 1$; $M_W$ denotes the mass of the vector meson; $4\pi/e^2 = 1/\alpha \simeq 137$. Assuming $M_W \sim 50$ GeV we get $M \sim 10$ GeV.

The magnetic charge is the topological invariant of the three-dimensional model. To prove that, we introduce the gauge-invariant electromagnetic field tensor

$$F_{\mu\nu} = \bar{\phi}^a F^a_{\mu\nu} - \frac{1}{e}\varepsilon_{abc}\bar{\phi}^a(\nabla_\mu\bar{\phi}^b)(\nabla_\nu\bar{\phi}^c)\,, \qquad \bar{\phi}^a = \frac{\phi^a}{|\phi|}\,, \tag{15.3.14}$$

which can be rewritten in a different form

$$F_{\mu\nu} = \partial_\mu P_\nu - \partial_\nu P_\mu - \frac{1}{e}\varepsilon_{abc}\bar{\phi}^a(\partial_\mu\bar{\phi}^b)(\partial_\nu\bar{\phi}^c)\,, \tag{15.3.15}$$

where $P_\mu = \bar{\phi}^a A^a_\mu$.

Substituting (15.3.12) into (15.3.14) or into (15.3.15) we find for the magnetic field

$$B_k = x_k \frac{1}{er^3}\,. \tag{15.3.16}$$

This formula does, indeed, coincide with the expression (15.3.8) for the magnetic field of the point-like monopole with magnetic charge $Q_\mathrm{m} = 1/e$.

In order to find the topological charges of the model, we consider the magnetic current which, by (15.3.5) and (15.3.14), is determined by

$$\tilde{j}_\mu = \partial_\nu \tilde{F}_{\mu\nu} = \frac{1}{2e}\varepsilon_{\mu\nu\alpha\beta}\varepsilon_{abc}\partial_\nu(\bar{\phi}^a\partial_\alpha\bar{\phi}^b\partial_\beta\bar{\phi}^c)\,. \tag{15.3.17}$$

This is a conserved quantity ($\partial_\mu\tilde{j}_\mu = 0$) which does not follow from the Noether theorem. The following magnetic flux $\Phi$ and the magnetic charge $Q_\mathrm{m}$ correspond to the current (15.3.17):

$$\Phi = 4\pi Q_\mathrm{m} = \int dx\,\tilde{j}_0 = \frac{1}{2e}\oint_{S^2_R}(d\sigma)_i\varepsilon_{ijk}\varepsilon_{abc}\bar{\phi}^a\partial_j\bar{\phi}^b\partial_k\bar{\phi}^c\,. \tag{15.3.18}$$

Here $S^2_R$ stands for the sphere of radius $R$ (where $R\to\infty$). Since the sphere can be parametrized with the aid of two coordinates $\xi_i$ ($i = 1, 2$), (15.3.18) is rewritten as

$$4\pi Q_\mathrm{m} = \frac{1}{e}\int_{S^2_R} d^2\xi \frac{1}{2}\varepsilon_{\alpha\beta}\varepsilon_{abc}\bar{\phi}^a\partial_\alpha\bar{\phi}^b\partial_\beta\bar{\phi}^c = \frac{1}{e}\int d^2\xi\sqrt{g} = \frac{4\pi}{e}d\,, \tag{15.3.19}$$

where $g = \det(\partial_\alpha\bar{\phi}^a\partial_\beta\bar{\phi}^a)$. The number $d$ entering (15.3.19) is called the *Kronecker index* of the mapping between the two $S^2$-spheres. This index is a topological invariant and, consequently, the magnetic charge is a topological invariant, too. The Kronecker index indicates the number of turns of winding of the $S^2$-sphere around the $S^2$-sphere. It can assume integer values only. Hence,

$$Q_\mathrm{m} = d/e = n/e\,, \quad \text{and} \tag{15.3.20}$$

$$\pi_2(SO_3/SO_2) = Z = d\,.$$

It is seen that the magnetic charge has no relation to the Noether theorem but is purely of topological nature. It also follows from (15.3.20) that the Dirac quantization condition (15.3.7) is satisfied.

Thus, the three-dimensional model with the Lagrangian (15.3.9) has a soliton solution which is the monopole.

Let us briefly list some other results pertaining to the monopole.

1) Setting in (15.3.12)

$$A_0^k = x^k J(r) \frac{1}{er^2}$$

(where $J(r)$ is an unknown function) instead of $A_0^k = 0$, leads to *dyonic solutions* with finite energy. Such solutions describe the "particles" which have both a continuous magnetic charge and a continuous electric charge.

2) A generalization of the model to the groups of higher rank leads to new magnetic monopoles with a wider range of charge values.

3) In the limit of $f \to 0$, Prasad and Sommerfield have found an exact solution of the set of Eqs. (15.3.13):

$$g(r) = Cr/\sinh(Cr)\,, \qquad h(r) = Cr \coth(Cr) - 1\,.$$

## 15.4 Instantons

Now we turn to finding non-trivial solutions for the four-dimensional Euclidean space (4 coordinates without time: 4+0). Such solutions are called *instantons.* As could already be seen in Sect. 15.2, in this case the model may include the gauge fields only. For this reason, we concern ourselves with the gauge-field Lagrangian that is invariant, e.g., under the group $SU_2$:

$$\mathscr{L} = -\tfrac{1}{4} F_{\mu\nu}^k F_{\mu\nu}^k\,. \tag{15.4.1}$$

Here

$$F_{\mu\nu}^k = \partial_\mu A_\nu^k - \partial_\nu A_\mu^k - g\varepsilon_{klm} A_\mu^l A_\nu^m\,.$$

Let us introduce the notations

$$\mathscr{A}_\mu = \tfrac{1}{2} A_\mu^k \tau_k\,, \qquad \mathscr{F}_{\mu\nu} = \tfrac{1}{2} F_{\mu\nu}^k \tau_k\,,$$

where $\tau_k$ denotes the Pauli matrices. In these notations, the field equations associated with the Lagrangian (15.4.1) read

$$\nabla_\mu \mathscr{F}_{\mu\nu} = \partial_\mu \mathscr{F}_{\mu\nu} - \mathrm{i}g[\mathscr{A}_\mu, \mathscr{F}_{\mu\nu}]_- = 0\,. \tag{15.4.2}$$

The requirement of finiteness of the action results, according to (15.4.1), in that the tensor $\mathscr{F}_{\mu\nu}$ ought to vanish at infinity. The field $\mathscr{A}_\mu(x)$ then becomes pure gauge, so that

$$\mathscr{A}_\mu(x) = \mathrm{i}\omega^{-1}(x)\,\partial_\mu \omega(x)\,, \tag{15.4.3}$$

where $\omega(x)$ is an element of the gauge group and $x$ belongs to the sphere $S^3$. It follows from (15.4.3) that the model has an infinitely degenerate vacuum corresponding to the minimum action.

In the four-dimensional model we are dealing with, the vector $\boldsymbol{k}$ which enters (15.1.26) varies on a three-dimensional sphere $S^3$. From the topological point of view, the problem thus consists in studying the mappings between a spatial $S^3$-sphere in the limit of infinite radius and the manifold of the vacuum values of the group $SU_2$. According to (15.2.1), together with Table 15.3, the number of soliton solutions of the model is

$$\pi_3(SU_2) = Z,$$

i.e., the model admits of an infinite discrete set of solutions.

Let us introduce the current density

$$J_0 = \frac{1}{16\pi^2} \operatorname{Tr}\{\mathscr{F}_{\mu\nu} \tilde{\mathscr{F}}_{\mu\nu}\} = \partial_\mu J'_\mu , \tag{15.4.4}$$

where

$$J'_\mu = 2\varepsilon_{\mu\nu\alpha\beta} \operatorname{Tr}\left\{\mathscr{A}_\nu \partial_\alpha \mathscr{A}_\beta + \frac{2g}{3i} \mathscr{A}_\nu \mathscr{A}_\alpha \mathscr{A}_\beta\right\}. \tag{15.4.5}$$

To the current density (15.4.4) there corresponds the charge

$$q = \int dx\, J_0 = \frac{1}{16\pi^2} \int dx \operatorname{Tr}\{\mathscr{F}_{\mu\nu} \tilde{\mathscr{F}}_{\mu\nu}\} \quad \text{or} \tag{15.4.6}$$

$$q = \frac{1}{16\pi^2} \int_{S^3} d^3\sigma_\mu J'_\mu , \tag{15.4.7}$$

where $S^3$ is a three-dimensional sphere with $R \to \infty$.

By virtue of (15.4.3), the last formula is rewritten as

$$q = \frac{1}{24\pi^2} \varepsilon_{\lambda\delta\alpha\beta} \oint_{S^3} d^3\sigma_\lambda \operatorname{Tr}\{\omega^{-1}\partial_\delta\omega\,\omega^{-1}\partial_\alpha\omega\,\omega^{-1}\partial_\beta\omega\}. \tag{15.4.8}$$

The integrand in (15.4.8) is referred to as the *Jacobian of the mapping* of $S^3$ onto $SU_2$. This Jacobian is a topological invariant which implies that $q$ is a topological invariant as well. The integral in (15.4.8) can take only integer values, i.e.,

$$q = n, \tag{15.4.9}$$

upon which $\pi_3(SU_2) = Z = q$.

The number $q$ is referred to as the *Pontryagin number.* It indicates how many times the spatial sphere is "wound" around the group $SU_2$. This can be put as follows: it is the Pontryagin number that is the topological number characterizing the topological charge of an instanton.

We shall not solve the field equations (15.4.2) directly. Instead, we make use of certain relations which are equivalent to the field equations. To find these relations, we consider the inequality

$$\int dx\,\mathrm{Tr}\{\mathscr{F}_{\mu\nu} \pm \tilde{\mathscr{F}}_{\mu\nu}\}^2 \geqslant 0\,. \tag{15.4.10}$$

Combining it with (15.4.6) we obtain

$$E = \tfrac{1}{2}\int dx\,\mathrm{Tr}\{\mathscr{F}_{\mu\nu}\,\tilde{\mathscr{F}}_{\mu\nu}\} \geqslant \tfrac{1}{2}\left|\int dx\,\mathrm{Tr}\{\mathscr{F}\tilde{\mathscr{F}}\}\right| = 8\pi^2|q|\,.$$

In the case of equality, it follows from (15.4.10) that

$$\mathscr{F}_{\mu\nu} = \pm\,\tilde{\mathscr{F}}_{\mu\nu}\,. \tag{15.4.11}$$

If this holds, the field equations are automatically satisfied, since

$$\nabla_\mu\,\mathscr{F}_{\mu\nu} = \pm\,\nabla_\mu\,\tilde{\mathscr{F}}_{\mu\nu} = 0\,.$$

Let us look for the functions which satisfy (15.4.11) (where the plus sign is taken) in the following spherically symmetric form:

$$\mathscr{A}_\mu = \mathrm{i}h(r)\,\omega^{-1}(x)\,\partial_\mu\,\omega(x)\,,$$

$$\omega(r) = \frac{x_4 - \mathrm{i}x_k\tau_k}{r}\,, \qquad r^2 = x_1^2 + x_2^2 + x_3^2 + x_4^2\,. \tag{15.4.12}$$

Substituting (15.4.12) into (15.4.11) we find

$$rh' = 2h(1-h)\,.$$

With $h(r) \underset{r\to\infty}{\longrightarrow} 1$, this equation has a solution

$$h(r) = \frac{r^2}{r^2+\lambda^2}\,. \tag{15.4.13}$$

Here $\lambda$ is an arbitrary parameter with the dimension of length. It can be considered as a size of the instanton.

Substitution of (15.4.13) into (15.4.12) yields the instanton solution

$$\mathscr{A}_\mu = \frac{\mathrm{i}r^2}{r^2+\lambda^2}\,\omega^{-1}\,\partial_\mu\,\omega\,. \tag{15.4.14}$$

With the aid of this formula we obtain the expression for $\mathscr{F}_{\mu\nu}$:

$$\mathscr{F}_{\mu\nu} = \frac{4\lambda^2}{(r^2+\lambda^2)^2}\,\sigma_{\mu\nu}\,, \qquad \text{where}$$

$$\sigma_{ij} = \frac{1}{4\mathrm{i}}\,[\tau_i, \tau_j]_-\,, \qquad \sigma_{i4} = \frac{1}{2}\,\tau_i = -\sigma_{4i}\,.$$

The Pontryagin number of the instanton solution found is equal to unity, $q = 1$.

Besides (15.4.14), there is another solution of (15.4.2); this corresponds to the topological charge $q = -1$:

$$\mathscr{F}_{\mu\nu} = -\tilde{\mathscr{F}}_{\mu\nu} = \frac{4\lambda^2}{(r^2+\lambda^2)^2}\bar{\sigma}_{\mu\nu}, \quad \text{where} \tag{15.4.15}$$

$$\bar{\sigma}_{ij} = \sigma_{ij}, \qquad \bar{\sigma}_{i4} = -\sigma_{i4}.$$

The so called anti-instanton can be attributed to this solution. There are probably no other solutions of (15.4.11) with $|q| = 1$. However, there exist solutions with larger values of $q$, and these have been found.

A similar analysis for models invariant under the groups of higher ranks has yielded the corresponding solutions.

## 15.5 Quantum Theory of Solitons

For constructing the quantum theory of solitons, in particular perturbation theory, one can utilize the path-integral formalism which was successfully applied for the models with ordinary particles (Chaps. 4, 5). One only ought to take into account certain specifics of the quantum theory of solitons. An example illustrating the technique of constructing the quantum theory of solitons is provided by the one-dimensional model described by the Lagrangian (15.1.1).

### 15.5.1 The Generating Functional for the Green's Functions

To find the amplitude of a process in perturbation theory, the form of the Green's functions has to be known (Chap. 5). These are determined by the generating functional $W$. For a system with the Lagrangian (15.1.1), the generating functional $W$ is expressed as a path integral over the canonical variables $\phi(x)$ and $\pi(x)$ as follows:

$$W(J', K') = \int \mathscr{D}\phi\, \mathscr{D}\pi \exp\{i\int dt[\pi\dot{\phi} - H(\pi,\phi) + J'\phi + K'\pi]\}. \tag{15.5.1}$$

Here $H(\pi,\phi)$ is the Hamiltonian corresponding to the Lagrangian (15.1.1):

$$H(\pi,\phi) = \frac{1}{2}\int dx\left((\pi^2 + \phi'^2 - \phi^2 + \frac{1}{2f^2}\phi^4 + \frac{1}{2}f^2\right), \tag{15.5.2}$$

$\phi$ and $\pi$ are the generalized coordinate and momentum, respectively, $\phi' = \partial\phi/\partial x$, $J'$ and $K'$ are the currents. Let us assume that, for the case of one-soliton state, the zeroth-order contribution to the amplitude is determined by the classical kink solution. To calculate the corrections, we use an expansion of the field around this solution. By that, the quantum effects are introduced as corrections to the results based on the classical solution.

The kink solution (15.1.10) involves an arbitrary parameter $X$ which implies that the kink coordinate as well as its momentum are not strictly fixed. Let us regard $X$ as a time-dependent dynamical variable and let us introduce the momentum $\tilde{P}$ canonically conjugate to $X$.

The expansion of the generalized coordinates $\phi(x,t)$ and $\pi(x,t)$ in the vicinity of the classical solution $\phi_0(x-X(t))$ will be expressed as

$$\begin{aligned} \phi(x,t) &= \phi_0(x-X(t)) + \eta(x-X(t),t)\,, \\ \pi(x,t) &= \pi_0(x;\tilde{P}(t),X(t)) + \zeta(x-X(t),t)\,. \end{aligned} \tag{15.5.3}$$

We further single out $X$ and $\tilde{P}$ from $\phi$ and $\pi$, respectively. For this purpose we make use of the identity

$$\int \mathscr{D}X\, \mathscr{D}\tilde{P}\,\delta[F_1(X;\phi)]\,\delta[F_2(\tilde{P};\pi,\phi)]\,\frac{\partial F_1}{\partial X}\,\frac{\partial F_2}{\partial \tilde{P}} = 1\,. \tag{15.5.4}$$

We choose the functions $\pi_0$, $F_2$ and $F_1$ in such a way that the variables $\eta$ and $\zeta$, as well as $\tilde{P}$ and $X$, be canonically conjugate, i.e., that the condition

$$\int dx\,\pi\dot{\phi} = \tilde{P}\dot{X} + \int dx\,\zeta\dot{\eta} \tag{15.5.5}$$

be satisfied.

This choice can be, for instance, as follows:

$$\begin{aligned} \pi_0 &= \frac{\tilde{P}}{m_0\left(1+\dfrac{1}{m_0}\xi\right)}\,\frac{\partial\phi_0(x-X(t))}{\partial x}\,, \\ F_1 &= \int dx\,\frac{\partial\phi_0(x-X(t))}{\partial x}\,\phi(x,t) \\ &= \int dx\,\frac{\partial\phi_0(x-X(t))}{\partial x}\,\eta(x-X(t),t)\,, \\ F_2 &= \frac{-\tilde{P}}{1+\dfrac{1}{m_0}\xi} + \int dx\,\frac{\partial\phi_0(x-X(t))}{\partial x}\,\pi(x,t) \\ &= \int dx\,\frac{\partial\phi_0(x-X(t))}{\partial x}\,\xi(x-X(t),t)\,, \end{aligned} \tag{15.5.6}$$

where $m_0 = (2\sqrt{2}/3)f^2$ is the kink mass and

$$\xi(t) = \int dx \frac{\partial \phi_0(x - X(t))}{\partial x} \frac{\partial \eta(x - X(t), t)}{\partial x}.$$

Let us introduce the new variable

$$\varrho = x - X(t) \tag{15.5.7}$$

and re-define the rest of the variables in the following way

$$\chi(\varrho, t) = \eta(x - X(t), t), \; \tilde{\pi}(\varrho, t) = \zeta(x - X(t), t), \tag{15.5.8}$$

$$P = \tilde{P} - \int d\varrho \, \tilde{\pi} \chi', \tag{15.5.9}$$

where $\chi' = \partial \chi / \partial \varrho$.

Substituting (15.5.4 – 6) into (15.5.1) gives the desired expression for the generating functional

$$W(J, K) = \int \mathscr{D}X \, \mathscr{D}P \, \mathscr{D}\chi \, \mathscr{D}\tilde{\pi} \, \delta(\int \phi_0' \chi \, d\varrho) \, \delta(\int \phi_0' \tilde{\pi} \, d\varrho)$$
$$\times \exp\{i \int dt [P\dot{X} + \int d\varrho (\tilde{\pi}\dot{\chi} - H' + J\chi + K\tilde{\pi})]\}, \tag{15.5.10}$$

where

$$H' = m_0 + \frac{(P + \int d\varrho \, \tilde{\pi} \chi')^2}{2 m_0 (1 + \xi/m_0)^2}$$
$$+ \frac{1}{2} \int d\varrho \left[ \tilde{\pi}^2 + \chi'^2 + \left( \frac{3\phi_0^2}{f^2} - 1 \right) \chi^2 + \frac{2}{f^2} \phi_0 \chi^3 + \frac{1}{2f^2} \chi^4 \right]. \tag{15.5.11}$$

As can be seen, passing from (15.5.1) to (15.5.10) is just passing from the variables $\phi$, $\pi$ to the variables $X$, $P$, $\chi$, $\tilde{\pi}$.

Due to the presence of the arbitrary parameter $X$ in the solution (15.1.10) the Hamiltonian of the system depends on the arbitrary momentum $P$. The generating functional (15.5.10) involves integrating over $X$ and $P$. It is convenient to bring the path integral (15.5.10) to a form as encountered in models with ordinary particles. To do so, the integration over the variables $\chi$ and $\tilde{\pi}$ is performed first, $X$ and $P$ being fixed. Then perturbation theory is formulated for this case. Subsequently, the path integration over $X$ and $P$ is carried out in the results obtained.

### 15.5.2 Perturbation Theory

For calculating the propagators and the interaction vertices, we shall use the generating functional. According to (15.5.10), it is written as

$$W(J, K) = \int \mathscr{D}\chi \, \mathscr{D}\tilde{\pi} \, \delta(\int d\varrho \, \phi_0' \chi) \, \delta(\int d\varrho \, \phi_0' \tilde{\pi})$$
$$\times \exp\{i \int dt [d\varrho (\tilde{\pi}\dot{\chi} + J\chi + K\tilde{\pi}) - H']\}. \tag{15.5.12}$$

Here $H'$ is the Hamiltonian determined by (15.5.11).

Let us express the Hamiltonian as the sum of the free Hamiltonian $H_0$ and the interaction Hamiltonian $H_{\mathrm{I}}$,

$$H_0 = \int d\varrho \left[ \frac{1}{2} \tilde{\pi}^2 + \frac{1}{2} \chi'^2 - \frac{1}{2} \left( 1 - \frac{3\phi_0^2}{f^2} \right) \chi^2 \right], \tag{15.5.13}$$

$$H_{\mathrm{I}} = \frac{(P + \int d\varrho\, \tilde{\pi} \chi')^2}{2 m_0 (1 + \xi/m_0)^2} - \int d\varrho \left( \frac{\phi_0}{f^2} \chi^3 + \frac{1}{4f^2} \chi^4 \right), \tag{15.5.14}$$

where $\xi = \int d\varrho\, \phi_0' \chi'$.

Substituting (15.5.13, 14) into (15.5.12) and taking the differential operator out of the integrand we find

$$W(J,K) = \exp\left[ -\mathrm{i} \int dt\, H_{\mathrm{I}} \left( \frac{1}{\mathrm{i}} \frac{\partial}{\partial J}, \frac{1}{\mathrm{i}} \frac{\partial}{\partial K} \right) \right] W_0(J,K), \tag{15.5.15}$$

where $W_0(J,K)$ is the generating functional corresponding to the free Hamiltonian:

$$\begin{aligned} W_0(J,K) = \int \mathscr{D}\chi\, \mathscr{D}\tilde{\pi}\, \delta(\int d\varrho\, \phi_0' \chi)\, \delta(\int d\varrho\, \phi_0' \tilde{\pi}) \\ \times \exp\{\mathrm{i} \int dt\, d\varrho [\tilde{\pi}\dot{\chi} - \tfrac{1}{2}\tilde{\pi}^2 - \tfrac{1}{2}\Omega^2 \chi^2 + J\chi + K\tilde{\pi}]\}, \end{aligned} \tag{15.5.16}$$

$$\Omega^2 = -\frac{d^2}{d\varrho^2} - 1 + \frac{3\phi_0^2}{f^2}.$$

Let us calculate the path integral figuring in (15.5.16). For that, we first find the eigenfunctions of the operator $\Omega^2$:

$$\left( -\frac{d^2}{d\varrho^2} - 1 + \frac{3\phi_0^2}{f^2} \right) \psi_n = \omega_n^2 \psi_n.$$

This operator has two discrete eigenvalues (labelled by $n = 0$ and $n = 1$); to the *zero-mode* eigenvalue $\omega_0 = 0$ corresponds the wave function $(1/m_0)\,\phi_0'$, i.e., a function contained among the $\delta$-functions of the expression (15.5.10). Furthermore, the operator has a continuous spectrum of eigenvalues ($\omega_k^2 = k^2 + 2$) to which correspond the normalized eigenfunctions of the form

$$\psi_1 = \sinh \frac{\varrho}{\sqrt{2}} \Big/ \cosh^2 \frac{\varrho}{\sqrt{2}},$$

$$\psi_k = \frac{1}{N_k} \mathrm{e}^{\mathrm{i}k\varrho} \left[ 3 \tanh^2 \frac{\varrho}{\sqrt{2}} - 3\mathrm{i}k\sqrt{2} \tanh \frac{\varrho}{\sqrt{2}} - 1 - 2k^2 \right]. \tag{15.5.17}$$

$$N_k = 2l(k^2 + 2)(2k^2 + 1) - 12\sqrt{2}(k^2 + 1).$$

Here the normalization of the wave function on the length $l$ and the periodicity of $\psi$ with respect to the spatial coordinate, with the period $l$, have been used.

Introducing the notations

$$\begin{aligned}\chi_n(t) &= \int d\varrho\, \chi(\varrho,t)\, \psi_n(\varrho) \equiv (\chi, \psi_n)\,, \\ \tilde{\pi}_n(t) &= \int d\varrho\, \tilde{\pi}(\varrho,t)\, \psi_n(\varrho) \equiv (\tilde{\pi}, \psi_n)\,,\end{aligned} \tag{15.5.18}$$

*we obtain*

$$\begin{aligned}\chi(\varrho,t) &= \sum_n \chi_n(t)\, \psi_n^*(\varrho)\,, \\ \tilde{\pi}(\varrho,t) &= \sum_n \tilde{\pi}_n(t)\, \psi_n^*(\varrho)\,.\end{aligned} \tag{15.5.19}$$

With the aid of the change $\tilde{\pi} \to \tilde{\pi} + \dot{\chi}$ the integral (15.5.16) is rewritten in the form where the variables $\tilde{\pi}$ and $\chi$ are separated

$$\begin{aligned}W_0(J,K) = \int \mathscr{D}\chi\, \mathscr{D}\tilde{\pi}\, \delta\Big(\int d\varrho\, \phi_0'\chi\Big)\, \delta\Big(\int d\varrho\, \phi_0'\tilde{\pi}\Big) \\ \times \exp\{\mathrm{i}\int dt\, d\varrho[-\tfrac{1}{2}\tilde{\pi}^2 - \tfrac{1}{2}\dot{\chi}^2 - \tfrac{1}{2}\Omega^2\chi^2 + J\chi + K\tilde{\pi} + K\dot{\chi}]\}\,.\end{aligned} \tag{15.5.20}$$

The integral over $\tilde{\pi}$ in the last expression equals

$$\int \mathscr{D}\tilde{\pi}\, \delta\Big(\int d\varrho\, \phi_0'\tilde{\pi}\Big) \exp\{\mathrm{i}\int dt[d\varrho(-\tfrac{1}{2}\tilde{\pi}^2 - K\tilde{\pi})]\}$$

or, upon substituting (15.5.18),

$$\begin{aligned}\int \prod_n \mathscr{D}\tilde{\pi}_n \delta(\tilde{\pi}_0) \exp\left\{\mathrm{i}\int dt \sum_n [-\tfrac{1}{2}\tilde{\pi}_n^*\tilde{\pi}_n + \tilde{\pi}_n^* K_n]\right\} \\ = \exp[\tfrac{1}{2}\int dt \sum_n{}' K_n^* K_n]\,.\end{aligned} \tag{15.5.21}$$

In the sum $\sum'$, the zero mode is excluded.

The integral over $\chi$ that enters (15.5.20) is calculated likewise:

$$\begin{aligned}\int \prod_n \mathscr{D}\chi_n \delta(\chi_0) \exp\left\{\mathrm{i}\int \sum_n [\tfrac{1}{2}\chi_n^* \mathrm{i} G_n^{-1}\chi_n + (J_n - \dot{K}_n)\chi_n^*]\right\} \\ = \exp\left[\mathrm{i}\int \tfrac{1}{2}\sum_n{}'(J_n^* - \dot{K}_n^*)\mathrm{i}G_n(J_n - \dot{K}_n)\right],\end{aligned} \tag{15.5.22}$$

where

$$\mathrm{i}G_n^{-1} = (-\partial_t - \omega_n^2 + \mathrm{i}\varepsilon)\,\delta(t-t')\,.$$

Combining (15.5.20, 21) and (15.5.22) provides the desired expression for the generating functional:

$$\begin{aligned}W_0(J,K) = \exp(-\int dt\, d\varrho \int dt'\, d\varrho' \{\tfrac{1}{2}[J(\varrho,t) - \dot{K}(\varrho,t)] \\ \times G(\varrho,\varrho';t-t')[J(\varrho',t') - \dot{K}(\varrho',t')] \\ + \tfrac{1}{2}K(\varrho,t)\Delta(\varrho,\varrho';t-t')K(\varrho',t')\})\,,\end{aligned} \tag{15.5.23}$$

where

$$G(\varrho,\varrho';t-t') = \sum_n{}' \psi_n(\varrho) \int \frac{d\omega}{2\pi} \mathrm{e}^{\mathrm{i}\omega(t-t')} \frac{1}{\omega^2-\omega_n^2+\mathrm{i}\varepsilon} \psi_n^*(\varrho'),$$

$$\Delta(\varrho,\varrho';t-t') = -\mathrm{i}\delta(t-t') \sum_n{}' \psi_n(\varrho)\psi_n^*(\varrho').$$

With the aid of (15.5.23) we find for the $\chi\chi$-propagator

$$\frac{1}{\mathrm{i}} \frac{\delta}{\delta J(\varrho,t)} \frac{\delta}{\delta J(\varrho',t')} W_0(J,K)\Big|_{J=K=0} = G(\varrho,\varrho';t-t'),$$

for the $\chi\pi$-propagator

$$\frac{1}{\mathrm{i}} \frac{\delta}{\delta J(\varrho,t)} \frac{\delta}{\delta K(\varrho',t')} W_0(J,K)\Big|_{J=K=0} = \partial_{t'} G(\varrho,\varrho';t-t'),$$

and for the $\pi\pi$-propagator

$$\frac{1}{\mathrm{i}} \frac{\delta}{\delta K(\varrho,t)} \frac{\delta}{\delta K(\varrho',t')} W_0(J,K)\Big|_{J=K=0} = \partial_t\partial_{t'} G(\varrho,\varrho';t-t') + \Delta(\varrho,\varrho';t-t').$$

These propagators are graphically presented in Fig. 15.6.

With regard to the interaction vertices, it should be noted that, besides the usual ones (corresponding to the terms $(\phi_0/f^2)\chi^3$ and $(1/4f^2)\chi^4$ in the Lagrangian, Fig. 15.7a, b), there emerge an infinite number of interaction vertices associated with the term

$$\frac{(P+\int d\varrho\, \pi\chi')^2}{2m_0(1+\xi/m_0)^2}. \tag{15.5.24}$$

Since $\xi/m_0 \sim 1/f$, the expression (15.5.24) is subject to series expansion with respect to the quantity $1/f$. The expansion of $(1+\xi/m_0)^{-2}$ yields a set of interaction vertices proportional to $P^2$:

$$-\frac{P^2}{2m_0}\left(-2\frac{\xi}{m_0}+3\frac{\xi^2}{m_0^2}-4\frac{\xi^3}{m_0^3}+\dots\right) \tag{15.5.25}$$

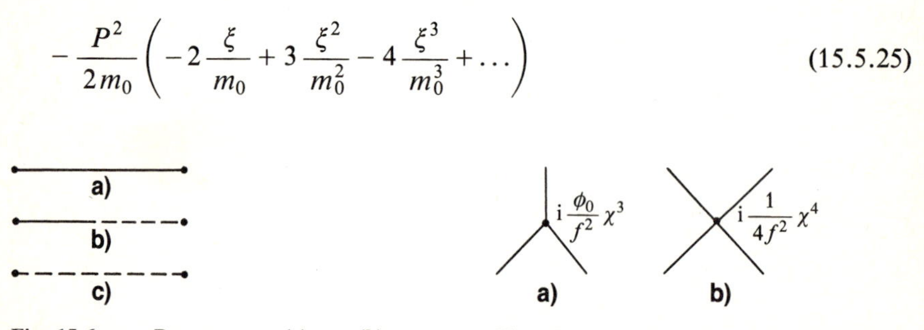


**Fig. 15.6a–c.** Propagators: **(a)** $\chi\chi$; **(b)** $\chi\pi$; **(c)** $\pi\pi$

**Fig. 15.7a, b.** Meson interaction vertices

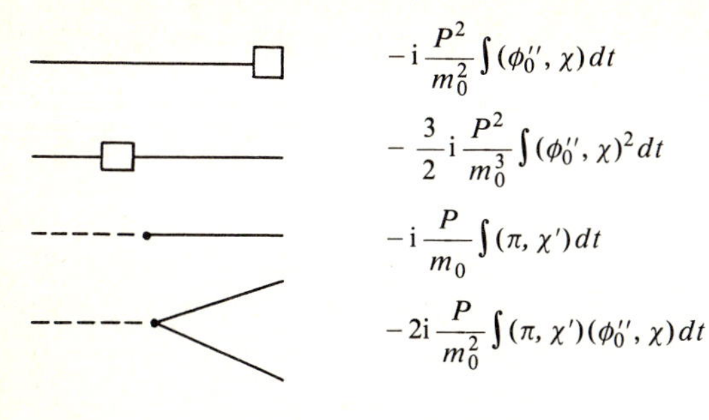


**Fig. 15.8.** Meson-soliton interaction vertices

whose order of magnitude is given by: $1/f^3$, $1/f^4$, etc. A second set of interaction vertices is proportional to $P$:

$$-\frac{P}{m_0}\int d\varrho\,\pi\chi'\left(1-2\frac{\xi}{m_0}+3\frac{\xi^2}{m_0^2}-4\frac{\xi^3}{m_0^3}+\dots\right). \tag{15.5.26}$$

The first two terms of the series (15.5.25, 26) are depicted in Fig. 15.8. By virtue of the obtained rules one can find expressions for the Green's functions which are to be subsequently integrated over $X$ and $P$.

Thus, constructing the quantum theory of solitons consists basically in performing an expansion around the classical solution, subsequently singling out and eliminating the zero-mode contributions followed by calculating the Green's functions at fixed $X$ and $P$ and, eventually, integrating over these variables.

The quantum theory of solitons in the case of larger dimensions can be constructed along the same lines.

# 16. Conclusion

The introduction of the gauge fields has made it possible to solve a number of problems of the field theory and of the physics of elementary particles.

1) The gauge fields are asymptotically free. This permits constructing asymptotically free models which include, besides the gauge fields, a certain set of the matter fields.
2) After introducing the gauge fields, non-renormalizable theories can become renormalizable (as is the case, for example, with the theory of weak interactions).
3) All three fundamental types of interactions – strong, electromagnetic and weak – have been reduced to exchange of gauge vector fields. The electromagnetic interactions are mediated by photons, the weak interactions by intermediate vector bosons, and the strong interactions by gluons.
4) There has been developed quantum chromodynamics which appears to be a prospective candidate for the theory of strong interactions.
5) Unified models for weak and electromagnetic interactions have been constructed. Theoretical results obtained are in a very good agreement with experimental data.
6) A possibility has been opened to construct models unifying all three types of interactions, in particular, models with a single coupling constant (grand unification).

On the other hand, some important problems are still unsolved. Intensive investigations are being conducted presently in the following main directions.

## 16.1 Quark and Gluon Confinement

The treatment of this problem in the framework of quantum chromodynamics on a lattice has been already discussed in Chap. 13.

To solve this problem, the properties of quantum chromodynamics in the infrared region are subject to intensive study.

There are also attempts to advance other treatments where the hadrons are considered as some extended objects, e.g., the bag model and the string model.

The bag model has various modifications. In one of them a hadron is thought of as a sphere containing quarks and gluons which are described by quantum chromodynamics. To make the system stable, the volume pressure and the surface tension are introduced, both quantities being free parameters of the theory. It is assumed that the quark and gluon currents are zero on the surface of the sphere, i.e., that the quark and gluon confinement takes place. For the main characteristics of hadrons (masses, magnetic moments, etc.) the model then gives values which are close to the experimental ones. Unfortunately, the model has too much of a phenomenological character.

The string model is based on the fact that at large distances (of the order of the hadron size) between the quarks the gauge fields are concentrated within a narrow region of the space in proximity to the line which connects the quarks, thus forming a string. Hence, a hadron may be considered as a string. The classical as well as the quantum theories of a relativistic string, which possesses the quantum numbers of hadron (spatial spin, isospin, etc.) have been put forward. However, the string model has a number of drawbacks (existence of tachyon solutions, difficulties in quantization, renormalization, etc.).

There have also been attempts to explain the confinement of quarks and gluons by invoking the instantons and their contributions.

## 16.2 Potential Approach

Another model which uses confinement as an ingredient, is the potential approach. The Hamiltonian of the system usually contains a Coulomb-like potential (derived from the one-gluon exchange between quark-(anti)quark, compatible with asymptotic freedom at short distances), and as well a confining potential (which is usually taken to rise linearly with the distance). A potential model is expected to work better for non-relativistic systems. Such models give excellent qualitative and rather good quantitative results for the spectroscopy of hadrons which contain heavy quarks.

## 16.3 QCD Sum Rules

A more basic approach with wider applicability is the QCD sum rules method. It generalizes the short-distance expansion of the field operator products to the case with the presence of a non-trivial vacuum. The vacuum structure is represented by condensates like the chiral one $\langle \bar{\psi}\psi \rangle$ for quarks, the gluon condensate $\langle F^a_{\mu\nu} F^a_{\mu\nu} \rangle$ and the higher-dimensional ones. This method is quite successfull in hadron spectroscopy of mesons and baryons. It applies to hadrons containing light and heavy quarks as well.

## 16.4 Theory of Loop Functionals

There is some hope that additional possibilities for describing the strong interactions – in particular, the confinement – can be provided by reformulating the local gauge theory as a non-local theory of loop functionals. In such a theory the gauge fields are no longer functions of the coordinates but become loop functionals $\phi(C)$ (the loops being the continuum analogue of the loops on a lattice considered in Chap. 13):

$$\phi(C) = P\exp\left\{\int_C dx_\mu A_\mu(x)\right\} .$$

Here $C$ is a closed loop and $P$ is the loop ordering operator. The quantum chromodynamics on the loops has been formulated. Unfortunately, it is still unclear how to handle the loop variables which is of a hindrance in deriving physical results.

## 16.5 Unified Gauge Models

There are also a number of unresolved problems pertaining to unified gauge models of particle interactions. There is still no unequivocal choice of the model unifying the strong interactions and the weak and electromagnetic interactions. There remain questions concerning the difference in the masses of the gauge fields corresponding to different interactions (gauge hierarchy), the explanation of the possible number of flavours, the existence of the Cabbibo angle, the $CP$ violation, etc.

Besides the Higgs mechanism, i.e., spontaneous symmetry breaking, discussed in this book, there exists a dynamical way of symmetry breaking. One of the patterns of dynamical symmetry breaking is technicolour. In the technicolour model, the Higgs bosons are considered as compound particles made out of techniquarks. The interaction between the techniquarks is mediated through the exchange of technigluons. The model aims at accounting for the appearance of the Higgs bosons with their masses in a dynamical way. No success has been achieved as yet in constructing a consistent technicolour model.

## 16.6 Unified Supersymmetric Models

Much attention is being paid to develop unified gauge models in which the global space-time group is the supersymmetric group, instead of the Lorentz group. The former has a number of specific features (in particular, its multiplets contain both bosons and fermions), which is a ground for the hope that unified supersymmetric gauge models might prove to be more perfect than the Lorentz ones.

## 16.7 Gauge Theory of Gravitation

In localizing the space-time symmetry groups, it is the gravitation fields which play the role of the gauge fields. Thus, localization of the Lorentz group leads to the Einstein theory of gravitation. Localizing other space-time symmetry groups (conformal, supersymmetric, etc.) gives rise to the corresponding theories of gravitation. All of them, and especially the supergravity theory, are subject to a very intensive study.

## 16.8 Superunification

Localizing both the internal and the space-time symmetry groups gives rise to the Yang-Mills and the gravitational fields. This paves the way for superunification, i.e., to a unification of the gravitational interaction with the strong, the electromagnetic and the weak interactions. It has been known for a rather long time that non-trivial unification (i.e., not in form of the direct product) of the Lorentz group and the internal symmetry group is not possible (the O'Raifeartaigh theorem). However, non-trivial unification of the supersymmetric group with the internal symmetry group (the extended supersymmetric group) is possible. Therefore, superunification is conducted on the basis of the local extended supersymmetric group. Only first steps have been made in this direction, and no realistic superunification model has been constructed to date.

## 16.9 Composite Models

Since the number of elementary particles is large (Higgs scalars, leptons, quarks, photons, intermediate bosons, gluons, gravitons) and since these particles are quite diversiform, the idea has come up to introduce more fundamental particles, out of which all the mentioned elementary particles, or part of them, could be constructed. Quite a number of models have been propounded; in each a definite set of fundamental particles is used. These have been given various names (subquarks, prequarks, metaquarks, preons, rishons, etc.). However, no success in constructing a comprehensive model with fundamental particles can be reported as yet.

## 16.10 The Elementary Particles and the Cosmology

Some important information about the characteristics of the elementary particles can be extracted from the available astrophysical data. In this way it was possible to establish the upper bounds for the masses of photon and neutrinos of various kinds as well as the limitations on the masses and the life-times of heavy neutral leptons and the upper bound for the number of possible types of neutrinos. Astrophysical data also provide arguments supporting the quark confinement, the baryon excess over the anti-baryons, etc.

* *

*

A more detailed account of the problems mentioned can be found in the surveys and the review articles cited in the References to the Conclusion.

# Bibliography

In the bibliography we cite only books and, as far as possible, recent review articles in which more details and references can be found. Unfortunately, with such a choice of references many important works are missing in our list which in no way claims to be complete.

#### On Gauge Field Theories in General

Abers, E. S., Lee, B. W.: Gauge Theories. Phys. Rep. **9**, 1 (1973)
Becher, P., Böhm, M., Joos, H.: *Eichtheorien der starken und elektroschwachen Wechselwirkung* (Teubner, Stuttgart 1983)
Faddeev, L. D., Slavnov, A. A.: *Gauge Fields: Introduction to Quantum Field Theory* (Benjamin/Cummings, Reading, MA 1980)
Itzykson, C., Zuber, J.-B.: *Quantum Field Theory* (McGraw-Hill, New York 1980)
Okun, L. B.: *Leptons and Quarks* (North-Holland, Amsterdam 1981)
Taylor, J. C.: *Gauge Theories of Weak Interactions* (Cambridge Univ. Press, Cambridge 1976)
Zinn-Justin, J.: *Trends in Elementary Particle Theory,* ed. by H. Rollnik, K. Dietz (Springer, Berlin, Heidelberg, New York 1975)

## Part I. **Invariant Lagrangians**

#### Quantum Field Theory

Bjorken, T. D., Drell, S. D.: *Relativistic Quantum Field Theory* (McGraw-Hill, New York 1964)
Bogoliubov, N. N., Shirkov, D. V.: *Introduction to the Theory of Quantized Field* (Wiley, New York 1980)
Itzykson, C., Zuber, J.-B.: *Quantum Field Theory* (McGraw-Hill, New York 1980)
Pauli, W.: Relativistic field theory of elementary particles. Rev. Mod. Phys. **13**, 203 (1941)
Ramond, P.: *Field Theory. A Modern Primer* (Benjamin/Cummings, Reading, MA 1981)
Schweber, S. S.: *An Introduction to Relativistic Quantum Field Theory* (Row, Peterson and Co., New York 1962)
Wentzel, G.: *Quantum Theory of Fields* (Interscience, New York 1949)

#### Elementary Particle Physics

Aitchison, I. J. R., Hey, A. J. G.: *Gauge Theories in Particle Physics. A Practical Introduction* (Adam Hilger, Bristol 1982)
Gasiorowicz, S.: *Elementary Particle Physics* (Wiley, New York 1966)
Källén, G.: *Elementary Particle Physics* (Addison-Wesley, Reading, MA 1964)
Lee, T. D.: *Particle Physics and Introduction to Field Theory* (Harwood, Chur 1981)
Nelipa, N. F.: *Physique des Particules Élémentaires* (Edition MIR, Moscow 1981)
Nishijima, K.: *Fields and Particles* (Benjamin, New York 1969)

#### Group Theory. Unitary Symmetry

Carruthers, P.: *Introduction to Unitary Symmetry* (Interscience, New York 1966)
Gel'fand, I. M., Minlos, R. A., Shapiro, Z. Ya.: *Representation of the Rotation Group and of the Lorentz Group and Their Applications* (MacMillan, New York 1963)
Lipkin, H. J.: *Lie Groups for Pedestrians* (North-Holland, Amsterdam 1965)

Slansky, R.: Group theory for unified model building. Phys. Rep. **79C**, 1 (1981)
Zhelobenko, D. P.: Compact Lie groups and their applications. Am. Math. transl. **40** (1973)

**Spontaneous Symmetry Breaking**

Bernstein, J.: Spontaneous symmetry breaking, gauge theories, the Higgs mechanism and all that. Rev. Mod. Phys. **46**, 7 (1974)
Coleman, S.: *"Secret Symmetry": Laws of Hadronic Matter*, Proc. of the 11th Course of the "Ettore Majorana", Intern. School of Subnuclear Physics, ed. by A. Zichichi (Academic, New York 1975)

## Part II. Quantum Theory of Gauge Fields

**Path Integrals and Covariant Perturbation Theory**

Berezin, F. A.: *Methods of Second Quantization* (Academic, New York 1966)
Dirac, P. A. M.: *Lectures on Quantum Mechanics* (Yeshiva University, New York 1964)
Faddeev, L. D.: *"Introduction to Functional Methods"*, in *Methods in Field Theory,* Les Houches Summer School 1975, ed. by R. Balian, J. Zinn-Justin (North-Holland, Amsterdam 1976)
Feynman, R. P., Hibbs, A. R.: *Quantum Mechanics and Path Integrals* (McGraw-Hill, New York 1965)
Klauder, J. R.: *"Path Integral"* in *Field and Strong Interactions,* Proc. XIX Intern. Universitätswochen für Kernphysik, Schladming, 1980, ed. by P. Urban (Springer, Wien, New York 1980) p. 3

## Part III. Gauge Theory of Electroweak Interactions

**Electromagnetic and Weak Interactions**

Akhiezer, A. I., Berestetski, V. B.: *Quantum Electrodynamics* (Interscience, New York 1965)
Berestetski, V. B., Lifshitz, E. M., Pitayevski, L. P.: *Relativistic Quantum Theory*, Part I (Pergamon, London 1971)
Commins, E. D.: *Weak Interactions* (McGraw-Hill, New York 1973)
Feynman, R. P.: *Quantum Electrodynamics* (Benjamin, New York 1961)
Lifshitz, E. M., Pitayevski, L. P.: *Relativistic Quantum Theory*, Part II (Pergamon, London 1972)
Okun, L. B.: *Leptons and Quarks* (North-Holland, Amsterdam 1981)
Taylor, J. C.: *Gauge Theories of Weak Interactions* (Cambridge University Press, Cambridge 1976)

**Leptons, Quarks. Quark Model**

Close, F. E.: *An Introduction to Quarks and Partons* (Academic, London 1979)
Greenberg, O. W., Nelson, C. A.: Colour models of hadrons. Phys. Rep. **32C**, 69 (1977)
Huang, K.: *Quarks, Leptons and Gauge Fields* (World Scientific, Singapore 1983)
Kalmus, G.: "Weak Decays of New Particles", in Proc. 21 Intern. Conf. on High Energy Physics, Paris 1982, ed. by P. Petian, M. Porneuf (Les Editions de Physique, Paris 1982), p. 431
Kokkedee, J.: *The Quark Model* (Benjamin, New York 1969)
Maiani, L.: *Theoretical Ideas on Heavy Flavor Weak Decays*, in Proc. 21 Intern. Conf. on High Energy Physics, Paris 1982, ed. by P. Petian, M. Porneuf (Les Editions de Physique, Paris 1982), p. 631
Trilling, G. H.: The properties of charmed particles. Phys. Rep. **75C**, 57 (1981)

**Standard Model. Neutral Currents**

Beg, M. A. B., Sirlin, A.: Gauge theories of weak interaction. Phys. Rep. **88C**, 1 (1982)
Dalitz, R. M.: "CP-Nonconservation", in *Electroweak Interactions*, Proc. XXI Intern. Universitätswochen für Kernphysik, Schladming 1982, ed. by H. Mitter (Springer, Wien, New York 1982) p. 393

Davier, M.: "Electroweak Neutral Current", in Proc. 21 Intern. Conf. on High Energy Physics, ed. by P. Petian, M. Porneuf (Les Editions de Physique, Paris 1982) p. 471

Ecker, G.: "Introduction to Gauge Theories of Electroweak Interactions", in *Electroweak Interactions*, Proc. XXI Intern. Universitätswochen für Kernphysik, Schladming 1982, ed. by H. Mitter (Springer, Wien, New York 1982) p. 3

Fritzsch, H., Minkowski, P.: Flavordynamics of quarks and leptons. Phys. Rep. **73C**, 67 (1981)

Harari, H.: Quarks and leptons. Phys. Rep. **42C**, 235 (1978)

Kim, J. E., Langacker, P., Levine, M., Williams, H. H.: A theoretical and experimental review of the weak neutral current. Rev. Mod. Phys. **53**, 211 (1981)

#### Renormalization

Faddeev, L. D., Slavnov, A. A.: *Gauge Fields: Introduction to Quantum Theory* (Benjamin/Cummings, Reading, MA 1980)

Fradkin, E. S., Tyutin, I. V.: Renormalizable theory of massive vector particles. Riv. Nuovo Cim. **4**, 1 (1974)

Leibbrandt, G.: Introduction to the technique of dimensional regularization. Rev. Mod. Phys. **47**, 849 (1975)

Narison, S.: Techniques of dimensional regularization and the two-point functions of QCD and QED. Phys. Rep. **84C**, 263 (1982)

Zinn-Justin, J.: "Renormalization of Gauge Theories", in *Trends in Elementary Particle Theory*, Lecture Notes Phys. Vol. 37, ed. by H. Rollnik, K. Dietz (Springer, Berlin, Heidelberg, New York 1975) p. 2

## Part IV. **Gauge Theory of Strong Interactions**

#### Strong Interactions

de Alfaro, V., Fubini, S., Furlan, G., Rossetti, C.: *Currents in Hadron Physics* (North-Holland, Amsterdam 1973)

Collins, P. D. B., Squires, E. J.: *Regge Poles in Particle Physics*. Springer Tracts Mod. Phys., Vol. 45 (Springer, Berlin, Heidelberg, New York 1968)

Eden, R. J., Landshoff, P. V., Olive, D. I., Polkinghorne, J. C.: *The Analytic S-Matrix* (Cambridge University Press, Cambridge 1966)

Nelipa, N. F.: *Physique des Particules Élémentaires* (Edition MIR, Moscow 1981)

#### Asymptotic Freedom

Politzer, H. D.: Asymptotic freedom: an approach to strong interactions. Phys. Rep. **14C**, 129 (1974)

#### Perturbative Quantum Chromodynamics. Hard Processes

Altarelli, G.: Partons in QCD. Phys. Rep. **81C**, 1 (1982)

Buras, A. J.: Asymptotic freedom in deep inelastic processes in the leading order and beyond. Rev. Mod. Phys. **52**, 199 (1980);
"A Tour of Perturbative QCD", in Proc. Int. Symp. on Lepton and Photon Interactions at High Energies, ed. by W. Pfeil (Bonn University, Bonn 1981) p. 636

Dokshitzer, Y. L., Dyakonov, D. I.,Troyan, S. I.: Hard processes in QCD. Phys. Rep. **58C**, 269 (1980)

Feynman, R. P.: *Photon-Hadron Interactions* (Benjamin, New York 1972)

't Hooft, G.: How well do we understand QCD? Status report of a theory, Proc. Intern. Conf. on High Energy Physics, ed. by J. Dias de Deus, J. Soffer (Lisbon 1981) p. 641

Marciano, W., Pagels, H.: Quantum chromodynamics. Phys. Rep. **36C**, 137 (1978)

Nachtmann, O.: "The Classical Tests of Quantum Chromodynamics", in *Field Theory and Strong Interactions*, Proc. of the XIX Intern. Universitätswochen für Kernphysik, Schladming 1980, ed. by P. Urban (Springer, Wien, New York 1980) p.101

Petermann, A.: Renormalization group and the deep structure of the proton. Phys. Rep. **53C**, 157 (1979)

Reya, E.: Perturbative quantum chromodynamics. Phys. Rep. **69C**, 195 (1981)

Walsh, T. F.: "Phenomenology of Jets", in *Field Theory and Strong Interactions*, Proc. XIX Intern. Universitätswochen für Kernphysik, Schladming 1980, ed. by P. Urban (Springer, Wien, New York 1980) p. 439

Yndurâin, F. J.: *Quantum Chromodynamics* (Springer, Berlin, Heidelberg, New York, Tokyo 1983)

### Lattice Gauge Theories

Hasenfratz, P.: "Lattice Gauge Theories", Proc. Intern. Conf. on High Energy Physics, ed. by J. Dias de Deus, J. Soffer (Lisbon 1981) p. 619

Kogut, J. B.: The lattice gauge theory approach to quantum chromodynamics. Rev. Mod. Phys. **55**, 775 (1983)

Wilson, K. G.: Quark on a lattice, or, the coloured string model. Phys. Rep. **23C**, 331 (1976)

Wilson, K. G: "Quantum chromodynamics on a lattice", in *New Developments in Quantum Field Theory and Statistical Mechanics*, ed. by M. Levy, P. Mitter (Plenum, New York 1977) p. 143

### Grand Unification

Ellis, J.: "Phenomenology of Unified Gauge Theories", in *Gauge Theories in High Energy Physics*, Les Houches Summer School Proc. 37, ed. by M. K. Gaillard, R. Stora (North-Holland, Amsterdam 1983) p. 155

Georgi, H.: "Grand Unification", in Proc. 21 Intern. Conf. on High Energy Physics, Paris 1982, ed. by P. Petian, M. Porneuf (Les Editions de Physique, Paris 1982) p. 705

Langacker, P.: Grand unified theories and proton decay. Phys. Rep. **72C**, 185 (1981)

Ramond, P. (ed.): *Grand Unified Theories and Related Topics* (World Scientific, Singapore 1981)

Salam, A.: Gauge unification of fundamental forces. Rev. Mod. Phys. **52**, 525 (1980)

Zee, A.: Unity of Forces in the Universe, Vol. 1 (World Scientific, Singapore 1982)

### Solitons, Instantons

Faddeev, L. D., Korepin, V. E.: Quantum theory of solitons. Phys. Rep. **42C**, 1 (1978)

Marciano, W., Pagels, H.: Quantum chromodynamics. Phys. Rep. **36C**, Chap. 3 (1978)

Olive, D., Saito, S., Crewther, R. J.: Instanton in field theory. Riv. Nuovo Cim. **2**, No. 8 (1979)

Olive, D.: Magnetic monopoles. Phys. Rep. **49C**, 165 (1979)

Rajaraman, R.: *Solitons and Instantons* (North-Holland, Amsterdam 1982)

Rossi, P.: Exact results in the theory of non-abelian magnetic monopoles. Phys. Rep. **86C**, 317 (1982)

## Conclusion

### Bag and String Model

Gervais, J. L., Jacob, M. (ed.): *Non-linear and Collective Phenomena in Quantum Physics* (World Scientific, Singapore 1983)

Hasenfratz, P., Kuti, J.: The quark bag model. Phys. Rep. **40C**, 76 (1978)

Jaffe, R. L.: "Application of the Bag Model", in *Field Theory and Strong Interactions*, Proc. XIX Intern. Universitätswochen für Kernphysik, Schladming 1980, ed. by P. Urban (Springer, Wien, New York 1980) p. 269

Neveu, A.: "Revival of the String Model", in Proc. 21 Intern. Conf. on High Energy Physics, Paris 1982, ed. by P. Petian, M. Porneuf (Les Editions de Physique, Paris 1982) p. 260

### Potential Models

Chaichian, M., Kögerler, R.: Coupling constants and the nonrelativistic quark model with charmonium potential. Ann. Physics (New York) **124**, 61 (1980)

Gottfried, K.: "Hadronic spectroscopy", in *Proc. Intern. Europhysics Conf. on High Energy Physics,* Brighton 1983, ed. by J. Guy, C. Costain (Rutherford Appleton Laboratory, Chilton, Didcot 1983) p. 743

**QCD Sum Rules**

Novikov, V. A., Okun, L. B., Shifman, M. A., Vainshtein, A. I., Voloshin, M. B., Zakharov, V. I.: Charmonium and gluons. Phys. Rep. **41C**, 1 (1978)

de Rafael, E.: Current algebra quark masses in QCD, Proc. of NSF-CNRS Joint Seminar on Theoretical Aspects of Quantum Chromodynamics, ed. by J. W. Dash, CPT-81-Pub. 1345 (Marseille, 1981) p. 259

Rubinstein, H. R.: "Non-Perturbative Effects and QCD Sum Rules", Proc. 21 Intern. Conf. on High Energy Physics, Paris 1982, ed. by P. Petian, M. Porneuf (Les Editions de Physique, Paris 1982) p. 249

Shifman, M. A.: "Theory of Heavy Quark-Antiquark States", Proc. Intern. Symp. on Lepton and Photon Interactions at High Energies, ed. by W. Pfeil (Bonn University, Bonn 1981) p. 242

**Theory of Loop Functionals**

Migdal, A. A.: "Loop Dynamics", Intern. Conf. on High Energy Physics, ed. by J. Dias de Deus, J. Soffer (Lisbon, 1981) p. 581

**Technicolour**

Farhi, E., Jackiw, R. (eds.): *Dynamical Gauge Symmetry Breaking* (World Scientific, Singapore 1982)

Farhi, E., Susskind, L.: Technicolour. Phys. Rep. **74C**, 277 (1981)

**Supersymmetric Unified Models**

Barbieri, R.: "Supersymmetric Gauge Models of the Fundamental Interactions", in *Electroweak Interactions*, Proc. XXI Intern. Universitätswochen für Kernphysik, Schladming 1982, ed. by H. Mitter (Springer, Wien, New York 1982) p. 363

Fayet, P.: "Unified Models of Particles and Interactions", Proc. 21 Intern. Conf. on High Energy Physics, ed. by P Petian, M. Porneuf (Les Editions de Physique, Paris 1982) p. 673

Fayet, P., Ferrara, S.: Supersymmetry. Phys. Rep. **32C**, 251 (1977)

Nanopoulos, D.: "Grand Unified Models and Physical Supersymmetries", in *SU(3)×SU(2)×U(1) and Beyond*, Proc. XIII GIFT Intern. Seminar an Theor. Phys., Masella, Spain 1982, ed. by A. Ferrando, J. A. Grifols, A. Mendez (World Scientific, Singapore 1983)

Zumino, B.: "New Developments in Grand Unified Theories", Proc. Intern. Conf. on High Energy Physics, ed. by J. Dias de Deus, J. Soffer (Lisbon, 1981) p. 601

**Supergravity**

van Nieuwenhuizen, P.: Supergravity. Phys. Rep. **68C**, 189 (1981)

**Superunification**

Ellis, J.: "GUTs, Astrophysics and Superunification", Proc. Intern. Conf. Neutrino '82, Vol. I, ed. by A. Frenkel, L. Jenik (KFKI, Budapest 1982) p. 304

**Composite Models**

't Hooft, G.: "Naturalness, Chiral Symmetry and Spontaneous Chiral Symmetry Breaking", in *Recent Developments in Gauge Theories*, ed. by G. 't Hooft, C. Itzykson, A. Jaffe, H. Lehmann, P. K. Mitter, I. Singer (Plenum, New York 1980) p. 135

't Hooft, G.: "Theoretical Perspectives", in Proc. 21 Intern. Conf. on High Energy Physics, Paris 1982, ed. by P. Petian, M. Porneuf (Les Editions de Physique, Paris 1982) p. 755

Peccei, R. D.: "Composite Models of Quarks and Leptons", in *Gauge Theories of the Eighties*, Lecture Notes Phys., Vol. 181, ed. by R. Raitio, J. Lindfors (Springer, Berlin, Heidelberg, New York, Tokyo 1983) p. 355

Peskin, M. E.: "Compositeness of Quarks and Leptons", in Proc. Int. Symp. on Lepton and Photon Interactions at High Energies, ed. by W. Pfeil (Bonn University, Bonn 1981) p. 880

**Elementary Particles and Cosmology**

Dolgov, A. D., Zeldovich, Ya. B.: Cosmology and elementary particles. Rev. Mod. Phys. **53**, 1 (1981)

# List of Symbols

| | |
|---|---|
| Fields | |
| in the general case | $u_i(x)$ |
| scalar field | $\phi(x)$ |
| spinor field | $\psi(x),\ \bar{\psi}(x) = \psi^{\dagger}(x)\gamma_0$ |
| electromagnetic field | $A_\mu(x)$ |
| non-Abelian gauge field | $A_\mu^k(x)$ |
| ghost field | $c^k(x),\ \bar{c}^k(x)$ |
| gluon field | $V_\mu^m(x)$ |
| Electromagnetic field tensor | $F_{\mu\nu}(x)$ |
| Non-Abelian gauge field tensor | $F_{\mu\nu}^k(x)$ |
| Lagrangian | |
| globally invariant | $L(x)$ |
| locally invariant | $\mathscr{L}(x)$ |
| Action | $I(u_i)$ |
| Transition amplitude | $S$ |
| Derivative | |
| ordinary | $\partial_\mu$ |
| covariant | $\nabla_\mu$ |
| Path integral | $\int \mathscr{D}u_i(x)$ |
| Transformation parameters | |
| global | $\varepsilon_k$ |
| local | $\varepsilon_k(x)$ |
| Element of the gauge group | $\omega(x)$ |
| Structure constants | $f_{abc}$ |
| Generators of transformations | |
| in the general case | $T_{ab}^k$ |
| for the fermions $\psi^a(x)$ | $(t_k)_{ab}$ |
| for the scalars $\phi^a(x)$ | $(\theta_k)_{ab}$ |

| | |
|---|---|
| Generating functionals for the Green's functions | |
| complete | $W(J_i)$ |
| connected | $Z(J_i)$ |
| one-particle-irreducible | $\Gamma(\Phi_i)$ |
| Propagators (the capital italic letters refer to the full propagators and the script letters to the propagators in the tree approximation) | |
| for photons | $D_{\mu\nu}(x-y)$, $\mathscr{D}_{\mu\nu}(x-y)$ |
| for non-Abelian gauge fields | $D^{kl}_{\mu\nu}(x-y)$, $\mathscr{D}^{kl}_{\mu\nu}(x-y)$ |
| for scalar fields | $D^{ab}(x-y)$, $\mathscr{D}^{ab}(x-y)$ |
| for spinor fields | $G^{ab}(x-y)$, $\mathscr{G}^{ab}(x-y)$ |
| Left-handed spinor doublets of particles | $L^a(x)$ |
| Right-handed spinor singlets | $R(x)$ |
| Currents | |
| electromagnetic | $J_\mu(x)$ |
| neutral weak | $J^n_\mu(x)$ |
| charged weak | $J^{(\pm)}_\mu(x)$ |
| Structure functions | $W_i(x, q^2)$ |

# Subject Index